Going to See the Doctor for a Check-Up Visit

A Book to Help Manage Expectations for Children and Adults with Autism Spectrum Disorder

Kristine Kenny, MSN, RN, CPN, CAS
Illustrated by Madelyn Hess

Going to See the Doctor for a Check-Up Visit:
A Book to Help Manage Expectations for Children and Adults with Autism Spectrum Disorder

ISBN-13: 979-8-9870047-0-8

Printed in the United States of America

Illustrated by Madelyn Hess

Going to See the Doctor for a Check-Up Visit

A Book to Help Manage Expectations for Children and Adults with Autism Spectrum Disorder

Use this space to tape a picture of your own doctor's office.

We encourage you to tape a picture of your own doctor's office on the cover of this book as well.

Kristine Kenny

This book is about going to the doctor for a well visit.
The doctor wants to make sure you are healthy.

When you get there, you will go to a waiting room. You will have to wait until it is your turn to see the doctor.

When it is your turn, you will leave the waiting room with your mom, dad, or caregiver with a nurse.

The nurse will check your height and weight.

The nurse may check your blood pressure, temperature, and heart rate.

Next, you will get to go to your own exam room.

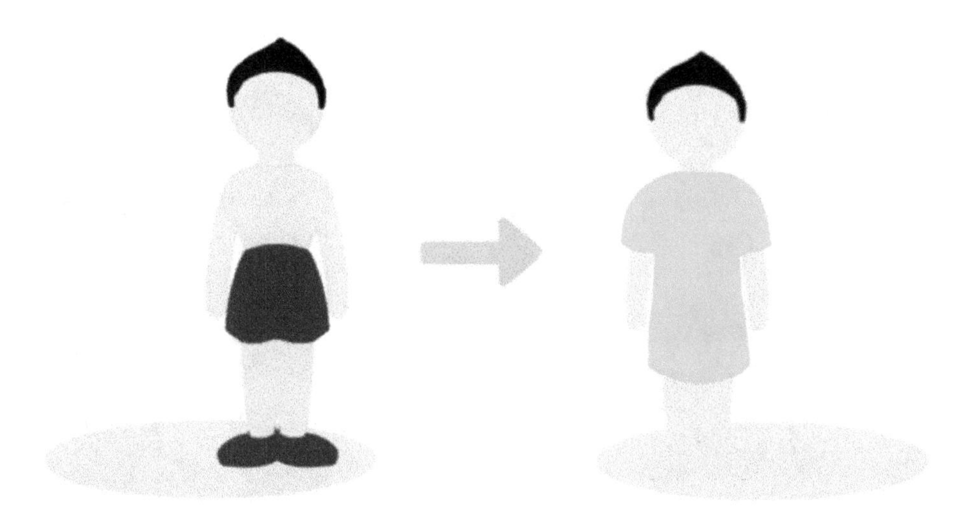

Here, you will get ready to see the doctor.

When the doctor comes in, they will talk with your mom, dad, or caregiver and you about how you have been doing.

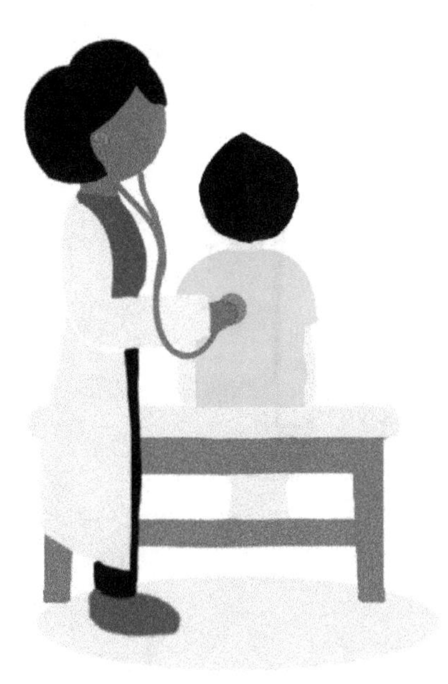

The doctor will listen to your heart. The doctor will listen to your lungs. You may need to take some deep breaths in and out to help the doctor listen.

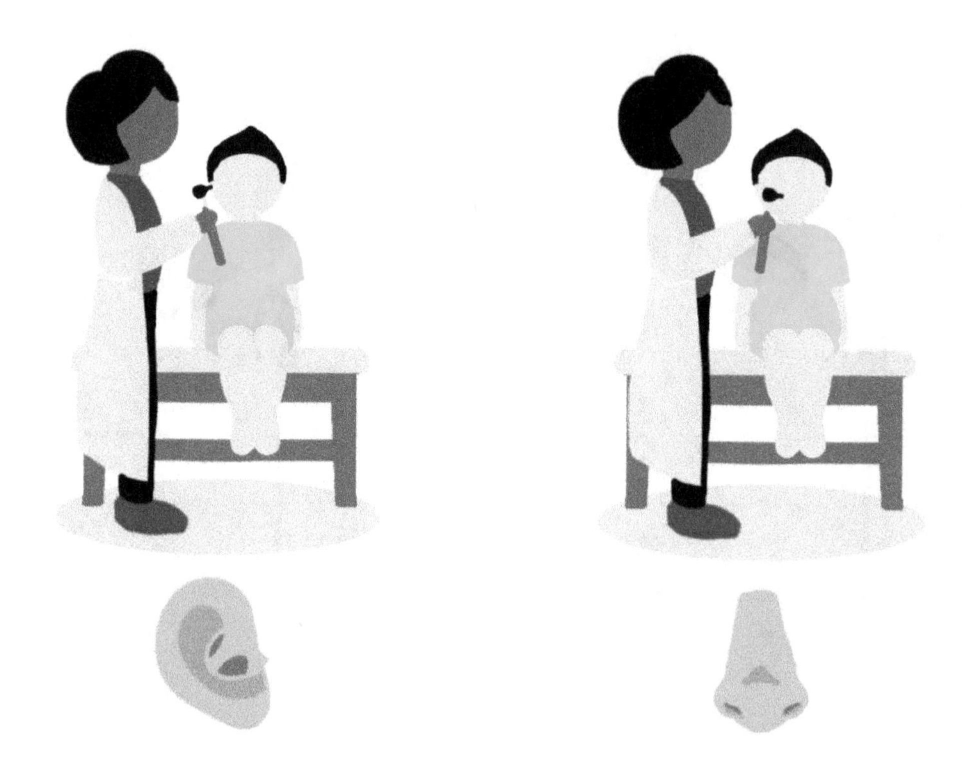

The doctor will look into your ears
and nose with a special tool.

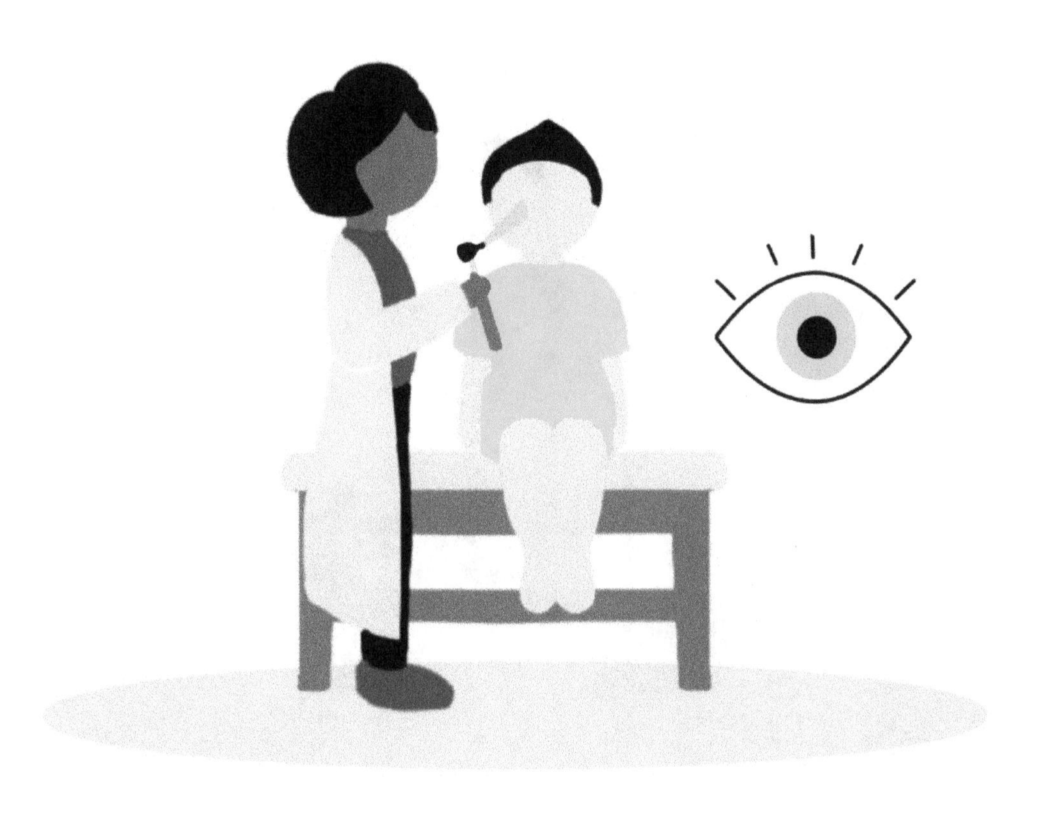

The doctor will look into each of your eyes with a small flashlight.

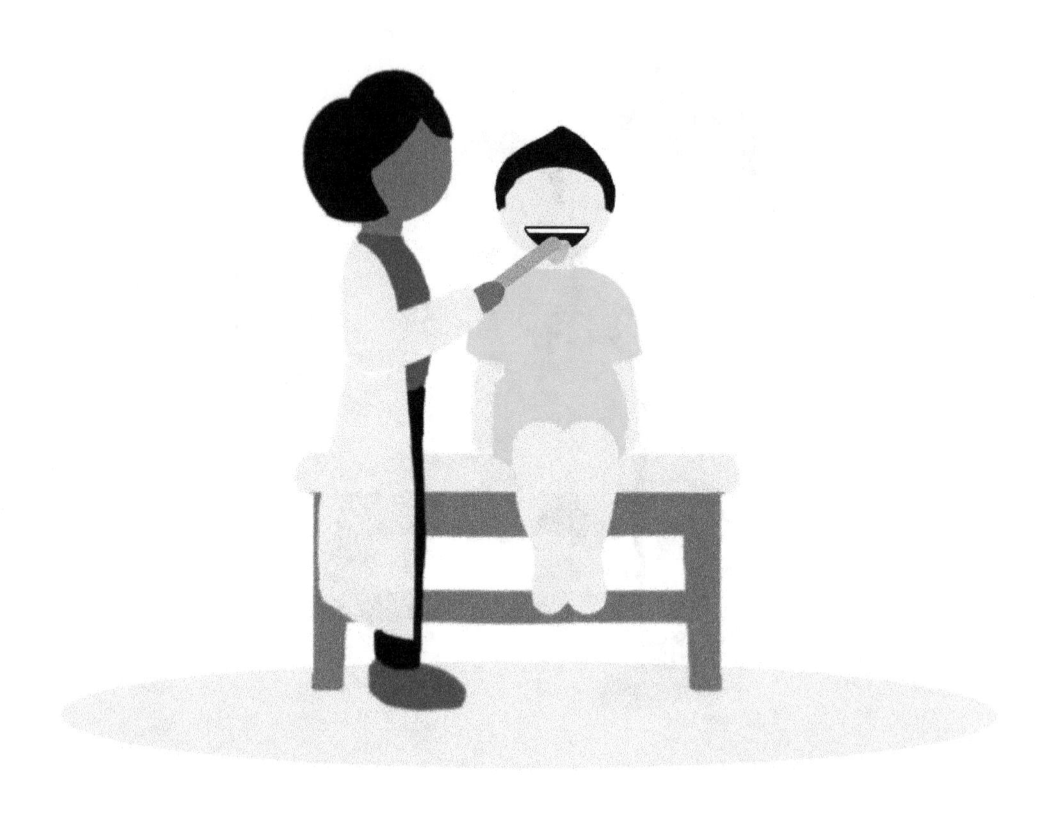

The doctor will look into your mouth. You can stick out your tongue so the doctor can see your teeth and inside your mouth.

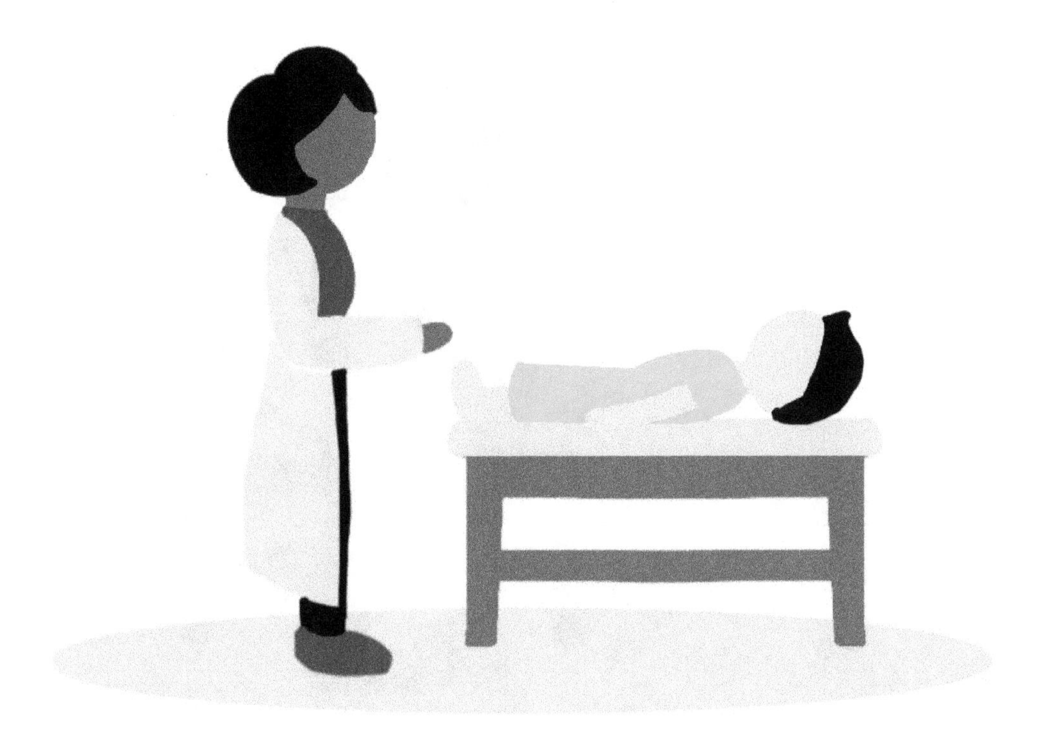

Next, the doctor may ask you to
lay down on the exam table.

The doctor will feel your neck.

The doctor will feel your belly.

The doctor will feel the skin on your legs and feet.

When the doctor says, “I am all done,” then you can put your clothes back on.

ALL DONE.
Great
Job!

You did a great job.
It is time to leave the doctor and go home.

Checklist

- [ ] 1. Arrive at doctor's office. 
- [ ] 2. Go inside and wait our turn. 
- [ ] 3. Nurse calls for us.  
- [ ] 4. Take vital signs. 
- [ ] 5. Go into room. 
- [ ] 6. Doctor comes in.

☐ 7. Doctor listens to heart and lungs.

☐ 8. Doctor looks at nose, ears, eyes.

☐ 9. Doctor touches skin.

☐ 10. Time to get dressed.

☐ 11. All done. Great job

Tape a photo of your reward here.

☐ 12. Go home.

ABOUT THE AUTHOR

Kristine Kenny

Kris has been a pediatric registered nurse for over 30 years, and is presently a nursing professor. Working as a registered nurse in a major pediatric hospital and being the mother of a special needs child, she noticed a huge gap in communication for children with special needs in healthcare settings. She has educated registered nurses at Children's Hospital of UPMC about the proper way to communicate with children with autism in a healthcare setting. She has written multiple stories for her daughter with autism to help guide her through medical experiences for many years. Her passion for education and empowerment is strong. Kris is committed to bringing quality health education for those with unique learning needs. This is the first of many books that will be available.

- Certified Pediatric Nurse and Certified Autism Specialist
- Presentations at Children Hospital of Pittsburgh-UPMC, National Conference- The Importance of Using Communication Tools for Children on the Autism Spectrum in a Healthcare Setting.
- Created an exercise class in her community for teenagers and young adults with special needs.
- Fundraised for Autism Speaks, TRY Special Needs, and TRAILS at Slippery Rock, PA.
- Nursing educator and faculty for 7+ years in RN program, specializing in Pediatric Nurse Care Education and Mental Health Nurse Care Education.
- Practicing RN for 27+ years, specializing in pediatric nursing.

CPSIA information can be obtained
at www.ICGtesting.com
Printed in the USA
LVHW010309241122
733904LV00002B/22

9 798987 004708

The Zynq Book

Embedded Processing with the
ARM® Cortex®-A9 on the Xilinx®
Zynq®-7000 All Programmable SoC

The Zynq Book

Embedded Processing with the ARM® Cortex®-A9 on the Xilinx® Zynq®-7000 All Programmable SoC

Louise H. Crockett

Ross A. Elliot

Martin A. Enderwitz

Robert W. Stewart

Department of Electronic and Electrical Engineering

University of Strathclyde

Glasgow, Scotland, UK

1st Edition

This edition first published July 2014 by Strathclyde Academic Media.

Tutorial Materials

Tutorial materials are distributed via the book's companion website: www.zynqbook.com.

The tutorials are provided with the same *Open Source License to Use and Reproduce*, and the same *Warning and Disclaimer*, as detailed elsewhere on this page in reference to the main book.

Contents

Foreword **xxi**

Acknowledgements **xxiii**

CHAPTER 1 Introduction **1**

1.1 System-on-Chip with Zynq 2
1.2 Simple Anatomy of an Embedded SoC 5
1.3 Design Reuse 7
1.4 Raising the Abstraction Level 7
1.5 SoC Design Flow 8
1.6 Practical Elements 10
1.7 About This Book 10
1.8 References 12

PART A Getting to Know Zynq **13**

CHAPTER 2 The Zynq Device ("What is it?") **15**

2.1 Processing System 16
2.1.1 Application Processing Unit (APU) 17
2.1.2 A Note on the ARM Model 20
2.1.3 Processing System External Interfaces 21
2.2 Programmable Logic 22

2.2.1 The Logic Fabric 23
2.2.2 Special Resources: DSP48E1s and Block RAMs 25
2.2.3 General Purpose Input/Output 28
2.2.4 Communications Interfaces 29
2.2.5 Other Programmable Logic External Interfaces 29
2.3 Processing System — Programmable Logic Interfaces 30
2.3.1 The AXI Standard 30
2.3.2 AXI Interconnects and Interfaces 31
2.3.3 EMIO Interfaces 34
2.3.4 Other PL-PS Signals 34
2.4 Security 35
2.4.1 Secure Boot 35
2.4.2 Hardware Support 36
2.4.3 Runtime Security 36
2.5 Zynq-7000 Family Members 39
2.6 Chapter Review 40
2.7 Architecture Reference Guide 41
2.8 References 44

CHAPTER 3 Designing with Zynq ("How do I work with it?") 47
3.1 Getting Started 48
3.1.1 Obtaining Design Tools 48
3.1.2 Design Tool Editions and Licensing 49
3.1.3 Design Tool Functionality 50
3.1.4 Third Party Tools 51
3.1.5 System Setup and Requirements 51
3.2 An Outline of the Design Flow 53
3.2.1 Requirements and Specification 54
3.2.2 System Design 54
3.2.3 Hardware Development and Testing 55
3.2.4 Software Development and Testing 58
3.2.5 System Integration and Testing 60
3.3 SoC Design Teams 60
3.4 System-Level IP-Focused Design with Vivado 62
3.5 The ISE and Vivado Design Suites 64
3.5.1 Features Comparison 64

3.5.2 Upgrading to Vivado ... 66
3.6 Development Boards ... 67
3.6.1 Zynq-7000 SoC ZC702 Evaluation Kit ... 67
3.6.2 Zynq-7000 SoC Video & Imaging Kit ... 69
3.6.3 Zynq-7000 ZC706 Evaluation Kit ... 69
3.6.4 ZedBoard ... 69
3.6.5 ZYBO ... 69
3.6.6 Third Party Boards ... 70
3.6.7 Accessories and Expansions ... 71
3.6.8 Working with Development Boards ... 72
3.7 Support and Documentation ... 72
3.8 Chapter Review ... 72
3.9 References ... 73

CHAPTER 4 Device Comparisons ("Why do I need Zynq?") 77
4.1 Device Selection Criteria ... 78
4.2 Comparison A: Zynq versus FPGA ... 80
4.2.1 MicroBlaze Processor ... 80
4.2.2 MicroBlaze MicroController System ... 84
4.2.3 PicoBlaze ... 85
4.2.4 ARM Cortex-M1 ... 85
4.2.5 Other Processor Types ... 85
4.2.6 Summary Comments ... 87
4.3 Comparison B: Zynq versus Standard Processor ... 89
4.3.1 Processor Operation ... 89
4.3.2 Execution Profiling ... 92
4.3.3 Summary Comments ... 94
4.4 Comparison C: Zynq versus a Discrete FPGA-Processor Combination ... 94
4.5 Exploiting the Zynq Architecture and Design Flow ... 96
4.6 Chapter Review ... 98
4.7 References ... 99

CHAPTER 5 Applications and Opportunities ("What can I do with it?") 101
5.1 An Overview of Applications ... 102

5.1.1 Automotive 102
5.1.2 Communications 102
5.1.3 Defence and Aerospace 103
5.1.4 Robotics, Control and Instrumentation 103
5.1.5 Image and Video Processing 104
5.1.6 Medical 105
5.1.7 High Performance Computing (HPC) 105
5.1.8 Others and Future Applications 105
5.2 When Can Zynq Really Help...? 106
5.3 Communications: Software Defined Radio (SDR) 107
5.3.1 Trends in Wireless Communications 107
5.3.2 Introducing Software Defined Radio (SDR) 108
5.3.3 SDR Implementation and Enabling Technologies 108
5.3.4 Cognitive Radio 110
5.4 Smart Systems and Smart Networks 111
5.4.1 What is a Smart System? 111
5.4.2 Examples of Smart Systems 112
5.4.3 Smart Networks: Communications for Smart Systems 114
5.4.4 Related Concepts 115
5.5 Image and Video Processing, and Computer Vision 115
5.5.1 Image and Video Processing 115
5.5.2 Computer Vision 116
5.5.3 Levels of Abstraction 117
5.5.4 Implementation of Image Processing Systems 118
5.5.5 Computer Vision on Zynq Example: Road Sign Recognition 120
5.6 Dynamic System-on-Chip 121
5.6.1 Run Time System Flexibility 121
5.6.2 Dynamic Partial Reconfiguration (DPR) 121
5.6.3 DPR Application Examples 122
5.6.4 Benefits of DPR 124
5.7 Further Opportunities: the Zynq 'EcoSystem' 125
5.7.1 What is the Ecosystem? 125
5.7.2 What is the Opportunity? 126
5.8 Chapter Review 128
5.9 References 128

CHAPTER 6 The ZedBoard 133

6.1 Introducing Zed 133
6.2 ZedBoard System Architecture 134
6.3 The Design Flow for ZedBoard 136
6.4 Getting Started with the ZedBoard 137
6.4.1 What's in the Box? 137
6.4.2 Hardware Setup 137
6.4.3 Programming the ZedBoard 138
6.5 MicroZed 142
6.6 Documentation, Tutorials and Support 142
6.6.1 Documentation about the ZedBoard 142
6.6.2 Demonstrations and Tutorials 143
6.6.3 Online Courseware 143
6.6.4 Other ZedBoard Resources and Support 144
6.7 ZedBoard.org Community 144
6.7.1 Community Projects 144
6.7.2 Blogs 145
6.7.3 Support Forums 145
6.8 Chapter Review 145
6.9 References 146

CHAPTER 7 Education, Research and Training 147

7.1 Technology Trends and SoC Education 148
7.2 University Teaching with Zynq 149
7.2.1 Teaching with Xilinx Tools and Boards 149
7.2.2 Digital Design and FPGA Teaching 150
7.2.3 Computer Science 150
7.2.4 Embedded Systems and SoC Design 150
7.2.5 Algorithm Implementation (e.g. Signal, Image, and Video Processing) 151
7.2.6 Design Reuse 152
7.2.7 New and Emerging Design Methods 153
7.2.8 Sensing, Robotics, and Prototyping 154
7.2.9 An Example Course 154
7.3 Projects and Competitions 155
7.4 Academic Research 156

7.5 The Xilinx University Program (XUP) 158
7.5.1 Introducing XUP 158
7.5.2 Software Support and Licenses 158
7.5.3 XUP Development and Teaching Boards 159
7.5.4 XUP Workshops and Training Materials 159
7.5.5 Technical Support for Universities 160
7.5.6 Eligibility 160
7.5.7 Getting in Touch with XUP 160
7.6 Training for Industry 160
7.6.1 Courses and Authorised Training Providers 160
7.6.2 Other Resources 161
7.6.3 Online Videos 161
7.7 Chapter Review 161
7.8 References 162

CHAPTER 8 First Designs on Zynq 165
8.1 Software Installation Guide 166
8.2 Aims and Outcomes 166
8.3 Overview of Exercise 1A 166
8.4 Overview of Exercise 1B 167
8.5 Overview of Exercise 1C 168
8.6 Possible Extensions 169
8.7 What Next? 169
8.8 References 169

PART B Zynq SoC & Hardware Design 171

CHAPTER 9 Embedded Systems and FPGAs 173
9.1 What is an Embedded System? 173
9.1.1 Applications 174
9.1.2 Generic Embedded System Architecture 175
9.2 Processors 176
9.2.1 Co-processors 177
9.2.2 Processor Cache 178
9.2.3 Execution Cycles 180
9.2.4 Interrupts 183

9.3 Buses ... 184
9.3.1 System and Peripheral Buses ... 185
9.3.2 Bus Masters and Slaves ... 186
9.3.3 Bus Arbitration ... 186
9.3.4 Memory Access ... 187
9.3.5 Bus Bandwidth ... 188
9.4 Chapter Review ... 189
9.5 References ... 189

CHAPTER 10 Zynq System-on-Chip Design Overview 191
10.1 Interfacing and Signals ... 192
10.1.1 PS-PL AXI Interfaces ... 192
10.1.2 PL Co-Processing Interfaces ... 193
10.1.3 Interrupt Interface ... 196
10.2 Interconnects ... 197
10.2.1 Interconnect Features ... 197
10.2.2 Interconnects, Masters and Slaves ... 198
10.2.3 Connectivity ... 199
10.2.4 AXI_HP Interfaces ... 200
10.2.5 AXI_ACP Interface ... 202
10.2.6 AXI_GP Interfaces ... 202
10.3 Memory ... 202
10.3.1 Memory Interfaces ... 203
10.3.2 On-Chip Memory (OCM) ... 208
10.3.3 Memory Map ... 210
10.4 Interrupts ... 211
10.4.1 Interrupt Signals ... 212
10.4.2 Generic Interrupt Controller (GIC) ... 212
10.4.3 Interrupt Sources ... 213
10.4.4 Interrupt Prioritisation and Handling ... 217
10.4.5 Further Reading ... 218
10.5 Chapter Review ... 219
10.6 References ... 219

CHAPTER 11 Zynq System-on-Chip Development 221
11.1 Hardware/Software Partitioning ... 221

11.2 Profiling 224
11.3 Software Development Tools 226
11.3.1 Software Tools 226
11.3.2 Hardware Configuration Tools 227
11.3.3 Software Development Kit (SDK) 228
11.3.4 Microprocessor Debugger 228
11.3.5 Sourcery CodeBench Lite Edition for Xilinx Cortex-A9 Compiler Toolchain 229
11.3.6 Logic Analysers 229
11.3.7 System Generator for DSP 229
11.4 Chapter Review 230
11.5 References 230

CHAPTER 12 Next Steps in Zynq SoC Design 231
12.1 Prerequisites 231
12.2 Aims and Outcomes 232
12.3 Overview of Exercise 2A 232
12.4 Overview of Exercise 2B 232
12.5 Overview of Exercise 2C 233
12.6 Overview of Exercise 2D 234
12.7 Possible Extensions 235
12.8 What Next? 236

CHAPTER 13 IP Block Design 237
13.1 Overview 237
13.2 Industry Trends and Philosophy 239
13.3 IP Core Design Methods 240
13.3.1 HDL 240
13.3.2 System Generator 241
13.3.3 HDL Coder 241
13.3.4 Vivado High-Level Synthesis 243
13.3.5 Choosing the Right IP Creation Method 244
13.4 Simulation and Documentation 244
13.4.1 Simulation 244
13.4.2 Documentation 249
13.5 Chapter Review 252

13.6 References 252

CHAPTER 14 **Spotlight on High-Level Synthesis** **255**
14.1 High-Level Synthesis Concepts 256
14.1.1 What is High-Level Synthesis (HLS)? 256
14.1.2 Motivations for High-Level Synthesis 257
14.1.3 Design Metrics and Hardware Architectures 259
14.2 Development of HLS Tools 260
14.3 HLS Source Languages 262
14.3.1 C 262
14.3.2 C++ 263
14.3.3 SystemC 263
14.3.4 Other Languages for High-Level Synthesis 264
14.4 Introducing Vivado HLS 264
14.4.1 What Does Vivado HLS Do? 264
14.4.2 Vivado HLS Design Flow 267
14.4.3 C Functional Verification and C/RTL Cosimulation 269
14.4.4 Implementation Metrics and Considerations 271
14.4.5 Overview of the High-Level Synthesis Process 272
14.4.6 Solutions: Exploring the Design Space 276
14.4.7 Vivado HLS Library Support 277
14.5 HLS in the Design Flow for Zynq 277
14.6 Chapter Review 278
14.7 References 278

CHAPTER 15 **Vivado HLS: A Closer Look** **281**
15.1 Anatomy of a Vivado HLS Project 282
15.2 Vivado HLS User Interfaces 283
15.2.1 Graphical User Interface 284
15.2.2 Command Line Interface (CLI) 286
15.3 Data Types 287
15.3.1 C and C++ Native Data Types 287
15.3.2 Vivado HLS Arbitrary Precision Data Types for C and C++ 289
15.3.3 Arbitrary Precision Types for SystemC 292
15.3.4 Floating Point Data Types and Operators 294

15.3.5 Validation of Arbitrary Precision Models 294
15.4 Interface Specification and Synthesis 295
15.4.1 C/C++ Function Definition 295
15.4.2 Synthesis of Port-Level Interfaces 296
15.4.3 Port Interface Protocol Types 298
15.4.4 Synthesis of Port Interface Protocols 300
15.4.5 Block-Level Interface Ports and Protocols 302
15.4.6 Interface Synthesis Directives 304
15.4.7 Manual Interface Specification 308
15.5 Algorithm Synthesis 309
15.5.1 Implementation Metrics and Constraints 309
15.5.2 Data Types 311
15.5.3 Pipelining 311
15.5.4 Dataflow 316
15.5.5 Algorithm Case Study: Loops 319
15.5.6 Arrays 327
15.6 Design Evaluation and Optimisation 328
15.6.1 Design Constraints 328
15.6.2 Synthesis Directives 329
15.6.3 Statistics and Reports 329
15.6.4 Design Iterations and Optimisation 329
15.7 Exporting from Vivado HLS 330
15.7.1 Vivado IP Catalog (IP-XACT Format) 330
15.7.2 System Generator for DSP 330
15.7.3 Pcore for XPS 330
15.8 Chapter Review 331
15.9 References 331

CHAPTER 16 Designing With Vivado High Level Synthesis 333

16.1 Prerequisites 333
16.2 Aims and Outcomes 333
16.3 Overview of Exercise 3A 334
16.4 Overview of Exercise 3B 334
16.5 Overview of Exercise 3C 334
16.6 Possible Extensions 335
16.7 What Next? 335

CHAPTER 17 IP Creation 337
17.1 Aims and Outcomes 337
17.2 Overview of Exercise 4A 338
17.3 Overview of Exercise 4B 338
17.4 Overview of Exercise 4C 339
17.5 Possible Extensions 340
17.6 What Next? 341

CHAPTER 18 IP Reuse and Integration 343
18.1 Overview 343
18.2 System Design — A System-Level Approach 344
18.3 IP-XACT 346
18.4 IP Libraries 346
18.4.1 Vivado IP Catalog 346
18.4.2 Third-Party 347
18.4.3 Custom IP 349
18.5 IP Integration 349
18.5.1 IP Integrator 349
18.5.2 IP Packager 349
18.6 Chapter Review 351
18.7 References 351

CHAPTER 19 AXI Interfacing 353
19.1 Development of AXI 353
19.2 Variations of AXI4 354
19.3 AXI Architecture 354
19.3.1 Address Channels 356
19.3.2 Write Data Channel 356
19.3.3 Read Data Channel 356
19.3.4 Write Response Channel 356
19.4 Examples of Applications 356
19.5 AXI Transactions 358
19.5.1 AXI Write-Burst Transaction 358
19.5.2 AXI Read-Burst Transaction 358
19.6 AXI in the Xilinx Toolflow 360
19.7 Summary 364

19.8 References ... 364

CHAPTER 20 **Adventures with IP Integrator** **365**
20.1 Aims and Outcomes ... 366
20.2 Exercise 4A ... 367
20.3 Exercise 4B ... 367
20.4 Exercise 4C ... 368
20.5 Possible Extensions ... 368
20.6 What Next? ... 368

PART C **Operating Systems & System Integration** **369**

CHAPTER 21 **Introduction to Operating Systems on Zynq** **371**
21.1 Why Use an Embedded Operating System? ... 371
21.1.1 Reducing Time to Market ... 371
21.1.2 Make Use of Existing Features ... 372
21.1.3 Reduce Maintenance and Development Costs ... 373
21.2 Choosing the Right Type of Operating System ... 373
21.2.1 Standalone Operating Systems ... 374
21.2.2 Real-Time Operating Systems (RTOS) ... 374
21.2.3 Other Embedded Operating Systems ... 375
21.2.4 Further Considerations ... 377
21.3 Applications ... 377
21.4 Multi-Processor Systems ... 378
21.5 Zynq Operating Systems ... 379
21.5.1 Linux ... 379
21.5.2 RTOS ... 382
21.5.3 Further Operating Systems ... 382
21.6 Chapter Review ... 383
21.7 References ... 383

CHAPTER 22 **Linux: An Overview** **385**
22.1 A Brief History ... 385
22.2 Linux System Overview ... 386
22.3 Licensing ... 387
22.3.1 GNU General Public licence ... 388

22.4 Development Tools and Resources 389
22.4.1 Virtual Machines 389
22.4.2 Version Control 391
22.4.3 Git 392
22.4.4 Debugging Linux 393
22.5 Chapter Review 395
22.6 References 396

CHAPTER 23 The Linux Kernel 397
23.1 Linux Kernel Hierarchy 397
23.2 System Call Interface 398
23.3 Memory Management 400
23.3.1 Virtual Memory 400
23.3.2 High and Low Memory 401
23.4 Process Management 401
23.4.1 Process Representation 402
23.4.2 Process Creation, Scheduling and Destruction 402
23.5 File System 404
23.5.1 Linux File Systems 404
23.5.2 Virtual File System 405
23.6 Architecture-Dependent Code 406
23.7 Linux Device Drivers 406
23.7.1 A Note on Mechanisms Vs. Policies 407
23.7.2 Module/Device Classification 407
23.8 Chapter Review 408
23.9 References 408

CHAPTER 24 Linux Booting 409
24.1 Overview 409
24.2 Stages of the Desktop Linux Boot Process 411
24.2.1 BIOS 411
24.2.2 First-Stage Bootloader (FSBL) 411
24.2.3 Second-Stage Bootloader (SSBL) 412
24.2.4 Kernel 413
24.2.5 Init 413
24.3 Booting Zynq 414

24.3.1 Zynq Boot Files 416
24.3.2 Stage-0 (Boot ROM) 417
24.3.3 Stage-1 (First-Stage Bootloader) 419
24.3.4 Stage-2 (Second-Stage Bootloader) 425
24.4 Chapter Review 425
24.5 References 426

Postscript 427

Glossary **429**

List of Acronyms **439**

Index **451**

Foreword

For over two years now, academics, industry professionals and "makers" worldwide have had access to development boards that use the Zynq®-7000 All Programmable SoC from Xilinx®. These boards including the ZedBoard, the Zc702, Zc706 and others have provided users heretofore unprecedented abilities to build their own customizable System on Chip (SoC) solutions. The Zynq SoC integrates an ARM® dual Cortex®-A9 based processor system with Xilinx 7-series FPGA fabric and by doing so it provides the power, performance and capacity benefits of an ASIC combined with the hardware programmability benefits of an FPGA.

The strong demand for these devices and the first of its kind nature goes hand-in-hand with a demand for supporting collateral in the form of documentation, training, tutorials and guidebooks. As the first book written in the English language, "The Zynq Book" does a remarkable job of filling a critical need, and the team from the University of Strathclyde have put together a very comprehensive book, which covers the essential information every Zynq user needs to know.

The book begins appropriately with an overview of the Zynq device, following up with a description of the ZedBoard. The book then moves very quickly into information needed to build designs targeted for the Zynq family, describing in depth both the development flow for these devices as well as the implications of various design choices. As this is a hybrid device which is both software and hardware programmable, the content comprehensively spans hardware design tools as well as the higher level software tools and flow. A special spotlight on Vivado® High Level Synthesis (HLS) is included, which showcases the productivity benefits offered by HLS as well as the synergy with the high level programming model offered by the Cortex-A9 processors. Of critical importance are the interfaces that connect the Processor System to the Programmable Logic or FPGA. The book does an excellent job of providing an overview of these interfaces and guidance on appropriately configuring them. Finally no book on an SoC is complete without a description of the embedded software run-time environment. The concluding chapter guides the reader through the nuances of booting Linux® on their very own custom SoC.

It is no coincidence that the first English language book on Zynq originates from the University of Strathclyde. Since 2005 Xilinx has worked closely with the Department of Electronic and Electrical Engineering at the University of Strathclyde in Scotland, UK. As the Xilinx Endowed Chair, Professor Bob Stewart played a leading role in developing and disseminating best practices for using Xilinx *All Programmable* technologies in academia and industry. He and his team have created outstanding pedagogical material that has enhanced the educational experience of tens of thousands of students and engineers around the world. The successful relationship undoubtedly reflects the University of Strathclyde's pragmatic, industry-driven ethos and its position as a leading technological institution with an excellent international reputation. Indeed the University of Strathclyde was founded in 1796 by Professor John Anderson, whose bequest was to create "a place of useful learning". It is very clear that the University continues to be a place of *useful learning* and a strong partner with Xilinx for technology education and research. It is also worth saying that we are particularly pleased to be associated with an institution recognized in the UK as University of the Year for 2013, and the UK Entrepreneurial University of the Year for 2014.

This book is indeed a must read item for the first time Zynq user!

Vidya Rajagopalan
Corporate Vice President of Processing, Systems, Software and Applications (PSSA),
Xilinx, Inc.

June 2014.

Acknowledgements

There are a number of people to whom we must extend our sincerest thanks for their support and practical help in producing The Zynq Book.

First of all, it simply would not have come to life without the vision of Patrick Lysaght, Senior Director at Xilinx, whose idea it was to create a book about Zynq. The project soon took on a life of its own, and we realised that there are many interesting things to write about Zynq! The resulting book is the product not just of our own efforts, but the many people who have helped us along the way.

Firstly, nobody could have been more helpful than Cathal McCabe, who has managed the project from Xilinx University Program. His knowledgeable support has been absolutely vital, and we are indebted to him for his patience, thoughtful input, and thoroughness. This book would not have turned out nearly so well without Cathal.

We also greatly appreciate the input of several colleagues from Xilinx: Sagheer Ahmad, Brian Gaide, Austin Lesea, Joshua Lu, Duncan Mackay, Daryl Nees, Stephen Neuendorffer, Parimal Patel, Fernando Martinez Vallina, Tim Vanevenhoven, and Y.C. Wang all gave of their time to read the book and provide constructive criticism. Some of these helpful people also took the trouble to work through the tutorials and share their feedback. It has been a great help to access their experience — thanks guys! We must also thank Barrie Mullins for his help and support.

Our colleagues at University of Strathclyde have been a great help as well, and we would like to thank them for their various inputs to the project. Several of them generously took the time to read and review chapters, or to work through and test tutorials, and provided much thoughtful feedback: many thanks to Douglas Allan, Dani Anderson, Dale Atkinson,

Kenneth Barlee, Iain Chalmers, Fraser Coutts, David Crawford, Sam Edwards, Poppy Harvey, Connor Hughes, Sarunas Kalade, Phil Karagiannakis, David Northcote and Kenneth Osborne. We also appreciate the contributions of all to the general "Zynq-chat" which has now become a feature of our working environment!

Last but not least, we must of course extend our thanks to our family and friends, who accepted with good grace our lack of participation in social activities, particularly as the various deadlines approached (!), and were most generous with their encouragement.

Louise Crockett, Ross Elliot, Martin Enderwitz, Bob Stewart.
June 2014.

1

Introduction

As you might have guessed from the title, this is a book about Zynq! That's Zynq as in the new generation of All-Programmable System-on-Chip (SoC) [10], not to be confused with zinc, or Zn, the chemical element.

In fact, there is a link between the two. Rumour has it that Xilinx gave their new device the name Zynq, because it represents a (processing) element that can be applied to anything. Zynq devices are intended to be flexible and form a compelling platform for a wide variety of applications, just as the metal zinc can be mixed with various other metals to form alloys with differing desirable properties.

The defining feature of Zynq is that it combines a dual-core ARM Cortex-A9 processor [1] with traditional Field Programmable Gate Array (FPGA) logic fabric. Although dedicated processors have been coupled with FPGAs before, it has never been quite the same proposition. In Zynq, the ARM Cortex-A9 is an application grade processor, capable of running full operating systems such as Linux, while the programmable logic is based on Xilinx 7-series FPGA architecture [5], [7]. The architecture is completed by industry standard AXI interfaces, which provide high bandwidth, low latency connections between the two parts of the device [8]. This means that the processor and logic can each be used for what they do best, without the overhead of interfacing between two physically separate devices. Meanwhile, benefits arising from simplifying the system to a single chip include reductions in physical size and overall cost.

As will be evident from the very size of this book, Zynq is about more than just the silicon. It is compelling because the software development tools, design flow, and standards-focussed integration methods are tailored to the requirements of Zynq-based system design [6]. The facility to create better designs more quickly is of interest to everyone! This book will introduce all of the topics necessary to get started in earnest, and also provide some practical tutorials to guide new users through the design flow and procedures.

We hope that you find the book informative and useful, whatever your level of prior experience. A particular target is to provide an accessible introduction for those new to the area.

1.1. System-on-Chip with Zynq

Given that we have already described Zynq as a System-on-Chip, an obvious first question would be "*What is an SoC?*".

As you may be aware, the concept has been around for a while; the implication is that a single silicon chip can be used to implement the functionality of an entire system, rather than several different physical chips being required. In the past, the term SoC has usually referred to an Application Specific Integrated Circuit (ASIC), which can include digital, analogue and radio frequency components, together with mixed signal blocks for implementing analogue-to-digital and digital-to-analogue converters (ADCs and DACs). Focussing on the digital aspect for a moment, an SoC can combine all aspects of a digital system: processing, high-speed logic, interfacing, memory, and so on. All of these functions might otherwise be realised using physically separate devices, and combined together into a system at the Printed Circuit Board (PCB) level. The SoC solution is lower cost, enables faster and more secure data transfers between the various system elements, has higher overall system speed, lower power consumption, smaller physical size, and better reliability. In fact there are a number of compelling reasons for preferring SoCs over discrete component equivalent systems! For a simple graphical comparison of the *system-on-a-board* and the *system-on-chip*, consider Figure 1.1.

The major disadvantages of ASIC-based SoCs are (i) development time and cost, and (ii) lack of flexibility. The non-recurring engineering effort (and cost) of developing an ASIC are significant, making this type of SoC suitable only for high-volume markets where there is no requirement for future upgrades. Representative examples of ASIC-based SoCs are the integrated processors found in PCs, tablets, and smartphones; these typically comprise at least two processor cores, memory, graphics, interfacing, and other functions [4], and are manufactured in high volume for products with a limited lifetime.

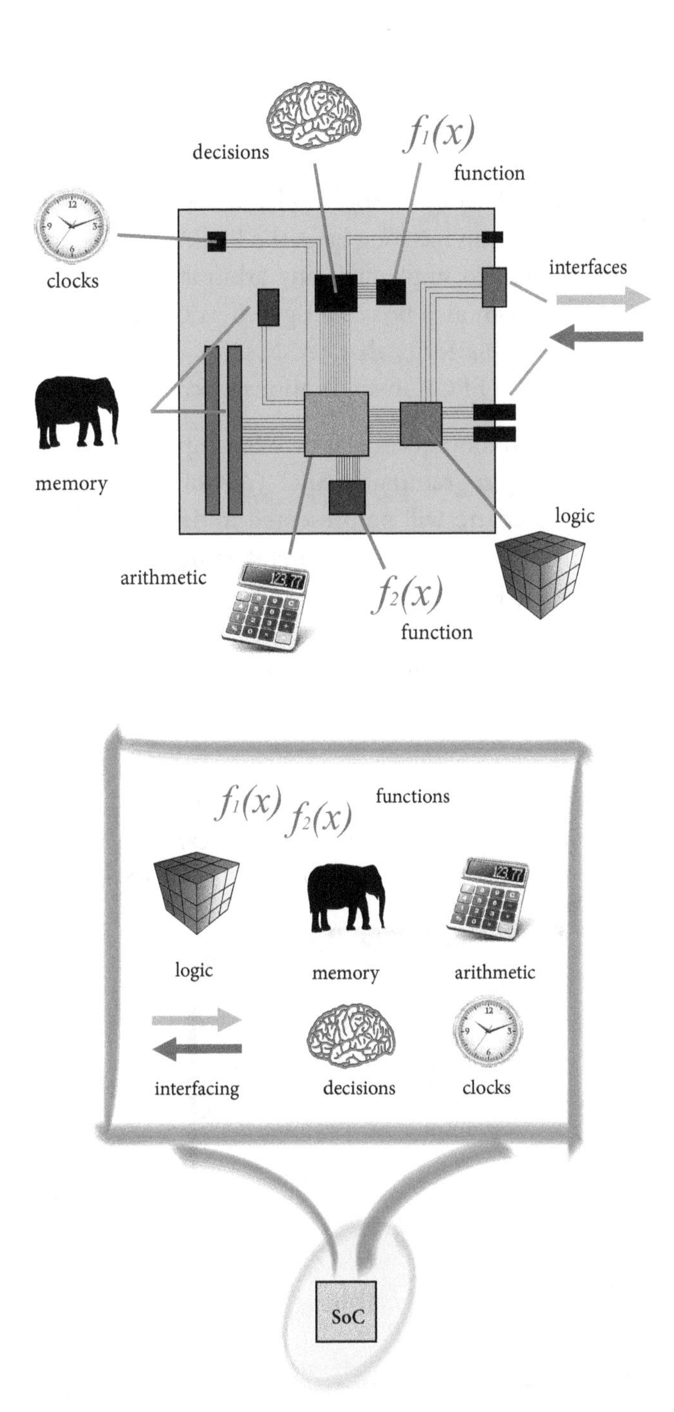

Figure 1.1: Comparison of System-on-a-Board (top) and System-on-Chip (bottom)

The limitations of ASIC SoCs render them incompatible with a significant number of applications, particularly where fast time-to-market, flexibility, and upgrade-ability are of key importance. They also constitute a poor solution for low or medium volume markets. There is a clear need for a more flexible solution, and this is what motivates the System-on-*Progammable*-Chip, a specific flavour of SoC implemented on a programmable, reconfigurable device. The natural solution has long been the FPGA. FPGAs are inherently flexible devices that can be configured to implement any arbitrary system, including embedded processors if needed. FPGAs can also be reconfigured as often as desired, thus offering a more fundamentally flexible platform than ASICs for implementing SoCs. There is virtually no risk in deploying an FPGA in applications where system upgrades are required.

Now, Zynq provides an even more ideal platform for implementing flexible SoCs: Xilinx markets the device as an 'All-Programmable SoC' (APSoC), which perfectly captures its capabilities. The Zynq architecture will be presented in detail in Chapter 2, but first it is useful to introduce a high level model of its architecture (Figure 1.2). Note that Zynq comprises two main parts: a Processing System (PS) formed around a dual-core ARM Cortex-A9 processor, and Programmable Logic (PL), which is equivalent to that of an FPGA. It also features integrated memory, a variety of peripherals, and high-speed communications interfaces.

The PL section is ideal for implementing high-speed logic, arithmetic and data flow subsystems, while the PS supports software routines and/or operating systems, meaning that the overall functionality of any designed system can be appropriately partitioned between hardware and software. Links between the PL and PS are made using industry

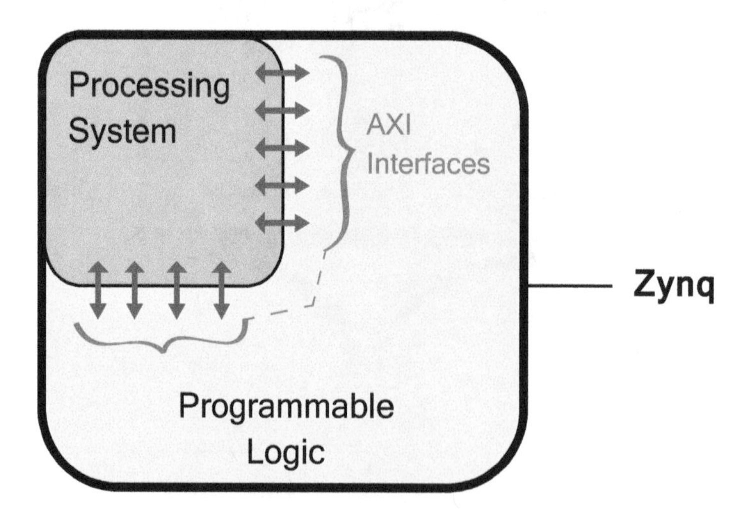

Figure 1.2: A simplified model of the Zynq architecture

standard Advanced eXtensible Interface (AXI) connections [8]. Further details of each of these features will be presented as the book progresses.

1.2. Simple Anatomy of an Embedded SoC

At this early stage, it is useful to set out a basic model for the types of digital systems to be discussed in this book. These will be systems incorporating a processor, memories, and peripherals, along with buses connecting the various elements together. (This represents the ***hardware system***, and here we consider a very simplified architecture — more details will be added in later chapters.) This model of the hardware system is shown in Figure 1.3.

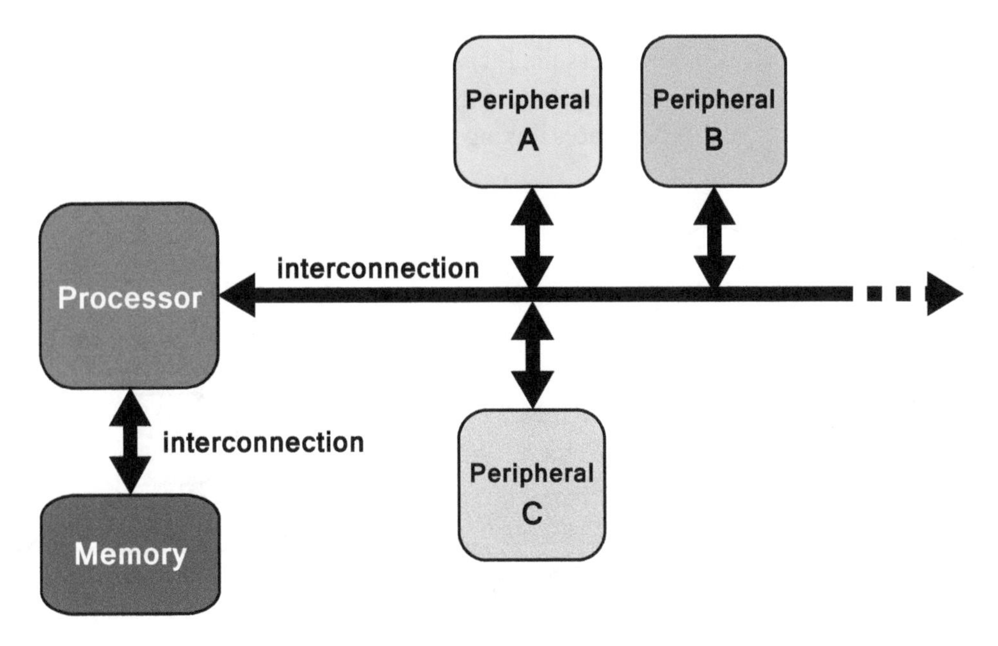

Figure 1.3: The hardware system architecture of an embedded SoC (simplified)

The processor can be regarded as the central element of the hardware system. The ***software system*** (a software 'stack') is run on the processor, comprising applications (usually based on an Operating System (OS)), and with a lower layer of software functionality for interfacing with the hardware system. Communication between system elements takes place via *interconnections*. These may be in the style of direct, point-to-point links, or buses serving multiple components. In the latter case, a protocol is required to manage access to the bus. Note that, although a single bus with connected peripherals is shown in Figure 1.3, a processor may serve several connected buses.

Peripherals are functional components residing away from the processor, and in general these perform one of three functions: (i) coprocessors — elements that supplement the primary processor, usually optimised for a certain task; (ii) cores for interacting with external interfaces, e.g. connecting to LEDs and switches, codecs, etc.; and (iii) additional memory elements. Later in the book, we will consider peripherals in more detail, but at this point it is useful to regard them as discrete functional blocks that can be designed, tested and integrated into a system, and also 'packaged' for later reuse.

Figure 1.4 provides a view of the hardware system shown in Figure 1.3, mapped to the Zynq device (depicted in Figure 1.2). The architectures of both have been substantially simplified, but the objective at this stage is to provide a high level clarification of how embedded SoCs map to Zynq devices. The PS has a fixed architecture and hosts the processor and system memory, whereas the PL is completely flexible, giving the designer a 'blank canvas' to create custom peripherals, or to reuse standard ones. The interconnections are implemented via AXI interfaces linking the PS and PL.

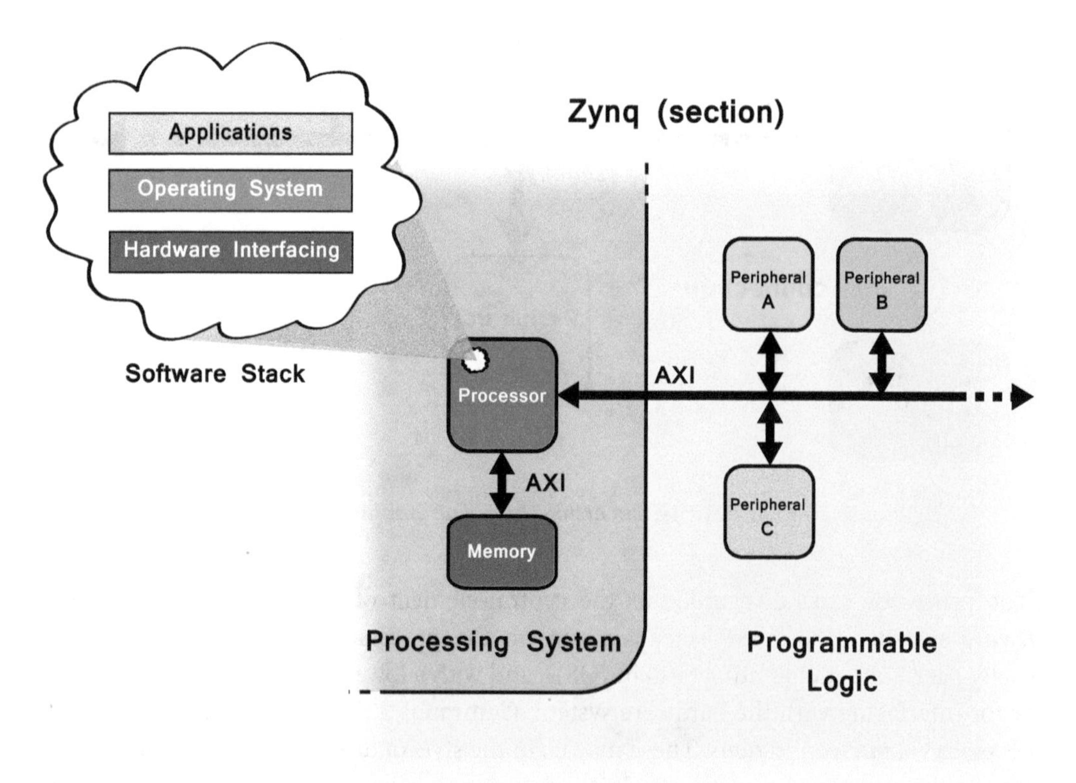

Figure 1.4: Relationship of the software system, hardware system, and Zynq architecture

The software system can also be seen on the left hand side of Figure 1.4. Software is hosted on the processor, which here is the ARM Cortex-A9, residing within the Zynq PS. It comprises a hierarchy of software elements, and this aspect will also be expanded upon later in the book.

1.3. Design Reuse

The development of a complete embedded system is a significant design task, and there are particular advantages to undertaking the design on a platform such as an FPGA or Zynq device, which make the process more straightforward. The underlying PL hardware is structured, and its performance characteristics are well known and integrated into the software development tools. Moreover, given this stable, common development platform, there is huge scope for design reuse. Intellectual Property (IP) functional blocks — corresponding to the peripheral components seen in Figure 1.4 — can be sourced from Xilinx libraries (provided with the design tools), reused from previous projects, or brought in from third parties or open source repositories, before being integrated together to form the system design.

Zynq is an SoC, and a wide variety of standard IP is available, meaning that there is no need to redesign these components. By raising the abstraction level in this way, and reusing components in the form of pre-tested and verified IP, development can be accelerated and costs can be lowered. As the popular saying goes, "*Why reinvent the wheel?*".

Given its importance to the SoC design philosophy, one of the key themes of this book will be design reuse. We will consider the various sources of IP, including Xilinx libraries, mechanisms for generating one's own IP, and sources of third party IP. Of course, it is not enough simply to attain these blocks — they must also be integrated into the system and appropriate connections and interactions established; therefore we also discuss the specific tools and methods required for this *IP integration* aspect of the design process. Finally, with a view to reuse and sharing of design elements, the 'packaging' of IP into the industry standard IP-XACT format will also be covered. These topics are the focus of Chapters 13 and 18, respectively.

1.4. Raising the Abstraction Level

A recurring trend in various software and hardware design processes is that of raising the level of abstraction. The motivation is clear: if the designer can effectively create systems with a lower requirement for explicit design input, while also supporting robust test processes, then there is great potential to accelerate the design process.

In terms of FPGA and Zynq design, advances in High Level Synthesis (HLS) mean that designers can create system components by specifying them using less detail than traditional, Register Transfer Level (RTL) methods, and instead rely upon the design tools to infer logic and optimise where possible, in accordance with user-supplied direction. Naturally this places a degree of trust in the development tools, which must be robust and produce repeatable and reliable designs. To meet this need, Xilinx has introduced the Vivado HLS tool, a high level synthesis development tool which specifically targets Xilinx devices. We will introduce Vivado HLS and associated design methods later in the book.

1.5. SoC Design Flow

A multitude of different models have been proposed for the SoC design flow with varying levels of complexity, but initially we aim to define the design flow for SoC development (as applied to Zynq) in very simple terms. The basic stages are shown in Figure 1.5.

Each of these will be expanded upon and discussed in greater detail later in the book, but a brief definition will suffice for now.

Naturally, as in any design project, the first stage is to define the desired behaviours of the system, i.e. to create an appropriate specification from a set of requirements. This is depicted as the starting point at the top of the diagram, and it forms the basis of the system design that is subsequently developed.

As mentioned earlier in this chapter, the Zynq architecture combines an ARM processor (for software elements of the designed system) with FPGA fabric (predominantly for hardware elements of the system, although additional processors can also be implemented here too, if desired). A key element of the system design stage, which comes next, is therefore to partition the intended functionality appropriately between software and hardware, and to define the interfaces between the two sections. Of course, it is possible that this partitioning will subsequently be adjusted as the designers iterate the system towards completion.

Having partitioned the system, software and hardware development can then progress in parallel, to a large extent. In terms of hardware development, the task is to identify the necessary functional blocks to achieve the design, and to thereafter assemble them through some combination of design reuse and new IP development, and make appropriate connections between the blocks. Similarly, the software aspect of the project can be realised through developing custom code or by reusing pre-existing software. Verification of both software and hardware will be required, and this forms an integral and important part of the process.

Lastly, the hardware and software elements of the system must be integrated according to the interfaces defined at the specification stage, and further 'whole system' testing undertaken.

The design flow will be discussed further in Chapter 3, and Zynq-based SoC design in Chapters 10 and 11.

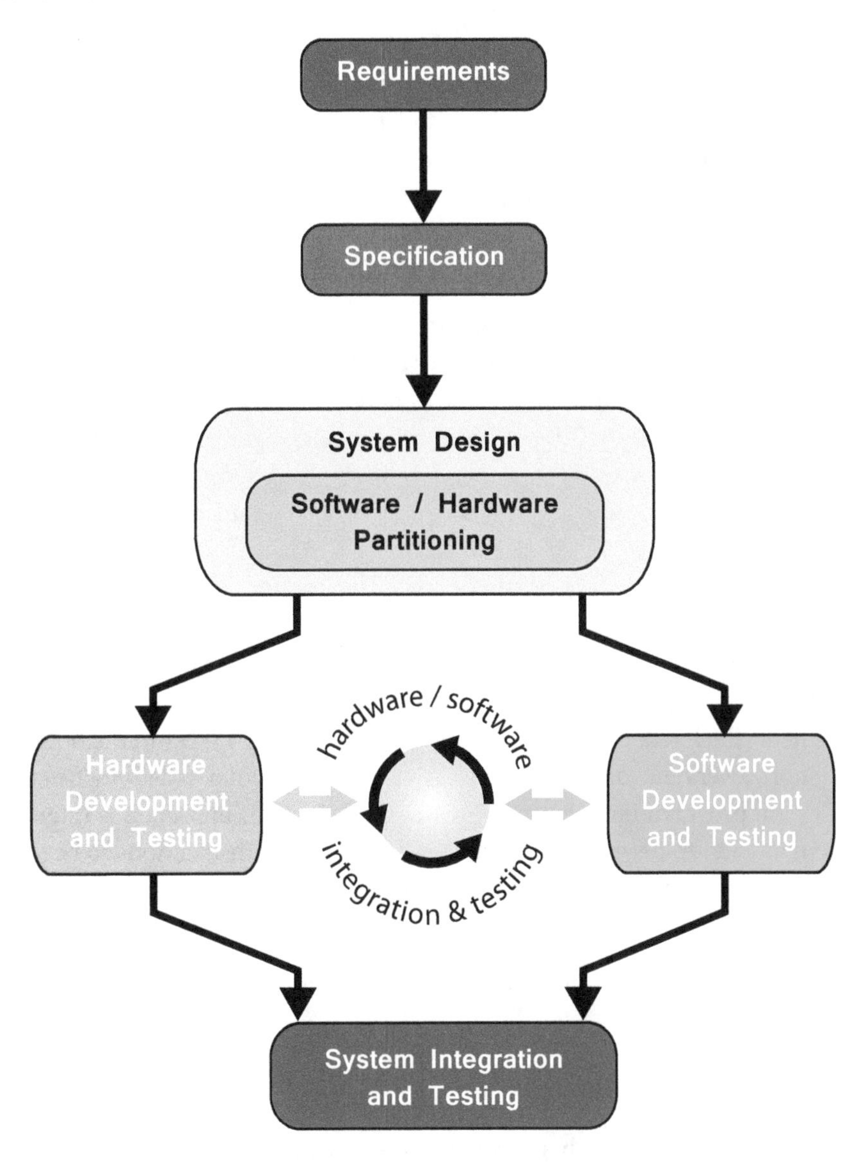

Figure 1.5: A basic model of the design flow for Zynq SoC

1.6. Practical Elements

This book indirectly features a set of practical exercises that can be followed alongside the main text. Detailed instructions and all necessary resource files for these are hosted on a companion website, while the book itself includes a concise overview of each exercise, confirming its purpose and the key points of note. The detailed instructions on the web are updated asynchronously to the book, in order to maintain compatibility with the latest releases of Xilinx development tools.

The exercises can be followed with the aid of the Xilinx Vivado Design Suite (the free WebPACK® edition or higher [9]), and the Zynq-based *ZedBoard* [11]. It is also possible to interpret the instructions for an alternative board, if you prefer. Should you already have them, the more fully specified Xilinx *Embedded* or *System* editions of Vivado would offer enhanced functionality, but WebPACK provides sufficient facilities to get started.

The ZedBoard is shown in Figure 1.6, and this is a good, representative example of a Zynq development board [11]. There are several Zynq boards available and, like this one, most provide a variety of peripheral interfaces and connectors including ethernet, audio and video. The availability of IP blocks and reference designs supporting these features makes it relatively quick and simple to create first interactive designs. Selected Zynq boards are introduced in Chapter 3, with further details of the ZedBoard provided in Chapter 6.

1.7. About This Book

This book provides both descriptive and hands-on introductions to working with the Zynq APSoC. Most of the chapters are standard chapters ("for reading") while there are also some others which relate to practical exercises ("for doing"). Within the book itself, we provide a short overview of each practical element, while detailed instructions and resource files are provided on a companion website [12]. This approach is intended to be more convenient, and to permit the detailed, tool-dependent instructions to be updated in response to revisions of the Xilinx software development tools, and asynchronously to the book.

The remainder of the book is organised into three, themed sections as follows:

Part A provides introductory information about the Zynq device, its associated toolflow, and the ZedBoard. Furthermore, Zynq is compared with alternative devices, and applications are explored. Part A also includes a dedicated section on Zynq and associated SoC concepts in the context of research, university-based teaching, and training catering for the general community.

Figure 1.6: A view of the ZedBoard

Part B is an in-depth review covering various aspects of the Xilinx SoC development using Zynq. It includes concepts of embedded systems design, IP block creation and integration, and software-hardware codesign. There is also a special 'spotlight' chapter, focussing on the increasingly important area of HLS for rapid development of IP.

Part C focuses on OSs for Zynq SoC development. Applications, motivations, trade-offs, OS alternatives, and features are all reviewed and discussed. Further, there is an in-depth look at the practical issues of deploying Linux on Zynq, and thus forming an embedded system in combination with PL-based elements.

There is also a **Glossary** and **List of Acronyms** at the back of this book. A large number of terms and acronyms are used, and these should provide useful references.

Next we continue to Chapter 2, in which a more detailed introduction to the Zynq device architecture is provided.

1.8. References

Note: All URLs last accessed June 2014.

[1] ARM, "The ARM Cortex-A9 Processors", White paper, v2.0, September 2009.
Available: http://www.arm.com/files/pdf/ARMCortexA-9Processors.pdf

[2] Avnet website.
Available: http://www.avnet.com

[3] Digilent website.
Available: http://www.digilentinc.com

[4] M. Dixon, P. Hammarlund, S. Jourdan and R. Singhal, "The Next Generation Intel Core Microarchitecture", *Intel Technology Journal*, Vol. 14, Issue 3, 2010. pp. 8 - 29.

[5] M. Santarini, "Xilinx Redefines State of the Art With New 7 Series FPGAs", *Xcell Journal*, Third Quarter 2010, pp. 6 - 11.
Available: http://www.xilinx.com/publications/archives/xcell/Xcell72.pdf

[6] M. Santarini, "Xilinx Unveils Vivado Design Suite for the Next Decade of 'All Programmable' Devices", *Xcell Journal*, Second Quarter 2012, pp. 8 - 13.
Available: http://www.xilinx.com/publications/archives/xcell/Xcell79.pdf

[7] Xilinx, Inc., "7 Series FPGAs Overview", Product Specification, DS180, v.1.15, February 2014.
Available: http://www.xilinx.com/support/documentation/data_sheets/ds180_7Series_Overview.pdf

[8] Xilinx, Inc., "AXI Reference Guide", UG761, v14.3, November 2012.
Available: http://www.xilinx.com/support/documentation/ip_documentation/axi_ref_guide/latest/ug761_axi_reference_guide.pdf

[9] Xilinx, Inc., "Vivado Design Suite", webpage.
Available: http://www.xilinx.com/products/design-tools/vivado/index.htm

[10] Xilinx, Inc., "Zynq 101" webpage.
Available: http://www.xilinx.com/products/silicon-devices/soc/zynq-7000/zynq-101.html

[11] ZedBoard website.
Available: http://www.zedboard.org/

[12] Zynq Book companion website.
Available: http://www.zynqbook.com

PART A

Getting to Know Zynq

2

The Zynq Device ("*What is it?*")

If you're reading this book, it is likely that you have some background in developing systems using FPGAs, or processors, or both. As set out at the start of the book, the Zynq is a new kind of device which combines both FPGA fabric and a capable applications processor, and therefore its features, capabilities, and potential applications are somewhat different to those of an FPGA or processor in isolation.

Over this and the next few chapters, we will take a more detailed look at the Zynq from a variety of perspectives, in the process addressing basic but important questions such as "*What is it?*", "*How do I work with it*?" and "*Why do I need Zynq?*". This chapter focuses on the first of these questions, and introduces the Zynq architecture.

Referring back to Figure 1.2 on page 4, the general architecture of the Zynq comprises two sections: the Processing System (PS), and the Programmable Logic (PL). These can be used independently or together, and in fact the power circuitry is configured with separate domains for each, enabling either the PS or PL to be powered down if not in use. However, the most compelling use model for Zynq is when both of its constituent parts are used in conjunction, and therefore it is important to appreciate the structure of both sections, as well as the interfaces between them.

The architecture of Zynq is reviewed over the remainder of this chapter, starting with the PS. Extended information can be found in the *Zynq-7000 Technical Reference Manual* [33].

2.1. Processing System

All Zynq devices have the same basic architecture, and all of them contain, as the basis of the processing system, a dual-core ARM Cortex-A9 processor. This is a 'hard' processor — it exists as a dedicated and optimised silicon element on the device.

For comparison purposes, the alternative to a hard processor is a 'soft' processor like the Xilinx MicroBlaze, which is formed by combining elements of the programmable logic fabric [27]. The implementation of a soft processor is therefore the equivalent of any other IP block deployed in the logic fabric of an FPGA. In general, the advantage of soft processors is that the number and precise implementation of processor instances is flexible. On the other hand, hard processors can achieve considerably higher performance, as is true of Zynq's ARM processor. This aspect will be discussed in more detail in Chapter 4.

It is worth noting that one or more MicroBlaze soft processors can be used within the PL portion of the Zynq, to operate in conjunction with the ARM processor. The MicroBlaze instances may have, for example, the role of co-ordinating specific low level functions within the system; less demanding tasks which can be delegated away from the main ARM Cortex-A9 processors to enhance overall performance. In other words, the presence of the ARM processor in the system does not preclude the use of soft processors, and indeed many applications may benefit from employing a processing model such as this, which uses both types.

Figure 2.1 confirms the positions of the ARM and MicroBlaze processors on the Zynq device; the ARM as a dedicated resource, and the MicroBlaze located in the logic fabric.

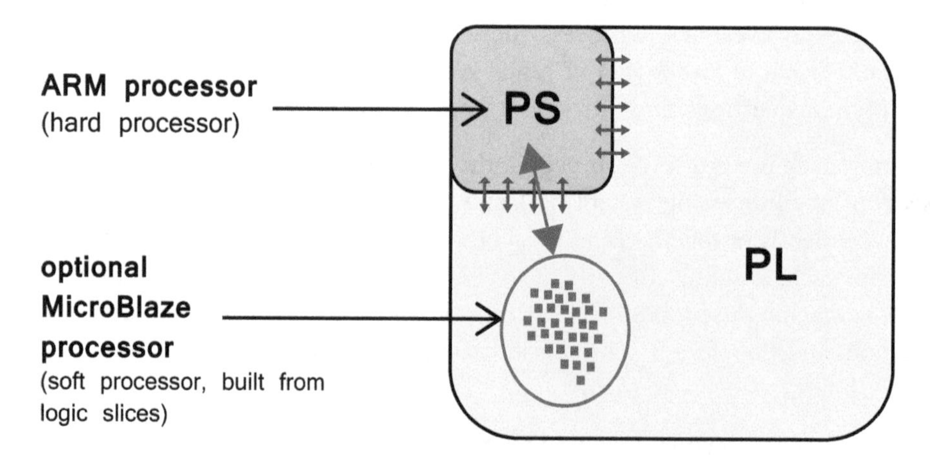

Figure 2.1: Locations of hard (ARM Cortex-A9) and soft (MicroBlaze) processors on a Zynq device

Importantly, the Zynq processing system encompasses not just the ARM processor, but a set of associated processing resources forming an Application Processing Unit (APU), and further peripheral interfaces, cache memory, memory interfaces, interconnect, and clock generation circuitry [8]. A block diagram showing the architecture of the PS is shown in Figure 2.2, where the APU is highlighted.

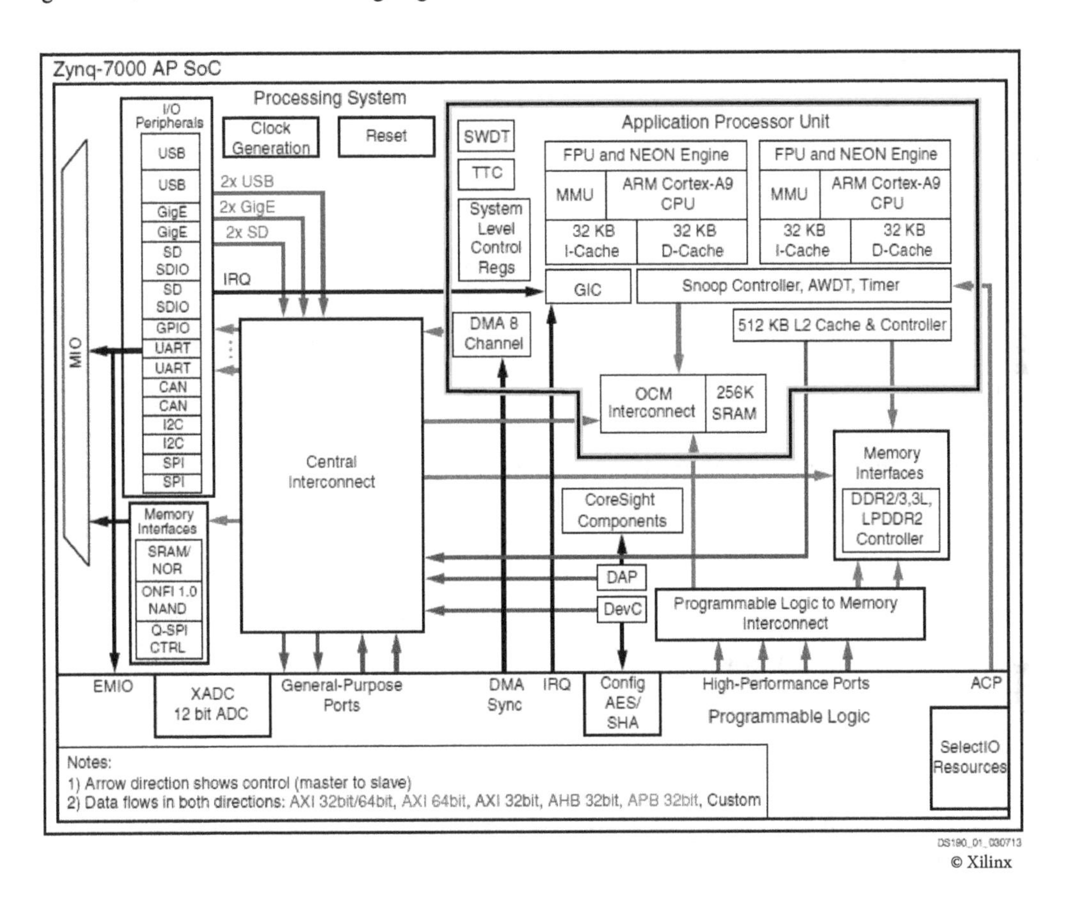

© Xilinx

Figure 2.2: The Zynq Processing System

2.1.1. Application Processing Unit (APU)

A simplified block diagram of the APU is shown in Figure 2.3. The APU is primarily comprised of two ARM processing cores, each with associated computational units: a NEON™ Media Processing Engine (MPE) and Floating Point Unit (FPU); a Memory Management Unit (MMU); and a Level 1 cache memory (in two sections for instructions and data). The APU also contains a Level 2 cache memory, and a further On Chip Memory

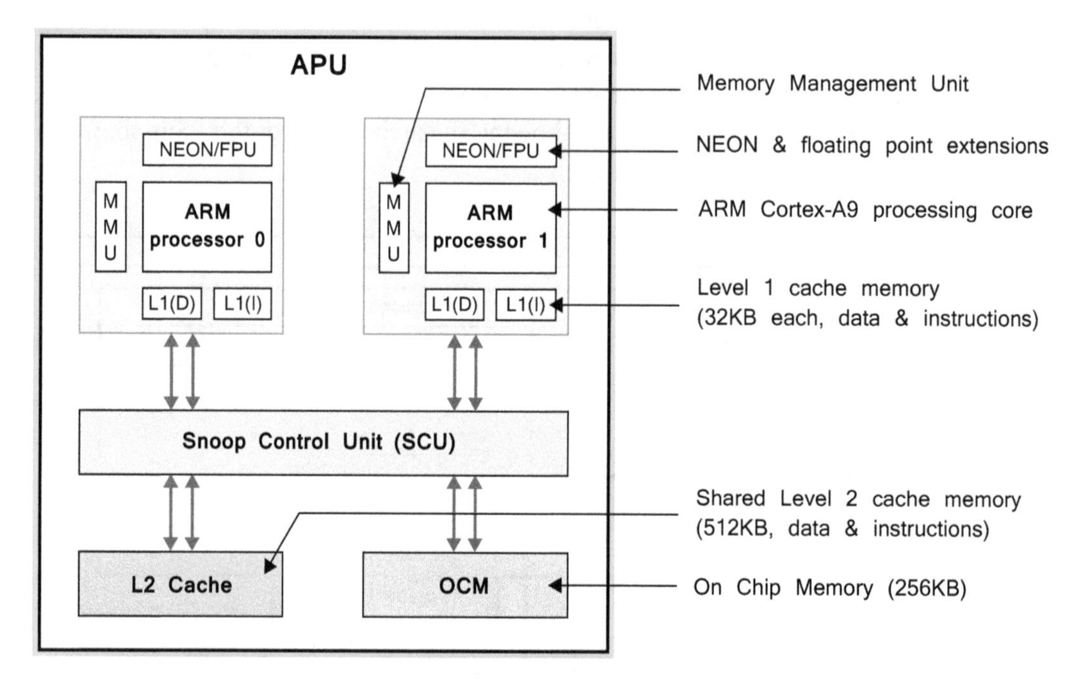

Figure 2.3: Block diagram of the application processing unit (simplified)

(OCM). Finally, a Snoop Control Unit (SCU) forms a bridge between the ARM cores and the Level 2 cache and OCM memories; this unit also has some responsibility for interfacing with the PL, which is not depicted here.

The ARM Cortex-A9 can operate at up to 1GHz, depending on the particular Zynq device (to be detailed in Section 2.5). Each of the two cores has separate Level 1 caches for data and instructions, both of which are 32KB; as in the general case, this permits local storage of frequently required data and instructions for fast access times and optimal processor performance. The two cores additionally share a larger Level 2 cache of 512KB for instructions and data, and there is a further 256KB of on-chip memory within the APU. The primary role of the MMU is to translate between virtual and physical addresses (refer to Section 23.3 for a discussion of memory management in the context of a Linux operating system).

The Snoop Control Unit undertakes several tasks relating to interfacing between the processors and Level 1 and 2 cache memories ('snooping' is one of several mechanisms for ensuring cache coherency, i.e. managing the consistency of data across shared cache resources [13]). The SCU is responsible for maintaining memory coherency between the processor data cache memories, which are marked as *L1(D)* in Figure 2.3, and the shared Level 2 cache memory. It also initiates and controls access to the Level 2 cache, arbitrating between requests from the two cores where necessary [8], [33]. The SCU additionally manages transactions that take place between the PS and PL via the Accelerator Coherency Port (ACP); not shown in the simplified APU diagram (Figure 2.3), but visible at the right hand side of Figure 2.2. Timers and an interrupt controller are further functional blocks located in the APU; more information about these can be found in [8].

From a programming perspective, support for ARM instructions is provided via the Xilinx *Software Development Kit* (SDK) which includes all necessary components to develop software for deployment on the ARM processor. The compiler supports the *ARM* and *Thumb*® instruction sets (16-bit or 32-bit), along with 8-bit Java bytecodes (used for Java Virtual Machines) when in the appropriate state. The reader is directed to [5] for further information about instruction set options and details.

As additional functionality to the main ARM processor, the NEON engine provides Single Instruction Multiple Data (SIMD) facilities to enable strategic acceleration of media and DSP type algorithms [9]. NEON instructions are an extension to the standard ARM instruction set, and can either be used explicitly, or by ensuring that the C code follows an expected form and thus allows NEON operations to be inferred by the compiler [15]. As the SIMD term suggests, the NEON engine can accept multiple sets of input vectors, upon which the same operation is performed simultaneously to provide a corresponding set of output vectors. This style of computation caters well to applications like image and video processing, which operate on a large number of data samples (pixels) simultaneously, and inherently parallel, generic signal processing functions such as Finite Impulse Response (FIR) filters and Fast Fourier Transforms (FFTs).

The computation of the NEON engine is depicted in Figure 2.4. There are two input registers, A and B, each of which contain a set of N individual input vectors. A single defined operation is performed between the N sets of input vectors to produce a corresponding set of output vectors which are written to the output register. The size of the vectors can vary, as can the number of vectors comprising each register; the important feature is that each 'lane' produces results arising from the same operation, which is performed on several different sets of inputs at the same time; hence the term *single instruction multiple data*.

NEON supports a variety of data types including signed and unsigned integers, single precision floating point, and half-precision floating point; however, double precision is not supported [9]. The floating point unit (which does not possess SIMD capabilities) is required if double precision computation is needed.

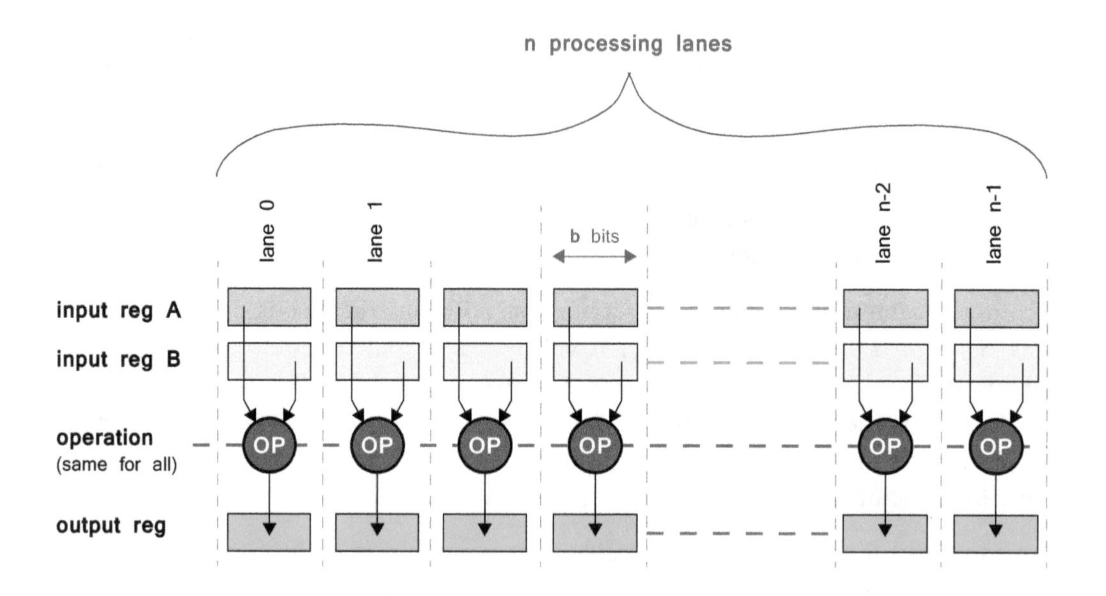

Figure 2.4: Single Instruction Multiple Data (SIMD) processing in the NEON MPE

In addition to NEON, there are also extensions for the Floating Point Unit (FPU). These are referred to as 'Floating Point Extensions', or sometimes 'VFP Extensions' (Vector Floating Point) for historical reasons. The unit provides hardware acceleration of floating point operations in compliance with the IEEE 754 standard [11], and supports single and double precision formats, with some additional support for half-precision and integer conversion. There are a small number of restrictions in terms of the functionality supported, as detailed in [7] and [5], but these do not affect compliance with the standard.

2.1.2. A Note on the ARM Model

It is useful to be aware of the processor licensing model used by ARM Limited, and the implications for you as a Zynq user when referring to Xilinx and ARM documentation.

ARM's business model is to license Original Equipment Manufacturers (OEMs), such as Xilinx, to utilise ARM processor IP within the devices they develop (in this case, Zynq). The Zynq includes the Cortex-A9, which is one of a range of available processors, and this is based on a specific profile (A) of a specific architecture (ARM v7). Even having selected the Cortex-A9 processor, the exact implementation can still be customised by the OEM for inclusion in the designed product, as was the case when Xilinx designed Zynq. The reader may wish to refer to [4], where a helpful overview of this structure and methodology is given.

As there is flexibility in the configuration of the processor IP ARM provides, there is not always a direct correspondence between the ARM and Xilinx documentation. For example, the ARM Cortex-A9 can have between 1 and 4 processing cores; when designing the Zynq device, Xilinx have specified a configuration which includes 2 cores. There are also other elements that can be configured (for example, the Level 1 cache memory size can be specified as 16KB, 32KB or 64 KB) and Xilinx have selected 32KB. Finally, there are optional extensions, and notably Xilinx have chosen to include the NEON and FPU extensions in the Zynq.

The ARM documentation describes the APU in detail, but with a level of generality, and with separate manuals for the Cortex-A9 core, optional extensions, and also the architecture on which it is based. Understanding the implementation in depth therefore may require consultation with several ARM manuals, and with knowledge of the parameters of the Zynq implementation; meanwhile, specifics relating to the Zynq configuration are confirmed in the Xilinx documentation.

Note that the Zynq-7000 specifically uses the *r3p0* revision of the ARM Cortex-A9, which is based on the ARM v7-A architecture. This is significant when referring to ARM documentation, as different versions of the manuals are available corresponding to the processor revisions.

2.1.3. Processing System External Interfaces

The Zynq PS features a variety of interfaces, both between the PS and PL, and between the PS and external components, as shown in Figure 2.2. In this section, we specifically cover the external interfaces; PS-PL interfacing will be covered later, in Section 2.3.

Communication between the PS and external interfaces is achieved primarily via the Multiplexed Input/Output (MIO), which provides 54 pins of flexible connectivity, meaning that the mapping between peripherals and pins can be defined as required. Certain connections can also be made via the Extended MIO (EMIO), which is not a direct path from the PS to external connections, but instead passes through and shares the I/O resources of the PL [30]. These are both shown at the left hand side of Figure 2.2. The EMIO can be used when extension beyond 54 pins is required, or as a method of interfacing the PS with an IP block implemented in the PL. EMIO is discussed further in Section 2.3.3.

The available I/O includes standard communications interfaces, and General Purpose Input/Output (GPIO) which can be used for a variety of purposes including simple buttons, switches, and LEDs. The complete set of I/O peripheral interfaces is reviewed in Table 2.1, where the names stated in the left hand column correspond to the shortened versions used within the Xilinx development tools. Note that there are two instances of each type of communications interface.

Extensive further information about each of these interfaces is available in the *Zynq-7000 Technical Reference Manual* [33].

Table 2.1: List of I/O Peripheral Interfaces

I/O Interface	Description
SPI (x2)	Serial Peripheral Interface [10] *De facto standard for serial communications based on a 4-pin interface. Can be used either in master or slave mode.*
I2C (x2)	I^2C bus [14] *Compliant with the I2C bus specification, version 2. Supports master and slave modes.*
CAN (x2)	Controller Area Network *Bus interface controller compliant with ISO 118980-1, CAN 2.0A and CAN 2.0B standards.*
UART (x2)	Universal Asynchronous Receiver Transmitter *Low rate data modem interface for serial communication. Often used for Terminal connections to a host PC.*
GPIO	General Purpose Input/Output *There are 4 banks GPIO, each of 32 bits.*
SD (x2)	*For interfacing with SD card memory.*
USB (x2)	Universal Serial Bus *Compliant with USB 2.0, and can be used as a host, device, or flexibly ("on-the-go" or OTG mode, meaning that it can switch between host and device modes).*
GigE (x2)	Ethernet *Ethernet MAC peripheral, supporting 10Mbps, 100Mbps and 1Gbps modes.*

2.2. Programmable Logic

The second principal part of the Zynq architecture is the programmable logic. This is based on the Artix®-7 and Kintex®-7 FPGA fabric, and is reviewed over the coming pages.

2.2.1. The Logic Fabric

The PL part of the Zynq device is depicted in Figure 2.5, with various features highlighted. The PL is predominantly composed of general purpose FPGA *logic fabric*, which is composed of *slices* and *Configurable Logic Blocks (CLBs)*, and there are also *Input/ Output Blocks (IOBs)* for interfacing. (Note that these are all Xilinx-specific terms.)

Next, each of the labelled features in Figure 2.5 will be explained. Note that there are also some other coloured blocks shown, and these will be reviewed in later sections.

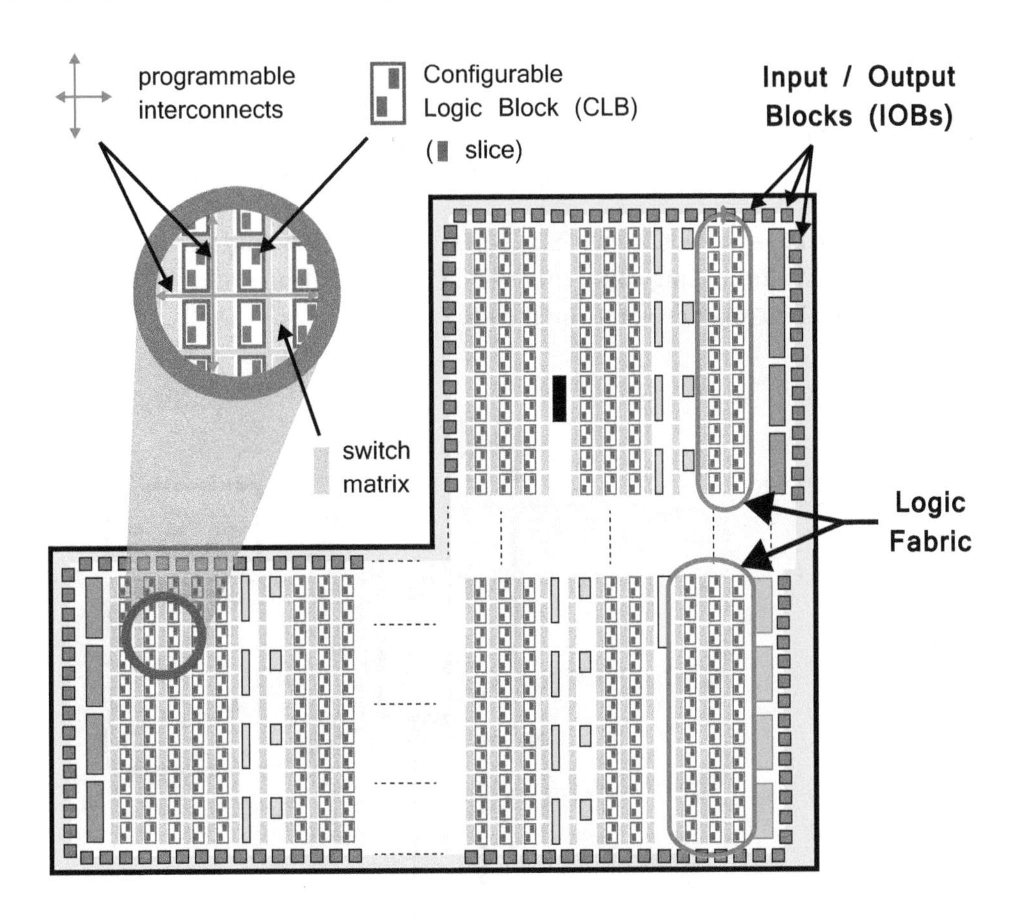

Figure 2.5: The logic fabric and its constituent elements

Features of the PL (shown in Figure 2.5) can be summarised as follows:

- ***Configurable Logic Block (CLB)*** — CLBs are small, regular groupings of logic elements that are laid out in a two-dimensional array on the PL, and connected to other similar resources via programmable interconnects. Each CLB is positioned next to a *switch matrix* and contains two logic *slices*, as shown in Figure 2.6.

- ***Slice*** — A sub-unit within the CLB, which contains resources for implementing combinatorial and sequential logic circuits. As indicated in Figure 2.6, Zynq slices are composed of 4 *Lookup Tables*, 8 *Flip-Flops*, and other logic.

- ***Lookup Table (LUT)*** — A flexible resource capable of implementing (i) a logic function of up to six inputs; (ii) a small Read Only Memory (ROM); (iii) a small Random Access Memory (RAM); or (iv) a shift register. LUTs can be combined together to form larger logic functions, memories, or shift registers, as required.

- ***Flip-flop (FF)*** — A sequential circuit element implementing a 1-bit register, with reset functionality. One of the FFs can optionally be used to implement a latch.

- ***Switch Matrix*** — A switch matrix sits next to each CLB, and provides a flexible routing facility for making connections (i) between elements within a CLB; and (ii) from one CLB to other resources on the PL.

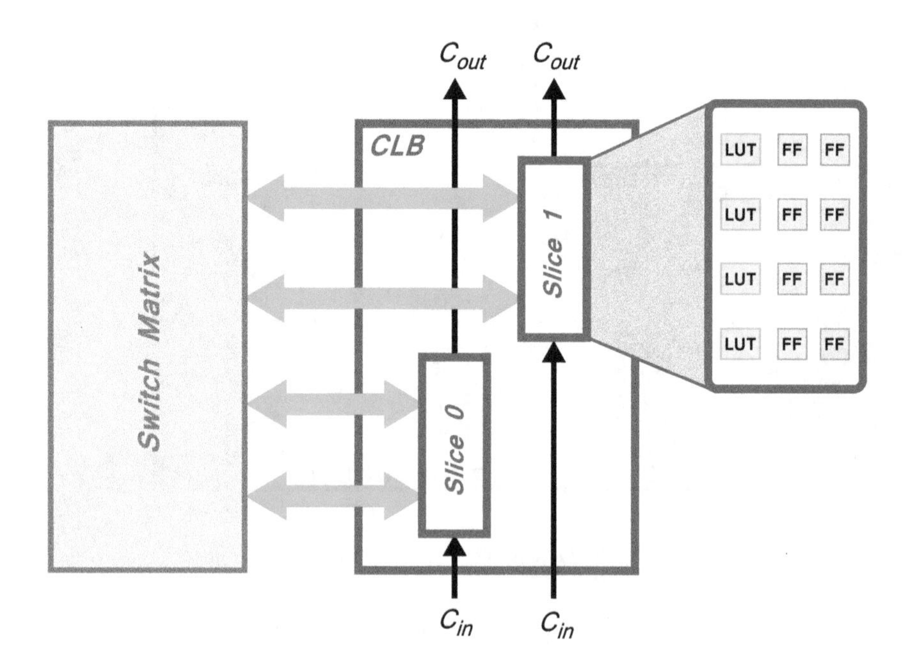

Figure 2.6: Composition of a Configurable Logic Block (CLB)

- ***Carry logic*** — Arithmetic circuits require intermediate signals to be propagated between adjacent slices, and this is achieved via carry logic. The carry logic comprises a chain of routes and multiplexers to link slices in a vertical column.

- ***Input / Output Blocks (IOBs)*** — IOBs are resources that provide interfacing between the PL logic resources, and the physical device 'pads' used to connect to external circuitry. Each IOB can handle a 1-bit input or output signal. IOBs are usually located around the perimeter of the device.

Extended information on the CLB resources within the logic fabric can be found in [20].

Although it is useful for the designer to have a knowledge of the underlying structure of the logic fabric, in most cases there is no need to specifically target these resources — the Xilinx tools will automatically infer the required LUTs, FFs, IOBs etc. from the design, and map them accordingly.

2.2.2. Special Resources: DSP48E1s and Block RAMs

In addition to the general fabric, there are two special purpose components: Block RAMs for dense memory requirements; and DSP48E1 slices for high-speed arithmetic. Both of these resources are integrated into the logic array in a column arrangement, embedded into the fabric logic and normally in proximity to each other (the reason being that intensive computation and storage of data in memory are often closely associated operations). This aspect of the device layout is indicated in Figure 2.5.

The Block RAMs in the Zynq-7000 are equivalent to those on Xilinx 7 series FPGAs, and they can implement Random Access Memory (RAM), Read Only Memory (ROM), and First In First Out (FIFO) buffers, while also supporting Error Correction Coding (ECC) [23].

Each Block RAM can store up to 36Kb of information, and may be configured either as one 36Kb RAM, or two independent 18Kb RAMs. The default word size is 18 bits, and in this configuration each RAM comprises 2048 memory elements. The RAM can also be 'reshaped' such that it contains more, smaller elements (for example 4096 elements x 9 bits, or 8192 x 4 bits), or alternatively, fewer, longer elements (e.g. 1024 elements x 36 bits, 512 x 72 bits). Larger capacity memories can be formed by combining two or more Block RAMs together.

Using a Block RAM means that a large amount of data can be stored in a small physical space on the device, within a dedicated and optimised memory element; the alternative is *Distributed RAM*, which is constructed from the LUTs within the logic fabric. A significant number of LUTs (spanned over a larger area) are required to form a memory of comparable size to a Block RAM, and the resulting implementation suffers from restricted timing performance due to the increased logic and routing delays. On the other hand, it is often advantageous to implement small memories using distributed RAM, both for resource efficiency, and because their placement is more flexible (distributed memories can be located close to the components that interact with them, which can result in fast timing performance too). Block RAMs can normally be clocked at the highest clock frequency supported by the device.

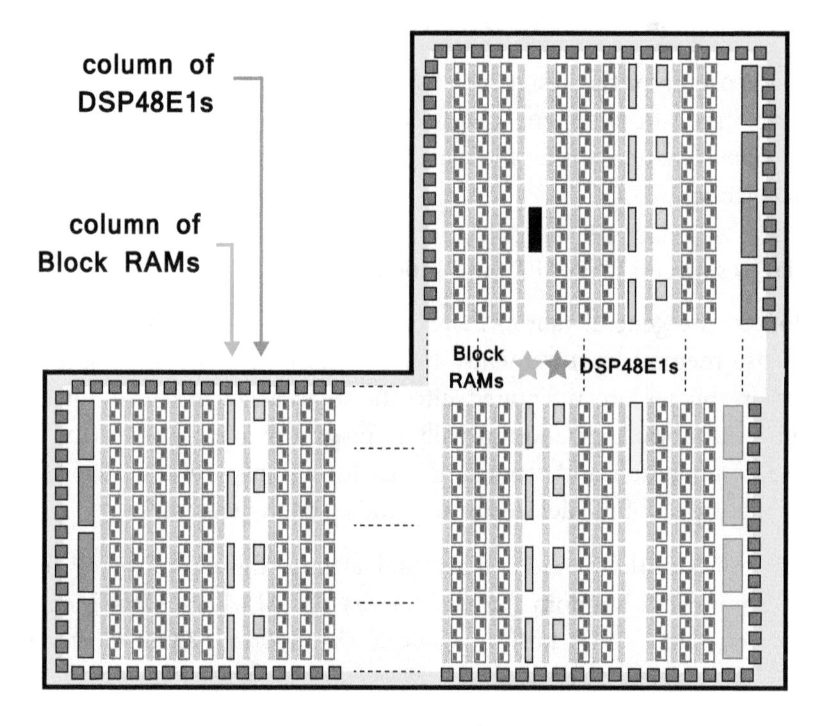

Figure 2.7: The logic fabric and its constituent elements

The LUTs in the logic fabric can be used to implement arithmetic operators of any arbitrary length, but are most suitable for arithmetic operators with short wordlengths (arithmetic circuits for long wordlengths can have a large footprint in slice logic, with placement and routing factors resulting in sub-optimal clock frequencies). DSP48E1s are specialist slices for implementing high-speed arithmetic on signals with medium to long arithmetic wordlengths. They are dedicated silicon resources, and primarily comprise a pre-adder/subtractor, multiplier, and post-adder/subtractor with logic unit, as shown in Figure 2.8, where maximum arithmetic wordlengths are marked.

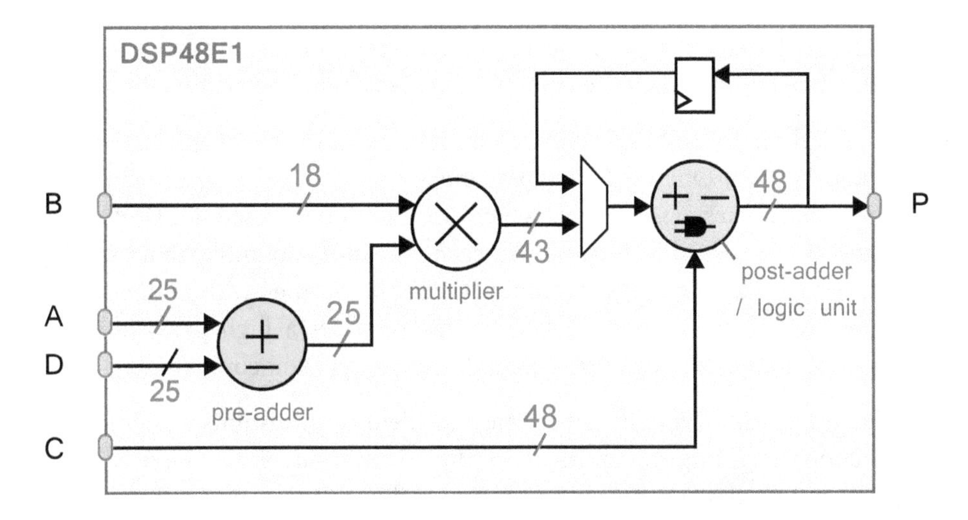

Figure 2.8: Arithmetic capabilities of the DSP48E1 slice

Figure 2.8 is a simplified diagram which omits low level implementation details. The DSP48E1 makes use of multiplexing circuitry to permit flexible usage of registers, and to support dynamic alteration of the computation (i.e. the function can be changed on a cycle-by-cycle basis as required). Various computations are possible, involving one, two or all of these arithmetic operators, and these are selected via an OPMODE input that configures the internal multiplexers (not fully shown in the diagram) and determines the arithmetic functionality implemented. Notice that the inputs are labelled A, B, C, and D, and that the output is labelled P. The unit can compute the functions P = (A+D)*B, or P = P' + C, or in fact many others. It is also capable of SIMD processing, implementing 2 or 4 shorter addition/subtraction/accumulation operations of 24 or 12 bits, respectively.

Notice also from Figure 2.8 that the post-adder has additional capabilities as a logic unit. When used in logic mode, it can perform logical functions instead of arithmetic, and supports all of the fundamental boolean operations: bit-wise NOT, AND, OR, NAND, NOR, XOR, and XNOR. It is also worth mentioning the pattern detector (not shown in the diagram) which adds the capability to detect overflow, perform rounding according to a selection of schemes, and undertake other related functions.

The standard arithmetic wordlengths marked on Figure 2.8 are adequate for most requirements, but they can also be extended by combining multiple DSP48E1s, if needed. Complex arithmetic can be undertaken, again by combining DSP48E1s, and the wordlengths are also suitable for implementing floating point arithmetic. Together with the advantage of high frequency operation (just like Block RAMs, DSP48E1s can be clocked at the maximum clock frequency of the device) and low power consumption, these DSP48E1 slices are attractive for implementing computationally demanding arithmetic circuits.

As a result of these properties, DSP48E1s are suited to a variety of applications in signal processing and beyond. One of their most compelling uses is to implement symmetric form Finite Impulse Response filters, which are commonly used in DSP and digital communications. The pre-adder ensures that each DSP48E1 can implement two filter taps, and entire filters can be formed by cascading DSP48E1s together, without the requirement to utilise any logic from the general fabric. This provides a high performance, highly efficient implementation for one of the fundamentally important computations in DSP.

When designing with Zynq, it makes sense to identify deterministic, computationally parallel functions and implement them in the PL section of the device, specifically targeting DSP and Block RAM resources where possible. In this way, the PL can be used to accelerate algorithms residing in the PS. There are many conceivable examples where the availability of PL directly adjacent to the processor, and the opportunity to allocate certain system functions to the PL, can bring significant benefits to the overall system implementation.

Further details about the DSP48E1 can be found in [17], and Block RAM memory resources in [23].

2.2.3. General Purpose Input/Output

The general purpose input / output facilities (IOBs) on the Zynq are collectively referred to as *SelectIO Resources*, and these are organised into banks of 50 IOBs each. Each IOB contains one pad, which provides the physical connection to the outside world for a single input or output signal.

The I/O banks are categorised as High Performance (HP) or High Range (HR), and these support a variety of I/O standards and voltages as detailed in [24]; the HP interfaces are limited to voltages of 1.8V and are typically used for high-speed interfaces to memory and other chips, while the HR interfaces permit voltages of up to 3.3V and cater for a wider variety of IO standards. Both single-ended and differential signalling are supported, requiring 1 IOB and 2 IOBs per connection, respectively.

Each IOB also includes an IOSERDES resource for programmable conversion between parallel and serial data formats (serialisation and deserialisation), of between 2 and 8 bits.

2.2.4. Communications Interfaces

The more highly specified Zynq devices include *GTX Transceivers*, high-speed communications interface blocks which are embedded into the logic fabric [21]. These are dedicated silicon blocks ("Hard IP" blocks), and they are capable of supporting a number of standard interfaces including PCI Express, Serial RapidIO, SCSI and SATA. Implementing PCI Express also requires a second Hard IP block (a PCI Express block [22], also present on the relevant Zynq devices) and Block RAMs, in addition to the GTX transceiver itself.

GTX Transceivers are implemented as 'quads', i.e. groups of 4 individual channels, each of which comprises a dedicated Phase Locked Loop (PLL) for that channel, a transmitter, and a receiver. Depending on the Zynq device and package chosen, rates of up to 12.5Gbps are supported. The interfaces can be used to create connections to independent external devices such as networking equipment, hard disks, and further FPGA or Zynq devices.

In terms of working with the GTX blocks, support is provided via a Wizard tool which automatically creates a core for the desired interface [26]. From the user perspective, this entails introducing a block into the system, choosing the desired protocol and hardware options, and setting parameters.

2.2.5. Other Programmable Logic External Interfaces

This section on I/O is completed by summarising the remaining external interfaces to the PL.

Analogue to Digital Conversion — The PL includes another hard IP component: the *XADC* block. This is a dedicated set of Analogue to Digital Converter (ADC) mixed signal hardware, which features two separate 12-bit ADCs both capable of sampling external analogue input signals at 1Msps. Control of the XADC is achieved using the PS-XADC interface block located within the PS, and the PS-XADC control block can itself be programmed from software executing on the APU. Extended information about the XADC and related resources can be found in [18].

Clocks — The PL receives four separate clock inputs from the PS, and additionally has the facilities to generate and distribute its own clock signals independently of the PS [33]. The independent PL resources are equivalent to those of a 7 series FPGA, further details of which can be found in [19].

Programming and Debug — A set of JTAG ports are provided in the PL section to facilitate configuration and debugging of the PL [33], [12]. Although more secure methods are normally preferred in deployment, JTAG configuration is often used during the development phase. The facilities offered via JTAG support debugging with both ARM and Xilinx tools.

2.3. Processing System — Programmable Logic Interfaces

As mentioned in the previous section, the appeal of Zynq lies not just in the properties of its constituent parts, the PS and the PL, but in the ability to use them in tandem to form complete, integrated systems. The key enabler in this regard is the set of highly specified AXI *interconnects* and *interfaces* forming the bridge between the two parts. There are also some other types of connections between the PS and PL, in particular EMIO.

This section examines the connections between the PS and PL and considers how they can be used. We begin by introducing the AXI standard, upon which most of these connections are based.

2.3.1. The AXI Standard

AXI stands for Advanced eXtensible Interface, and the current version is AXI4, which is part of the ARM AMBA® 3.0 open standard. Many devices and IP blocks produced by third party manufacturers and developers are based on this standard.

The AMBA standard was originally developed by ARM for use in microcontrollers, with the first version being released in 1996. Since then, the standard has been revised and extended, and it is now described by ARM as "*the de facto standard for on-chip communication*" [3]. The focus is now on System-on-Chip, including SoCs based on FPGAs or, in the case of Zynq, a device which includes FPGA fabric. In fact, Xilinx contributed strongly to defining AXI4 as an optimal interconnect technology for use within FPGA architectures [16], [29].

Support for AXI was first introduced into the Xilinx tool flow in release 12.3 of the ISE® Design Suite [25], and extensive support is now available in the Vivado Design Suite. AXI buses can be used flexibly, and in the general sense are used to connect the processor(s) and other IP blocks in an embedded system. In fact there are three flavours of AXI4, each of which represents a different bus protocol, as summarised below. The choice of AXI bus protocol for a particular connection depends on the desired properties of that connection.

- ***AXI4 [2]*** — For memory-mapped links, and providing the highest performance: an address is supplied followed by a data burst transfer of up to 256 data words (or 'data beats').

- ***AXI4-Lite [2]*** — A simplified link supporting only one data transfer per connection (no bursts). AXI4-Lite is also memory-mapped: in this case an address and single data word are transferred.

- ***AXI4-Stream [1]*** — For high-speed streaming data, supporting burst transfers of unrestricted size. There is no address mechanism; this bus type is best suited to direct data flow between source and destination (non memory mapped).

The term 'memory mapped' is used in the above descriptions, and it is useful to briefly confirm its meaning. If a protocol is memory mapped, an address is specified within the transaction issued by the master (read or write), which corresponds to an address in the system memory space. In the case of AXI4-Lite, which supports a single data transfer per transaction, data is then written to, or read from, the specified address; in the case of AXI4 bursts, the address specified is for the first data word to be transferred, and the slave must then calculate the addresses for the data words that follow.

2.3.2. AXI Interconnects and Interfaces

The primary interface between the PS and PL is via a set of nine AXI interfaces, each of which is composed of multiple channels. These make dedicated connections between the PL, and interconnects within the PS, as indicated in Figure 2.9. It is useful to briefly define these two important terms:

- ***Interconnect*** — An interconnect is effectively a switch which manages and directs traffic between attached AXI interfaces. There are several interconnects within the PS, some which are directly interfaced to the PL (as in Figure 2.9), and others which are for internal use only. The connections between these interconnects are also formed using AXI interfaces.

- ***Interface*** — A point-to-point connection for passing data, addresses, and handshaking signals between master and slave clients within the system.

Note from the diagram that all of the interfaces are specifically connected to AXI interconnects residing within the PS, with the exception of the ACP interface, which is connected directly to the Snoop Control Unit inside the APU.

Internally to the processing system, AXI interfaces are used within both the ARM APU (making connections between the processing cores and SCU, cache memory and OCM), and more generally to connect the various interconnects within the PS. These connections are in addition to those at the PS-PL boundary. In particular, the three interconnects shown in Figure 2.9 (the *Memory*, *Master* and *Slave* Interconnects) are internally connected to the

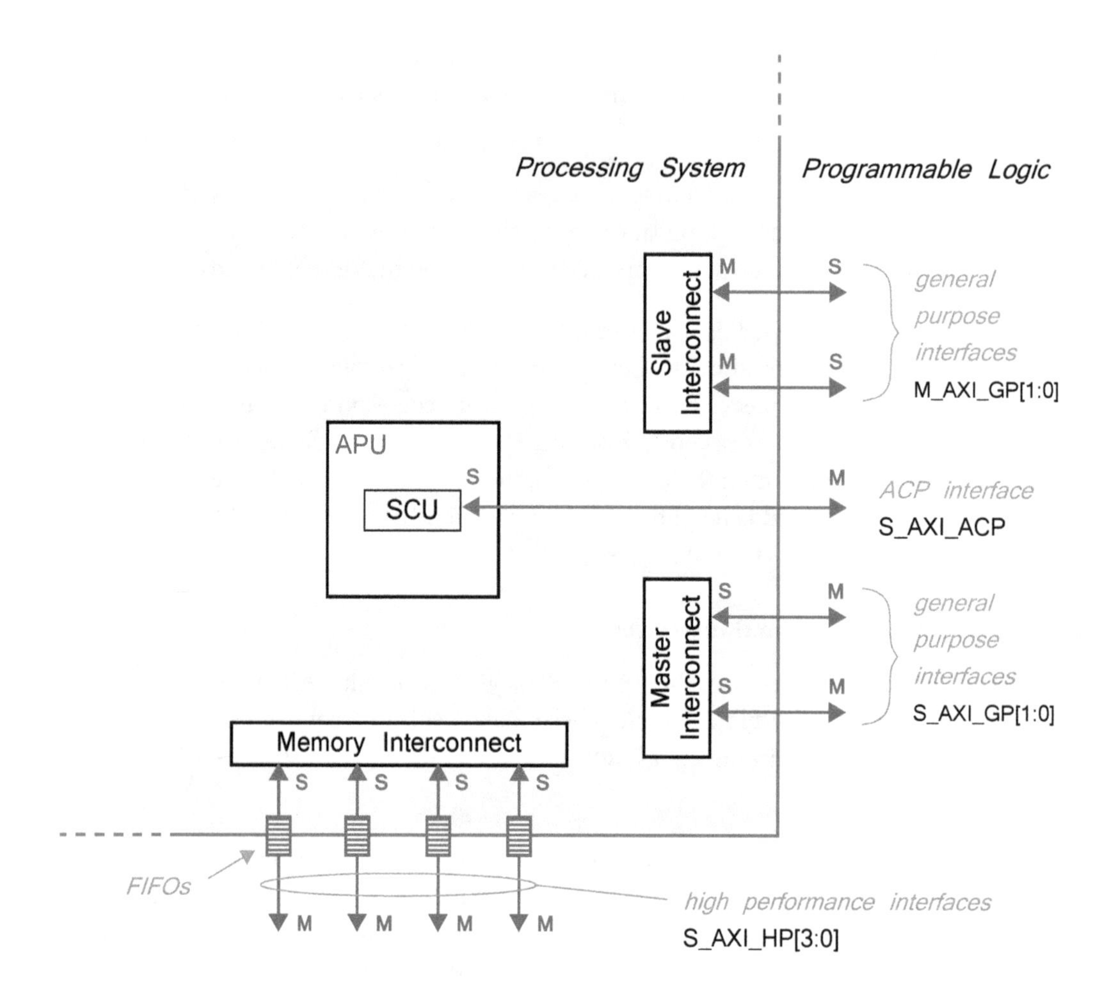

Figure 2.9: The structure of AXI interconnects and interfaces connecting the PS and PL

Central Interconnect, which is not shown here, but which is visible in Figure 2.2. Full details of PS internal connections, including a block diagram showing all AXI interconnects and interfaces, are available in [33].

Table 2.2 provides a summary of the interfaces shown by the arrows in Figure 2.9. A short description of each interface is given, and the master and slave are indicated (in accordance with convention, the master is in control of the bus, and initiates transactions, while the slave responds). Note that the interface naming convention (as in the left hand column of Table 2.2) is to indicate the role of the *PS*, i.e. "M" is the first letter where the PS is the master, and "S" is the first letter where the PS is the slave.

Table 2.2: Interfaces between PS and PL [33]

Interface Name	Interface Description	Master	Slave
M_AXI_GP0	General Purpose (AXI_GP)	PS	PL
M_AXI_GP1		PS	PL
S_AXI_GP0	General Purpose (AXI_GP)	PL	PS
S_AXI_GP1		PL	PS
S_AXI_ACP	Accelerator Coherency Port (ACP), cache coherent transaction	PL	PS
S_AXI_HP0	High Performance Ports (AXI_HP) with read/write FIFOs. (Note that AXI_HP interfaces are sometimes referred to as AXI Fifo Interfaces, or AFIs).	PL	PS
S_AXI_HP1		PL	PS
S_AXI_HP2		PL	PS
S_AXI_HP3		PL	PS

To further explain the roles of these different types of PS-PL AXI interfaces:

- ***General Purpose AXI*** — A 32-bit data bus, which is suitable for low and medium rate communications between the PL and PS. The interface is direct and does not include buffering. There are four general purpose interfaces in total: the PS is the master of two, and the PL is the master of the other two.

- ***Accelerator Coherency Port*** — A single asynchronous connection between the PL and the SCU within the APU, with a bus width of 64 bits. This port is used to achieve coherency between the APU caches and elements within the PL. The PL is the master.

- ***High Performance Ports*** — The four high performance AXI interfaces include FIFO buffers to accommodate "bursty" read and write behaviour, and support high rate communications between the PL and memory elements in the PS. The data width is either 32 or 64 bits, and the PL is the master of all four interfaces.

Each bus is made up of a collection of signals, and transactions on these buses take place in accordance with the defined bus standard, AXI4, which will be introduced next. It is beyond the scope of the current discussion to describe the AXI buses and transactions in depth, but we will return to this topic later in the book, in Chapter 19.

2.3.3. EMIO Interfaces

As mentioned in Section 2.1, several connections from the PS can be routed through the PL to external interfaces, and this is referred to as Extended MIO, or EMIO.

EMIO involves signal transfer between the two domains, and is achieved through a simple set of wire connections; consequently, not all MIO interfaces are supported in EMIO, and some of those which are supported have reduced capability [33]. The connections are arranged in two 32-bit banks.

Interfaces routed through the EMIO are in many cases connected directly to the desired external pins of the PL, as specified by the inclusion of appropriate entries in a constraints file. In this mode, EMIO can provide an additional 64 inputs, and 64 outputs with corresponding output enables. Another option is to use EMIO to interface the PS with a peripheral block in the PL. Figure 2.10 depicts both of these use models.

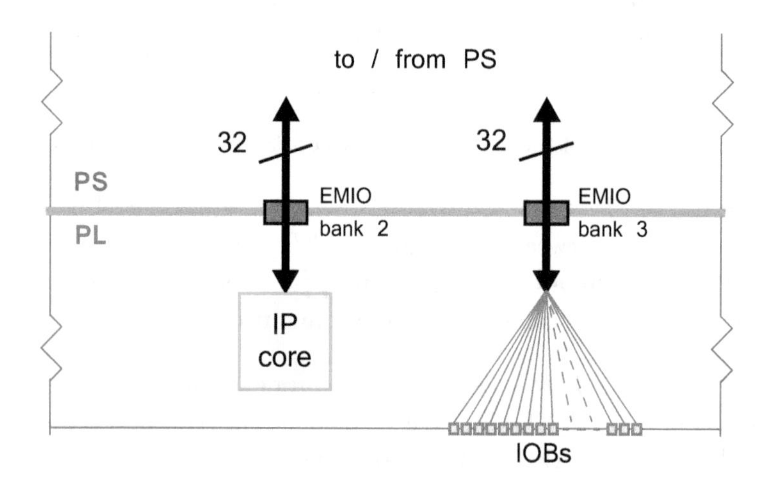

Figure 2.10: Using the EMIO to interface between PS and PL

2.3.4. Other PL-PS Signals

Other signals crossing the PS-PL boundary include watchdog timers, reset signals, interrupts, and DMA interfacing signals. These will not be covered here, but later in the context of embedded systems design using Zynq (Chapter 10), when their various functionalities will also be explained.

2.4. Security

Traditionally, the deployment of secured, tamper-proof devices has mainly been of concern to applications under the purview of defence or security. More recently, however, with the growing trend of development and marketing of custom IP resources, the ability to deploy devices which offer the ability to securely protect the internal software and hardware IP has become an ever-increasing concern to markets such as avionics, automotive, broadcast, industrial and wired/wireless networking and communications [28].

Zynq-7000 devices provide a wide range of security features which offer protection of the internal functionality of your system, ranging from dedicated hardware support for multiple encryption standards, secure system boot facilities, and software execution protection.

In this section, the security features of Zynq will be briefly introduced. It is outwith the scope of this chapter to provide any great detail on the individual security aspects of Zynq, but instead relevant features will be introduced. Further information on theses features can be found in the cited references.

2.4.1. Secure Boot

One of the main architectural points of note with Zynq-7000 devices is that the boot method is restricted to a single source — device boot must be driven by the processor. When the device is powered on or reset, the first core of the PS boots from external memories before going on to configure the PL [28]. By restricting the boot method to a single source, it ensures that the there is no manual way to load malicious software after the PL has been configured, and also no way to load a malicious image to the PL after the processor has initialised.

A number of features have been incorporated into Zynq-7000 devices which facilitate secure booting. One of these features is the boot ROM, which has been designed to handle various forms of security. Both asymmetric and symmetric authentication of the First-Stage Boot Loader (FSBL), U-Boot, PL bitstream and user software (OS and user applications) is supported. In the case of asymmetric authentication, RSA-2048 primary and secondary public keys are used, where as HMAC (SHA-256) is used for symmetric authentication. Further, encryption of the boot files mentioned above is supported with 256-bit AES/CBC key which can be either volatile (battery backed up) or non-volatile (eFuses).

One other feature which facilitates secure boot is the OCM, which has been provided to be large enough (256KB) to run the FSBL from an internal location which is immune to any external probing attack. The OCM is also large enough to securely store TrustZone® software routines (more on these in the *Zynq-7000 and ARM TrustZone Technology* subsection below).

2.4.2. Hardware Support

All Zynq-7000 devices benefit from a host of hardware security IP, which are implemented either as hard IP blocks within the PS, or as soft IP in the PL. The functionality of these security IPs includes anti-tamper, trust and information assurance, to protect the system from power-on and through runtime [28].

Further to the available security IP, Zynq-7000 devices have a number of embedded blocks which can support the creation of secured systems. Such blocks include authentication, decryption engines, key storage and unique device identification possibilities.

Some of the features of Zynq devices which relate to security are listed as follows [28]:

- ARM TrustZone support (PS and PL)
- AES-256 encryption (BBRAM key and eFUSE key)
- Secure Configuration and Boot (PS and PL)
- HMAC bitstream authentication
- FSBL RSA-2048 Authentication
- Hardened readback disable
- JTAG disable/monitor
- SEU checker

2.4.3. Runtime Security

The need for preventing unwanted access to the internal device data or memory doesn't end after the boot process has completed, and as such there is a need to provide runtime security. By not employing runtime security on a device, confidential user or system data might be compromised, along with the stability and operation of the system. In order to prevent such compromise to your system, any malicious access to internal data, memory or peripherals must be obviated.

Runtime security can be split into three areas of protection, which are outlined as follows:

- **Processing System to Programmable Logic** — The prevention of software running on the Zynq PS from accessing hardware-based IP and slaves running in the PL. Zynq devices have two methods of implementing such protection: (i) a Zynq-specific implementation of ARM TrustZone technology (see the dedicated *Zynq-7000 and ARM TrustZone Technology* subsection below), and (ii) based on the monitoring of AXI port transaction from the master, and the corresponding slave address.

- **Processing System to Processing System** — Previous generations of embedded systems were made up of an amalgamation of various independent subsystems, which each, in turn, comprised of dedicated hardware, operating systems and software. This architecture was inherently secure, in terms of runtime security, because each subsystem made use of its own dedicated hardware (CPU, buses, memory and peripherals.). With todays embedded systems, such as those based on Zynq-7000 devices, making use of shared resources, such as the PS, PL and configurable interconnects, runtime security is of greater concern. It is therefore important to ensure that shared resources have sufficient security in place.

  One such area which must have sufficient security is the MMU:

  - *Memory Management Unit Security* — By configuring MMU Page Tables in a way that is security-aware, system security is improved by restricting the access of unauthorised software applications and hardware drivers to specific memory regions, devices, configuration registers and IP cores [28].

    All members of the Zynq-7000 family have a dedicated MMU for each of the two Cortex-A9 processing cores. The Page Tables of each of the MMUs allows for fine-grained access to be controlled for the DDR Memory, OCM, system level control registers, memory mapped blocks in the PS and memory mapped IP blocks within the PL.

- **Programmable Logic to Processing System / Programmable Logic** — One of the main advantages of the Zynq PL is the ability to easily instantiate multiple IP blocks which can act as AXI masters (a MicroBlaze processor, for example). Such AXI masters are subject to various levels of restriction which limits their access to slave devices which are associated with the PS (such as CAN, Ethernet, GPIO and USB), as well as other soft IP slaves instantiated in the PL [28].
  Further to this, during system development, the developer has the freedom to

control which slave addresses are accessible to any one master IP within the PL. This functionality reduces the chances of a compromised master IP from accessing restricted hardware.

Zynq-7000 and ARM TrustZone Technology

One feature of Zynq devices which can prevent such venerabilities is the Zynq-specific implementation of ARM TrustZone technology [28]. The TrustZone architecture enables trusted computing within embedded systems by establishing a hardware architecture which is capable of spreading the security framework throughout the design of the system. This is accomplished by running specific subsystems in either a "normal world" or a "secure world", rather than protecting the entirety of the systems assets in a single, dedicated hardware resource [34]. By operating in this manner, and combined with software which is capable of making full use of the offered advantages, TrustZone establishes a security solution which can operate from one end of a system to the other.

For Zynq devices, a *normal world* is defined as a subset of hardware comprising of memory and L2 cache regions, and specific AXI devices [34]. Non-trusted software can run in a normal world, but its access and awareness of additional hardware will be limited, as it may be dedicated to the TrustZone architecture in the *secure world*. Software applications which are classified as trusted will execute in the secure world, which is a separate, trusted environment isolated from the main OS to prevent any malicious access to the embedded system.

2.5. Zynq-7000 Family Members

At the time of writing, the Zynq product range comprises six different general purpose Zynq-7000 devices, all with slightly different features and sizes. The prominent features of these are summarised in Table 2.3 (extended details can be found in [32]).

Table 2.3: Zynq-7000 family members

	Z-7010	Z-7015	Z-7020	Z-7030	Z-7045	Z-7100
Processor	Dual core ARM Cortex-A9 with NEON and FPU extensions					
Max. processor clock frequency	866MHz			1GHz		
Programmable Logic	Artix-7			Kintex-7		
No. of FlipFlops	35,200	96,400	106,400	157,200	437,200	554,800
No. of 6-input LUTs	17,600	46,200	53,200	78,600	218,600	277,400
No. of 36Kb Block RAMs	60	95	140	265	545	755
No. of DSP48 slices (18x25 bit)	80	160	220	400	900	2020
No. of SelectIO Input/Output Blocks[a]	HR: 100 HP: 0	HR: 150 HP: 0	HR: 200 HP: 0	HR: 100 HP: 150	HR: 212 HP: 150	HR: 250 HP: 150
No. of PCI Express Blocks	-	4	-	4	8	8
No. of serial transceivers	-	4	-	4	8 or 16[b]	16
Serial transceivers maximum rate	-	6.25Gbps	-	6.6Gbps / 12.5Gbps[c]	6.6Gbps / 12.5Gbps[b]	10.3Gbps

a. Depends on the package; maximum numbers are shown here. HR = High Range, HP = High Performance.
b. Depends on package chosen.
c. Depends on the speed grade of the device.

As shown in the table, the main differentiator between the specific devices within the Zynq family is the type and quantity (or 'density') of the programmable logic. Of the general purpose Zynq family members, the smaller devices are based on Xilinx' Artix-7 FPGA logic fabric, and the larger devices on the Kintex-7 logic fabric. Each of the six family members provides a different amount of general purpose logic, Block RAMs, and DSP48E1s, and naturally the overall processing capability of the PL section increases in proportion to its resources. The PS is standard across all family members, the only difference being that the maximum frequency of the ARM core differs: the PS on the Artix-7 based devices can be clocked at up to 866GHz, the Kintex-based devices up to 1GHz.

There are also some other differentiating factors. In particular, PCI Express blocks and high-speed communications interfaces are integrated within the PL section of select devices, providing transceivers capable of operating at multi-Gbps rates [31].

In addition to general purpose Zynq devices, there are also two automotive grade Zynqs (based on the Z-7010 and Z-7020), and three defence grade Zynqs (based on the Z-7020, Z-7030 and Z-7045). Both automotive and defence grade devices have an extended temperature range compared to mainstream equivalents, while the latter additionally has ruggedised packaging and enhanced security features.

Given all of these factors, there are a wide variety of device options to choose from, ensuring that Zynq is suitable for a number of applications.

2.6. Chapter Review

This chapter has reviewed the general architecture of the Zynq device, and detailed its two component parts, the PS and PL, together with resources for interfacing between them.

The processing system was shown to include a dual-core ARM Cortex-A9 processor with associated extensions for SIMD and floating point processing, grouped together in an APU unit along with memory resources. The capabilities of the APU and its interfacing to the rest of the processing system were highlighted. The internal structure of the PS as a whole, and its connections with the programmable logic, were also explained.

The structure and resources of the programmable logic section were outlined, including the fabric logic, Block RAM and DSP48E1 resources, and interfacing resources, and the important issue of interfacing between the PL and PS using AXI was reviewed. Lastly, a comparison was made between device members of the Zynq-7000 family.

2.7. Architecture Reference Guide

This section acts as a visual guide to the list of references provided in Section 2.8, and is included to help clarify helpful sources of detailed information relating to the basic architecture of Zynq as presented in this chapter.

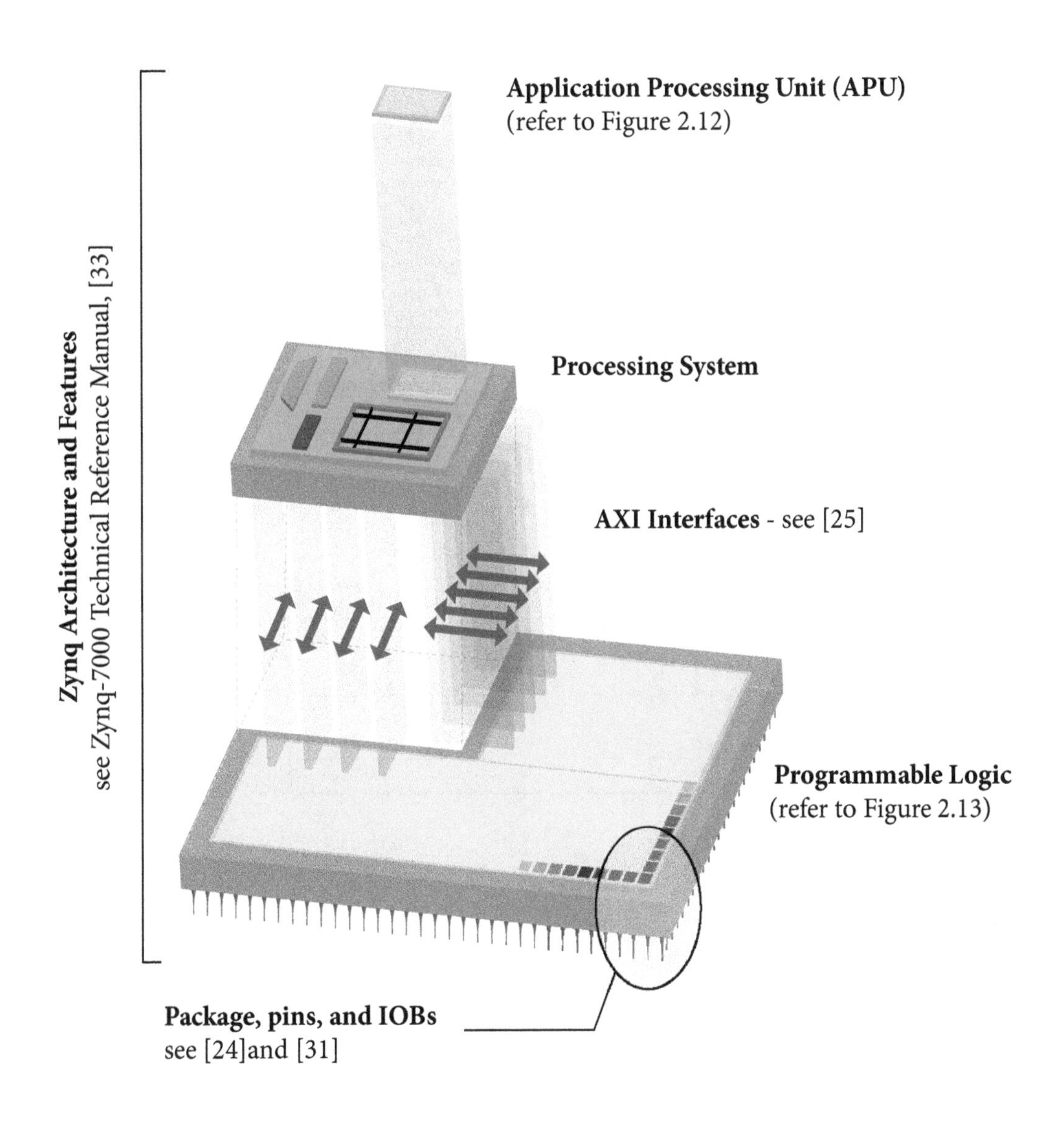

Figure 2.11: General Zynq architecture and references

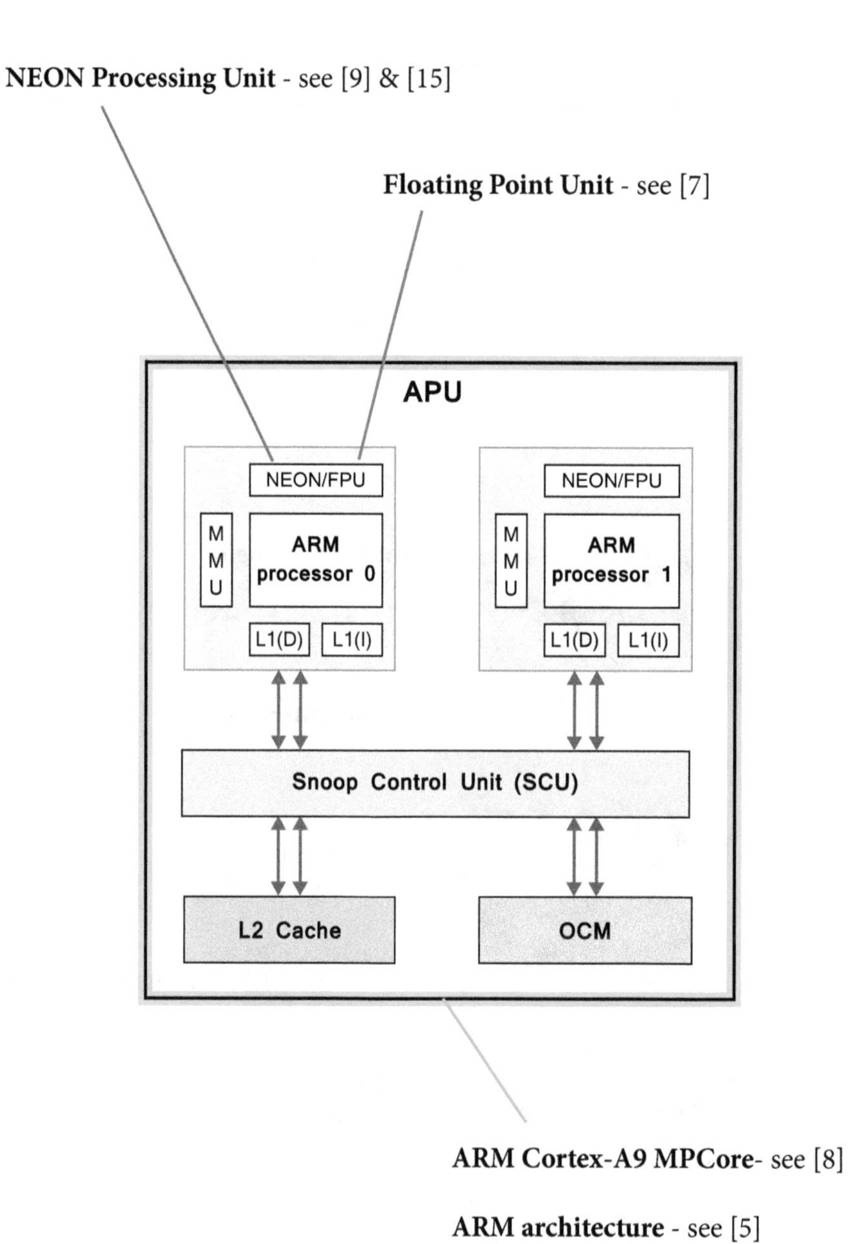

ARM Cortex-A9 MPCore- see [8]

ARM architecture - see [5]

Figure 2.12: Application Processing Unit - references

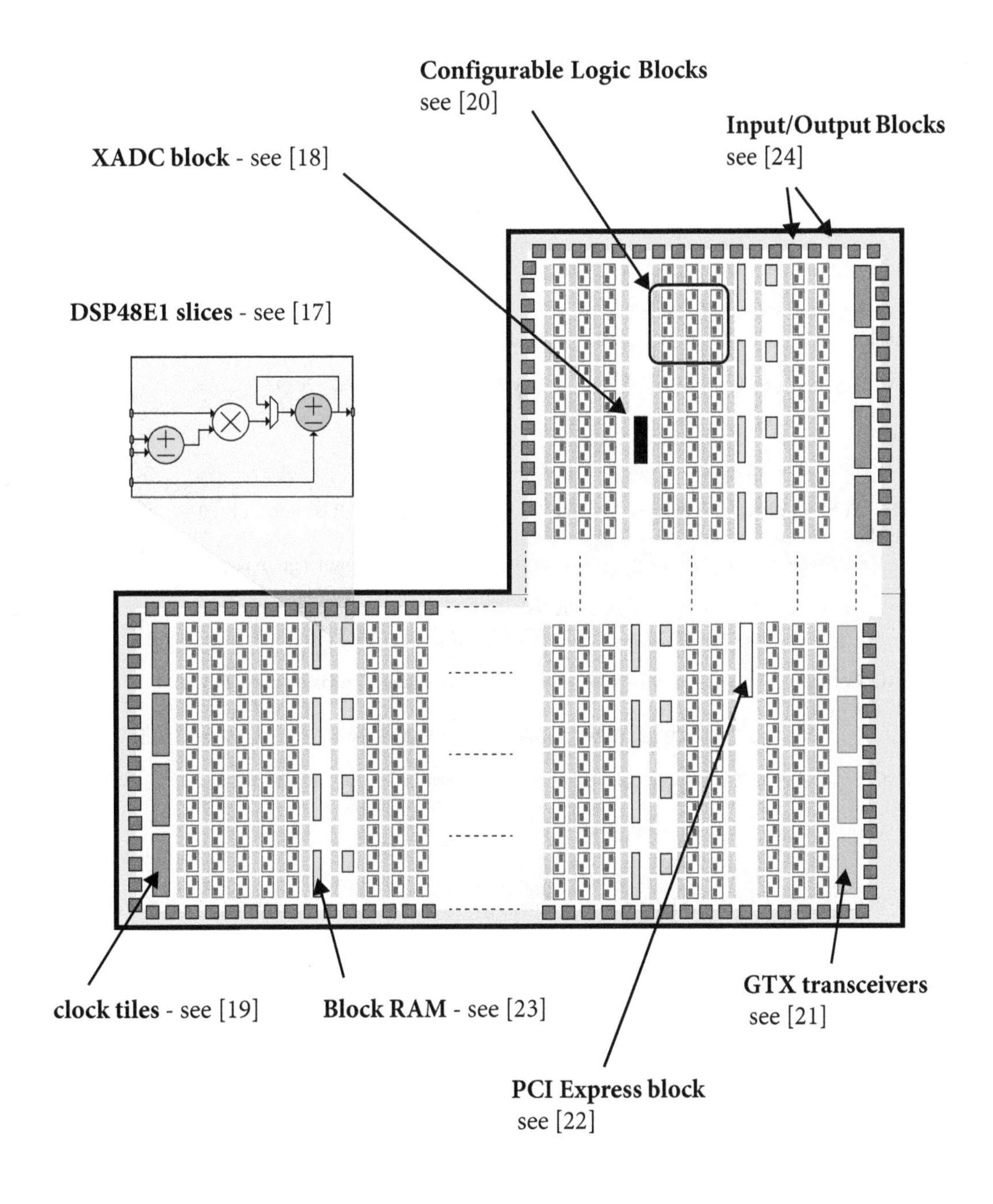

Figure 2.13: Programmable Logic - references

2.8. References

NOTE: all URLs last accessed June 2014.

[1] ARM, "AMBA 4 AXI4-Stream Protocol Specification", v1.0, March 2010.
Available: http://www.arm.com/products/system-ip/amba/ (then "Download Specifications").

[2] ARM, "AMBA AXI and ACE Protocol Specification: AXI3, AXI4, and AXI-Lite, ACE and ACE-Lite", February 2013.
Available: http://www.arm.com/products/system-ip/amba/ (then "Download Specifications").

[3] ARM, "AMBA Open Specifications", webpage.
Available: http://www.arm.com/products/system-ip/amba/amba-open-specifications.php

[4] ARM, "Architectures, Processors and Devices Development Article", May 2009.
Available:
http://infocenter.arm.com/help/topic/com.arm.doc.dht0001a/
DHT0001A_architecture_processors_and_devices.pdf

[5] ARM, "ARM Architecture Reference Manual: ARMv7-A and ARMv7-R edition", July 2012.
Available: https://silver.arm.com/download/ARM_and_AMBA_Architecture/AR570-DA-70000-r0p0-00rel1/DDI0406C_b_arm_architecture_reference_manual.pdf (account sign in required)

[6] ARM white paper, "The ARM Cortex-A9 Processors", v2.0, September 2009.
Available: http://www.arm.com/files/pdf/ARMCortexA-9Processors.pdf

[7] ARM, "Cortex-A9 Floating-Point Unit Technical Reference Manual", revision r3p0, July 2011.
Available:
http://infocenter.arm.com/help/topic/com.arm.doc.ddi0408g/DDI0408G_cortex_a9_fpu_r3p0_trm.pdf

[8] ARM, "Cortex-A9 MPCore Technical Reference Manual", revision r3p0, July 2011.
Available:
http://infocenter.arm.com/help/topic/com.arm.doc.ddi0407g/
DDI0407G_cortex_a9_mpcore_r3p0_trm.pdf

[9] ARM, "Cortex-A9 NEON Media Processing Engine Technical Reference Manual", revision r3p0, July 2011.
Available:
http://infocenter.arm.com/help/topic/com.arm.doc.ddi0409g/
DDI0409G_cortex_a9_neon_mpe_r3p0_trm.pdf

[10] Freescale Semiconductor, "Reference Manual - M68HC11" (*Section 8. Synchronous Serial Peripheral Interface*), Rev. 6.1, 2007.
Available: http://www.freescale.com/files/microcontrollers/doc/ref_manual/M68HC11RM.pdf

[11] "IEEE Standard for Floating-Point Arithmetic", IEEE Computer Society, revision IEEE Std. 754-2008, August 2008.

[12] "IEEE Standard Test Access Port and Boundary-Scan Architecture", IEEE Computer Society, revision IEEE Std. 1149.1-1990 including IEEE Std 1149.1a-1993, February 1990 and June 1993.

[13] David A. Patterson and John L. Hennessy, *Computer Architecture and Design: The Hardware / Software Interface*, 4th Ed., Morgan Kaufmann, 2012.

[14] Philips, "I2C Bus Specification and User Manual", UM10204, Rev. 5, October 2012.
Available: http://www.nxp.com/documents/user_manual/UM10204.pdf

[15] Qin, Leon, "Using NEON for Parallel Data Processing; Zynq-7000 Hardware Architecture", Xilinx presentation, October 2012.
Available: http://www.xilinx.com/Attachment/53775/Neon_Introduction_for_Avnet_training.pdf

[16] R. Wilson, "Truth About Xilinx Love Affair with AMBA", Electronics Weekly, 28th June 2010.
Available: http://www.electronicsweekly.com/articles/28/06/2010/48931/truth-about-xilinx-love-affair-with-amba.htm

[17] Xilinx, Inc., "7 Series DSP48E1 Slice User Guide", UG479, v1.7, May 2014.
Available: http://www.xilinx.com/support/documentation/user_guides/ug479_7Series_DSP48E1.pdf

[18] Xilinx, Inc., "7 Series FPGAs and Zynq-7000 All Programmable SoC XADC Dual 12-Bit 1 MSPS Analog-to-Digital Converter User Guide", UG480, v1.4, May 2014.
Available: http://www.xilinx.com/support/documentation/user_guides/ug480_7Series_XADC.pdf

[19] Xilinx, Inc., "7 Series FPGAs Clocking Resources User Guide", UG472, v1.10, May 2014.
Available: http://www.xilinx.com/support/documentation/user_guides/ug472_7Series_Clocking.pdf

[20] Xilinx, Inc., "7 Series FPGAs Configurable Logic Block User Guide", UG474, v1.5, August 2013.
Available: http://www.xilinx.com/support/documentation/user_guides/ug474_7Series_CLB.pdf

[21] Xilinx, Inc., "7 Series FPGAs GTX/GTH Transceivers User Guide", UG476, v1.10, February 2014.
Available: http://www.xilinx.com/support/documentation/user_guides/ug476_7Series_Transceivers.pdf

[22] Xilinx, Inc., "7 Series FPGAs Integrated Block for PCI Express Product Guide", PG054, v3.0, June 2014.
Available: http://www.xilinx.com/support/documentation/ip_documentation/pcie_7x/v3_0/pg054-7series-pcie.pdf

[23] Xilinx, Inc., "7 Series FPGAs Memory Resources User Guide", UG473, v1.10.1, May 2014.
Available:
http://www.xilinx.com/support/documentation/user_guides/ug473_7Series_Memory_Resources.pdf

[24] Xilinx, Inc., "7 Series FPGAs SelectIO Resources User Guide", UG471, v1.4, May 2014.
Available: http://www.xilinx.com/support/documentation/user_guides/ug471_7Series_SelectIO.pdf

[25] Xilinx, Inc., "AXI Reference Guide", UG761, v14.3, November 2012.
Available: http://www.xilinx.com/support/documentation/ip_documentation/axi_ref_guide/latest/ug761_axi_reference_guide.pdf

[26] Xilinx, Inc., "LogiCORE® IP 7 Series FPGAs Transceivers Wizard v2.6 User Guide", UG769, June 2013.
Available: http://www.xilinx.com/support/documentation/ip_documentation/gtwizard/v2_6/ug769_gtwizard.pdf

[27] Xilinx, Inc., "LogiCORE IP MicroBlaze Micro Controller System, Product Specification, DS865, v1.1, April 2012.
Available: http://www.xilinx.com/support/documentation/sw_manuals/xilinx14_1/ds865_microblaze_mcs.pdf

[28] Xilinx, Inc., "Security Solutions" webpage.
Available: http://www.xilinx.com/products/silicon-devices/soc/zynq-7000/security.html

[29] Xilinx, Inc., "Xilinx and ARM Announce Development Collaboration", press release, 19th October 2009.
Available: http://press.xilinx.com/2009-10-19-Xilinx-and-ARM-Announce-Development-Collaboration

[30] Xilinx, Inc., "Zynq-7000 All Programmable SoC Overview", *Preliminary Product Specification*, DS190, v1.6, December 2013.
Available: http://www.xilinx.com/support/documentation/data_sheets/ds190-Zynq-7000-Overview.pdf

[31] Xilinx, Inc., "Zynq-7000 All Programmable SoC Packaging and Pinout Product Specification", UG865, v1.3, November 2013.
Available: http://www.xilinx.com/support/documentation/user_guides/ug865-Zynq-7000-Pkg-Pinout.pdf

[32] Xilinx, Inc., "Zynq-7000 All Programmable SoCs Product Table", XMP087, v1.7.
Available: http://www.xilinx.com/publications/prod_mktg/zynq7000/Zynq-7000-combined-product-table.pdf

[33] Xilinx, Inc., "Zynq-7000 Technical Reference Manual", UG585, v1.7, February 2014.
Available: http://www.xilinx.com/support/documentation/user_guides/ug585-Zynq-7000-TRM.pdf

[34] Y. Gosain and P. Palanichamy, "TrustZone Technology Support in Zynq-7000 All Programmable SoCs", Xilinx White Paper, WP429, v1.0, May 2014.
Available: http://www.xilinx.com/support/documentation/white_papers/wp429-trustzone-zynq.pdf

3

Designing with Zynq (*"How do I work with it?"*)

This book aims to take a practical view, and therefore it is important to establish the general flow and procedures for developing Zynq systems, and to outline the software tools and hardware resources necessary to design for Zynq.

The chapter begins with a 'getting started' overview, which explains how to obtain and configure the required design tools, and the basic setup for working with a development board. Vivado, the suite of design tools provided by Xilinx for FPGA and Zynq design, is reviewed, and the functionality of its components explained. Licensing options for Xilinx software, and the role of third party tools in the design process are reviewed. The design flow itself is then discussed. In doing so, we cover conceptual and practical aspects of working with design tools to develop the hardware and software aspects of a Zynq system.

One of the key aspects brought out in this chapter is the system-oriented design philosophy and design flow supported by the development tools. Towards the end of the chapter, the discussion focuses on this fresh emphasis within the tools, and explains its advantages, as well as the evolution from the earlier design suite for FPGA design, ISE. A specific section is included to clarify the correspondence between tools in the two suites, for those who may be familiar with ISE but not Vivado.

Finally, a summary of currently available Zynq development boards is provided, and a brief overview of available support and documentation resources is included.

3.1. Getting Started

If you are new to Zynq, then you may be wondering what is required in order to get started with your first design, and where to look for support. This section provides a brief summary. More detail on each aspect will be provided during the remainder of the chapter.

3.1.1. Obtaining Design Tools

To start designing for Zynq, you will need to obtain the appropriate design tools from Xilinx. These can be ordered on DVD, or downloaded from the Xilinx website at the URL:

http://www.xilinx.com/support/download/index.htm

(Alternatively, the download page can be accessed by following the *Downloads* link from the Xilinx home page).

There are a number of design tools available, but you will need only the following:

- ***Vivado Design Suite*** (version 2014.1 or later)

Note that Vivado can be downloaded as a full version, or with customs selections of tools and target boards, based on user preferences. You may also note the presence of the *ISE Design Suite* in the list of available downloads, but this is *not* required (Vivado replaces ISE for newer devices like the Zynq). Further explanation is provided in Section 3.5, which compares elements of the Vivado and ISE Design Suites for readers who may be experienced with the older tool flow. However, please note that the ISE Design Suite is not recommended for new designs, and instead Vivado should be adopted.

Aside from the tools themselves, it is worth highlighting the *Documentation Navigator* utility, which is a very useful option. This provides an easy way to access all of the relevant user guides and other support documents:

Finally, if used in a network license configuration, then the following separate download will be required for installation on the license server machine:

- ***License Management Tools*** (2014.1 Utilities or later)

This download is not needed for standalone licensing methods.

Further guidance about downloading and installing design tools can be found in [27] (or check for the version of this document corresponding to your desired version of Vivado Design Suite).

3.1.2. Design Tool Editions and Licensing

The Vivado design tools are available in different editions, all of which may be installed from the same set of downloaded files (as outlined in Section 3.1.1), but which are differentiated by licence type. Licensing is administered online via a Xilinx user account.

There are three licence options for Vivado: *WebPACK, Design Edition*, and *System Edition*, the primary features of which are summarised in Table 3.1. The WebPACK edition is a free version, including all of the core software for creating first designs with Zynq: Vivado Integrated Development Environment (IDE) for hardware design, SDK and GNU Compiler Connection (GCC) compiler for software development, and Vivado Simulator for verification. A time-limited evaluation version of the System Edition can also be obtained.

Table 3.1: Versions of the Vivado Design Suite [23]

Design Phase	Feature	WebPACK[a]	Design Edition	System Edition
IP Integration and Implementation	Integrated Design Environment (IDE)	Included	Included	Included
	Software Development Kit (SDK)	Included	Included	Included
Verification and Debug	Vivado Simulator	Included	Included	Included
	Vivado Logic Analyser	-	Included	Included
	Vivado Serial I/O Analyser	-	Included	Included
Design Exploration and IP Generation	Vivado High-Level Synthesis (HLS)	-	-	Included
	System Generator for DSP	-	-	Included

a. The WebPACK edition is free, with features as listed in the table, and is device limited: it supports Zynq devices, and a subset of Artix and Kintex FPGAs. Refer to [27] for device limitation details.

Licenses can be configured either as server / floating licenses (i.e. a bank of licenses hosted on a network license server), or as individual licenses fixed to specific computers ('node-locked'). The WebPACK edition is free to obtain, and available as a node-locked license only.

3.1.3. Design Tool Functionality

The functionality of the various tools listed in Table 3.1 will be explained as required throughout the rest of the book, starting in Section 3.2 of this chapter when the design flow is discussed in detail. As the discussion will illustrate, ***Vivado IDE*** and ***SDK*** provide the bases for hardware and software system design, respectively, and therefore might be considered the two most important tools in the suite [4].

Vivado IDE is an integrated development environment for creating the hardware system part of the SoC design, i.e. the processor, memories, peripherals, external interfaces and bus connections. Vivado IDE interacts with other tools in the Vivado Design Suite, and also includes facilities for integrating and packaging IP, which enhances possibilities for design reuse.

SDK is a software design suite based on the popular Eclipse platform, which includes driver support for all Xilinx IPs, GCC library support for ARM and NEON extensions using the C and C++ languages, and tools for debugging and profiling [2], [29].

Of the verification and debugging tools, ***Vivado Simulator*** represents a Hardware Description Language (HDL) simulator environment for testing the hardware components within the system. The ***Vivado Logic Analyser*** provides facilities for 'in-system' verification of the design: specific additional cores are included in the hardware design and implemented on the Zynq device to enable probing of on-chip behaviour during run-time; the captured data is then transmitted back to the host PC where it may be viewed in a logic analyser [26]. ***Vivado Serial I/O Analyser*** is, as its name suggests, a similar tool catering specifically for the high speed communications interfaces.

The ***Vivado High-Level Synthesis (Vivado HLS)*** and ***System Generator for DSP*** tools are specifically used for creating, testing and managing IP for inclusion in the hardware system. ***System Generator*** is a block-based tool which is used for creating and simulating DSP designs, and it will be featured in Chapter 13 along with other methods of IP creation. ***Vivado HLS*** is a design tool for hardware synthesis from C-level descriptions, and is the subject of a special focus chapter later in the book (Chapter 15). This method of design creation is potentially a rapid one, allowing subsystems to be described at a high level of abstraction, and it is particularly relevant in today's electronics industry where time-to-market is an increasingly important factor.

3.1.4. Third Party Tools

The Vivado design suite supports certain third party tools, details of which may be found in [27] (or check the Xilinx website for the most recent version).

Of these, the majority are swappable replacements for certain tools within the Vivado design suite; for example, a particular HDL simulator (such as Aldec® *Active HDL®* or Mentor Graphics® *ModelSim®* or *Questa®*) may be preferred in place of the Vivado Simulator. Likewise there is potential to utilise a third party synthesis tool like Synopsys® *Synplify®* or *Synplify Pro®*.

A notable exception to the above relates to the *System Generator* tool. This is hosted in the MathWorks® MATLAB® / Simulink® environment, and therefore relies on the MATLAB and Simulink third party software products in order to function [6]. Two associated MathWorks components, the *Signal Processing Toolbox* and *DSP System Toolbox,* are additionally needed to support most DSP designs. One of the particular advantages of System Generator as a development environment is that the developer can exploit all of the capabilities of MATLAB and Simulink for scripting, simulation, visualisation, and file I/O, together with the specific facilities of any additional toolboxes that may be available (e.g. image processing, communications, etc.).

As a general point, it is important to note that particular versions of third party software may be required for compatibility with the installed version of the Vivado suite, as detailed in [27], hence it is recommended that these details are checked prior to installation.

3.1.5. System Setup and Requirements

The reader is again referred to [27], or the version of that user guide corresponding to their installation, for detailed information regarding supported operating systems, as this may change over time. As a general statement, recent versions of Windows and selected versions of Linux are supported. When using Vivado, it is important that the OS grants the user write permissions for all directories containing design files.

Naturally the hardware specification of the development computer is significant, as this influences execution times when running the design tools. Xilinx does not publish system recommendations, but does provide an indication of memory requirements by target device [21]. It is particularly notable that 32-bit operating systems are not suitable for targeting the two largest Zynq devices. At least 4GB of RAM is recommended for the three smaller devices, while the largest may require up to 12GB of RAM. Generally speaking, at least a dual-core processor is recommended, and some of the design tools have multi-core processing support, i.e. they can exploit processors with multiple cores.

The computer hardware configuration also requires a USB port for programming the Zynq over JTAG, and ideally another for PC-Zynq communication via the UART and Terminal application (this is a common and useful method for debugging designs).

Finally, as implied above, the user should have a Zynq development board on which to prototype and test their designs. A development board hosts a Zynq device, together with various other resources such as power circuitry, external memory, interfaces for programming and communication, simple user I/O such as buttons, LEDs, switches, and usually a number of other peripheral interfaces and connectors. During the debugging stage, designs developed on the computer using the Vivado design suite can be downloaded onto the development board using a Joint Test Action Group (JTAG) or ethernet connection, then tested in hardware, using peripherals and external interfaces if required. Debugging may include, for instance: using a debugger to interact with the processor and monitor its behaviour; user interaction with the design running on the chip via a USB-UART connection and the Terminal interface on the PC; and by executing hardware-in-the-loop simulations with the aid of an ethernet connection.

The topic of development boards will be returned to in Section 3.6, when a review of currently available boards will be presented (note that the selection may expand over time).

Taking all of the factors discussed in this chapter so far into consideration, Figure 3.1 provides a graphical summary of a typical setup for getting started with Zynq.

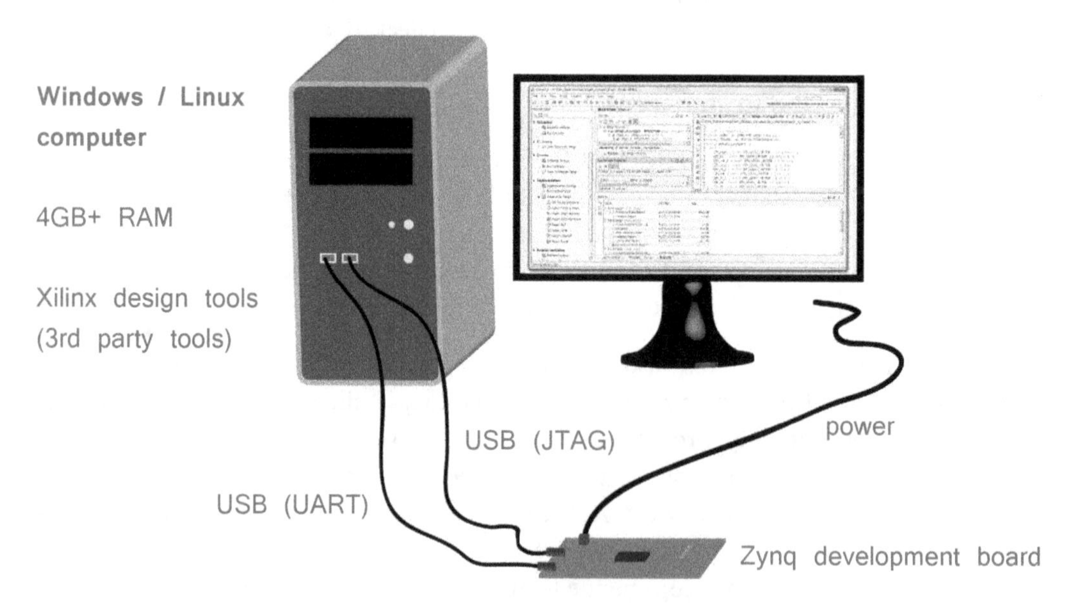

Figure 3.1: Zynq development setup

3.2. An Outline of the Design Flow

Having established the software and hardware requirements for starting out designing for Zynq, we next return to the topic of the design flow, which was first introduced in Chapter 1. Figure 3.2 represents an augmented diagram, including references to the relevant design tools. This will form the basis of our discussion over the next few pages.

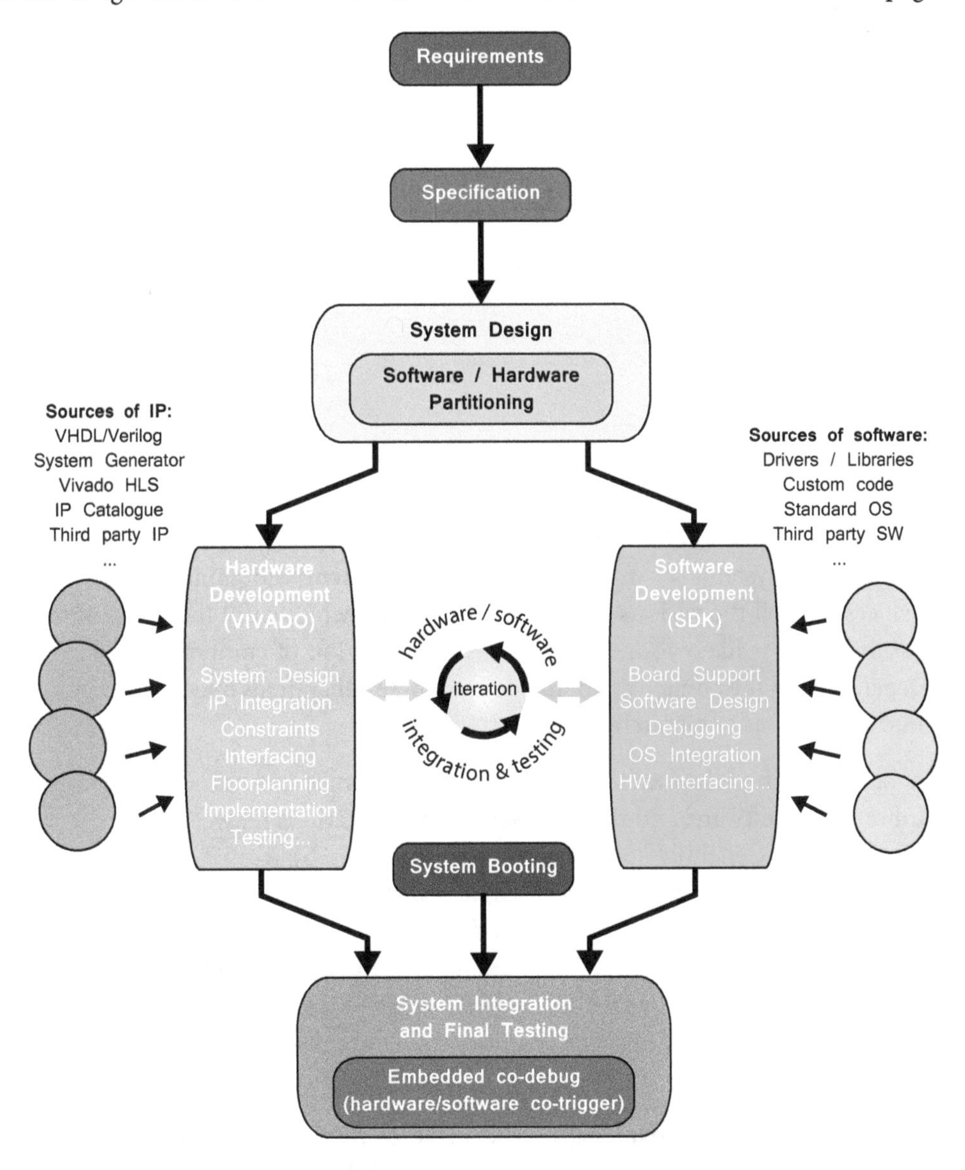

Figure 3.2: The design flow for Zynq SoC (expanded model)

This represents the flow that a single designer would undertake if completing the design by themselves, and is equivalent to following the practical examples. In Section 3.3, we will consider a team-based design flow, which is more representative of development in industry, and a task for which the Vivado Design Suite is naturally well-suited.

3.2.1. Requirements and Specification

As in any project, the starting point will be a specification of the system to be designed, based on an assessment of project requirements. It goes without saying that it is important to define the system parameters as fully and accurately as possible before embarking on practical design work.

The specification will include aspects such as the intended functionality of the design, its interfaces, performance criteria and the target device or platform. Although the initial specification is defined in as much detail as possible at the outset, it is likely to be refined, and the level of detail extended, as the project proceeds. The exact content of the specification will of course depend on the requirements and scope of the project.

3.2.2. System Design

Design of the system architecture is usually approached from a top-down perspective. This means defining the top-level system interface and parameters first, before identifying the individual subsystems or functional tasks. Subsystem functionality and required performances, as well the various interactions between them, would then be defined. The output of this stage is likely to be an abstract representation of components and transactions. Depending on the complexity of the design, these subsystems may be further decomposed into lower levels of hierarchy.

Then, having defined the functional units, the design must be appropriately partitioned into hardware and software, and the necessary communications between different parts of the system defined. Generally speaking, software (on the PS) will be used to implement general purpose sequential processing tasks, an operating system, user applications and GUIs, while computationally intensive data flow parts of the design are more suitably realised in the PL. Furthermore, any software algorithms which exhibit significant parallelism can be identified are strong candidates for implementation in PL; this corresponds to a 'coprocessor' model where computationally intensive but parallel tasks can be off-loaded from the processor into hardware to achieve an overall performance increase [15].

One of Zynq's particular advantages is the close coupling between the processor and programmable logic, which reside on the same physical device. The low latency, high performance AXI links between the PS and PL allow the differing properties of these two resources to be maximally exploited when partitioning the system into software and hardware elements. This is due to the reduced overhead of communicating between them as compared to a discrete-component alternative.

Knowledge of the Zynq architecture also enables the designer to make appropriate use of special resources like the NEON and FPU in the PS, and the BlockRAMs and DSP48E1 slices in the PL. The high-speed interfaces for external communications that are integrated into the device can also be utilised as needed.

It is significant that the Vivado flow places a particular emphasis on system design. Vivado IDE is the natural starting point, and acts as the 'cockpit' within which the top-level design is created. The integration between this tool and others within the suite (in particular System Generator and Vivado HLS) then supports population of the subsystem blocks with functional design units. Once the hardware design is completed, it is exported to SDK for software development, with design iterations undertaken by returning to Vivado IDE and then SDK if required. In a team-based development scenario, software development may be progressed in parallel with the hardware system.

3.2.3. Hardware Development and Testing

Development of the hardware system involves designing the peripheral blocks and other logic to be implemented in the PL, creating the appropriate connections between these blocks and the PS, and suitably configuring the PS. For example, a hardware system might include interfaces to a CAN bus, a UART for debugging, and GPIO, along with hardware coprocessors to support software running on the ARM. The system is visualised in Figure 3.3.

The development of the hardware system is undertaken in the Xilinx Vivado IDE development suite. The designer can sketch out their desired system with blocks originating from an IP library, parameterise the blocks, and design appropriate internal connections and external ports. This process is undertaken using the ***IP Integrator*** component of Vivado, which will be presented in greater detail later in the book, in Chapter 18.

Figure 3.4 shows an example screenshot of an IP Integrator design. There is a block representing the Zynq PS (seen here at the bottom left), a further block associated with resetting the PS, a peripheral, and an AXI interconnect block. The major connections between blocks are made using AXI interfaces. External ports are shown at the right hand side of the diagram.

There are various mechanisms available for testing the hardware system. Firstly, while building it in IP Integrator as a block diagram, a set of Design Rule Checks (DRCs) are applied. These ensure the fundamental completeness and integrity of the design by, for example, checking that all required connections are made correctly. Once satisfied with the diagram, the system is then synthesised and implemented. Each of these stages involves more detailed processing and integrity checking of the design, and errors are flagged by Vivado if there are problems requiring attention.

The original source of IP blocks implemented in the PL may be:

- The Xilinx library
- Third party sources
- Designed by the User, or by colleagues within the same organisation ('in-house')

Depending on the source of the block, it may already have been verified as an independent entity. IP from the Xilinx library is provided on a pre-verified basis, but blocks from third parties may not be. Any IP you design in-house, together with third party blocks not already verified, should be validated as standalone blocks before integration into

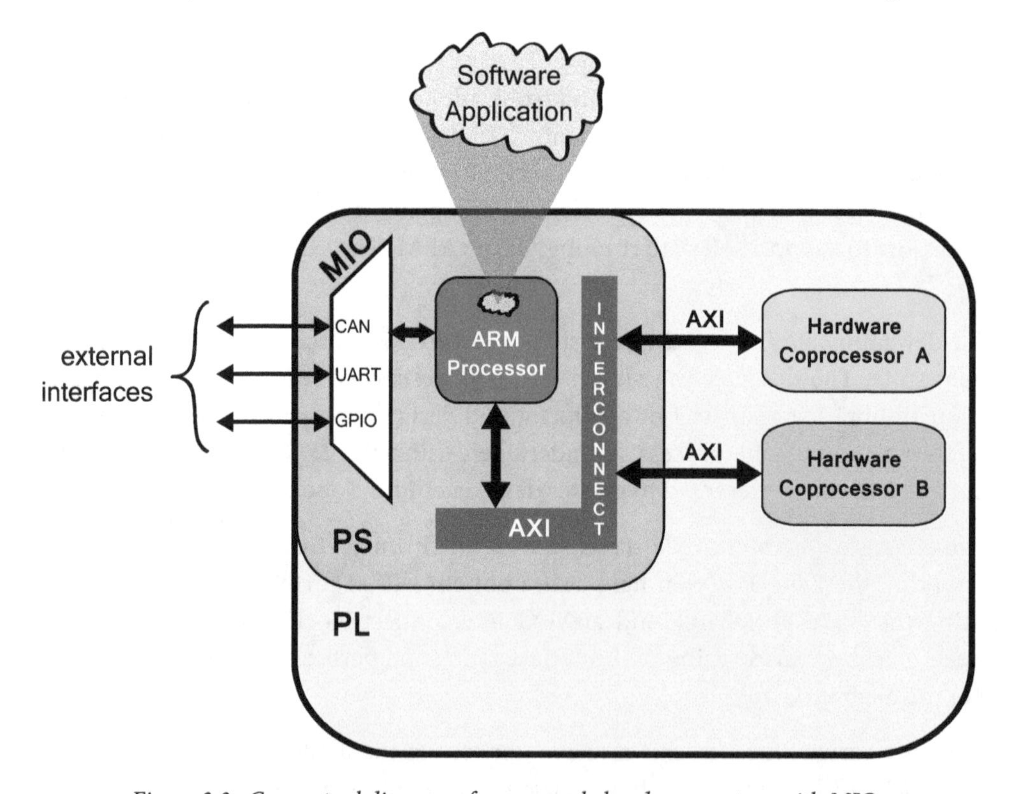

Figure 3.3: Conceptual diagram of an example hardware system with MIO

an embedded system. Depending on the design method, this may take place using HDL simulation, or simulation in MATLAB / Simulink, for example. Further discussion of the creation, testing and management of IP will appear in Chapters 13 to 18.

The hardware system is usually tested in conjunction with the software that is deployed upon it, i.e. while the hardware is operational in the context of the integrated system. Certain desired signals can be *marked for debug* in the IP Integrator block diagram, i.e. specified for inclusion in the hardware testing session. Later, at the point of running the software upon the hardware system, these signals can be inspected in a waveform viewer on the host PC.

Another powerful method of testing is Hardware In the Loop (HIL). Using this method, part of the design runs on a development board and signals are returned to the simulation environment for inspection. For instance, Xilinx provide a tutorial wherein the PS runs on the board, while a PL component is exercised in simulation [19]. This provides an opportunity to inspect the PL signal behaviour resulting from real PS operation in detail.

Once an iteration of the hardware system has been completed, the project is exported to SDK for work on the software portion of the design. Further design iterations may follow,

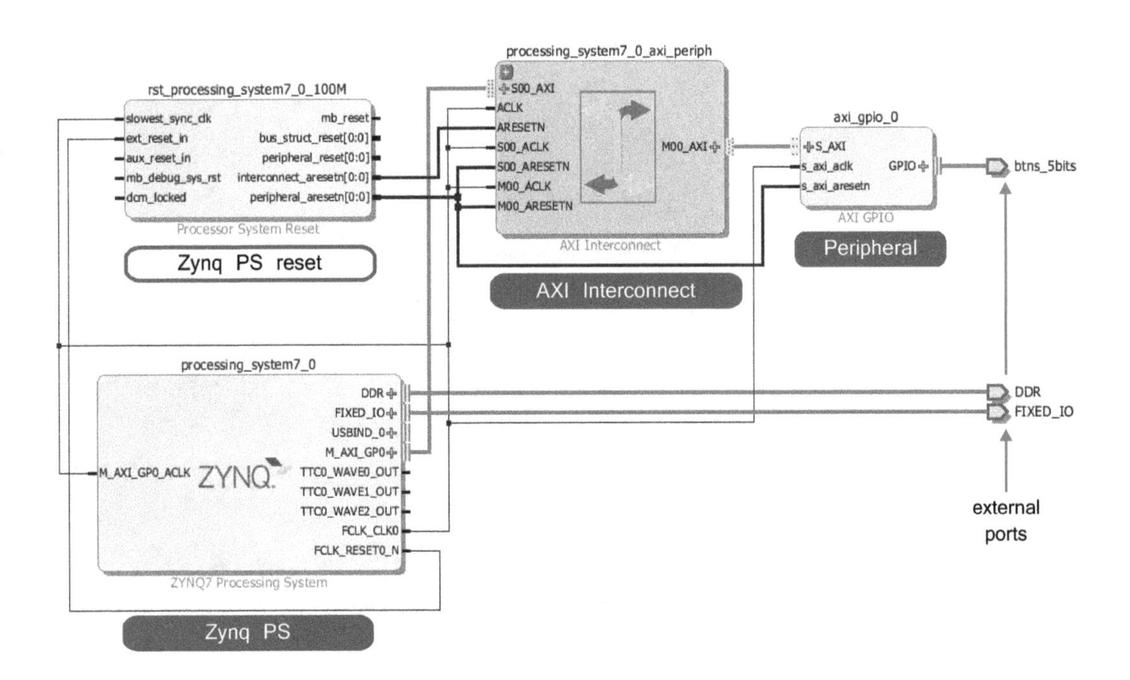

Figure 3.4: An example IP Integrator block diagram

involving a return to Vivado and changes to the IP Integrator design or further partitioning of software to hardware. Design teams are likely to adopt an approach wherein the software and hardware sub-teams progress and iterate the two aspects in parallel.

3.2.4. Software Development and Testing

Given that Zynq is a flexible platform, the hardware system on which the software operates can vary. The project exported from Vivado into SDK represents the customised hardware upon which the software is based; usually this is referred to as the 'hardware base system' or 'hardware platform'. It corresponds to the configuration designed in IP Integrator, such as the system shown in Figure 3.4.

The software system can be thought of as a stack, or set of layers, which is built upon the ***Hardware Base System*** as shown in Figure 3.5.

The layer directly above the Hardware Base System is the ***Board Support Package (BSP)***, the set of low-level drivers and functions that are used by the next layer up (the ***Operating System***) to communicate with the hardware. ***Software Applications*** run on top of the Operating System — these collectively represent the uppermost layer in the software stack, and the highest level of abstraction from hardware.

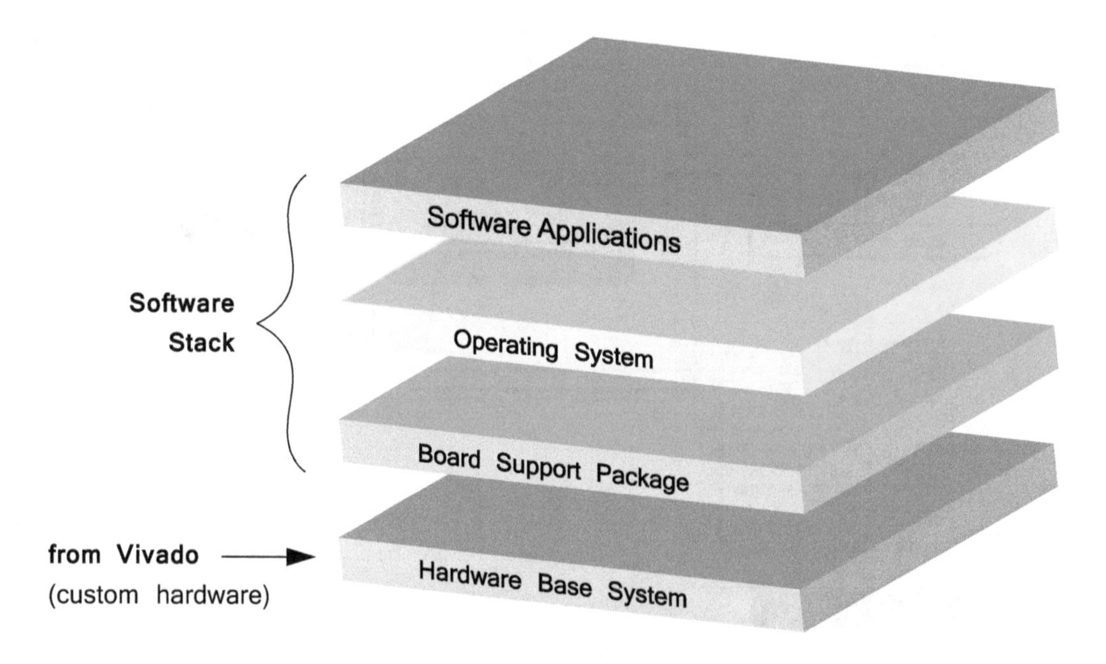

Figure 3.5: Hardware and software layers of a Zynq design

When creating the software stack, the first design choice is usually to decide on the OS to deploy: this can be a fully-fledged OS such as Linux or Android; an embedded OS; a Real-Time Operating System (RTOS) for deterministic, time-critical applications; or *Standalone*, a 'light' OS including only the most basic functions [22]. It is also possible for applications to communicate directly with hardware, and this is often referred to as a 'bare-metal' application. With two processor cores available, there is also the possibility of deploying two different types of OS, one on each core. This theme will be explored in Chapter 21, and is also outlined in the introductory part of the *Zynq-7000 All Programmable SoC Software Developers Guide* [33].

The BSP is tuned to the hardware base system, allowing an OS to operate efficiently on the given hardware. The BSP is customised to the combination of base system and operating system, and includes hardware parameters, device drivers, and low-level OS functions. Therefore, in terms of Vivado / SDK development, the BSP should be refreshed if changes are made to the hardware base system. SDK provides the environment for creating BSPs, and developing and testing software for deployment in the upper layers. It also supports the creation of BSPs for use in third party development tools such as ARM Development Studio 5 (DS-5), which may be used in place of Xilinx SDK if desired [33], [36].

In terms of testing, SDK contains the Xilinx Microprocessor Debugger (XMD) and System Debugger (TCF) utilities, which are features for testing software while it is running on the target hardware platform, i.e. a form of HIL. This process can be undertaken by programming the PL part of the Zynq with a bitstream (*.bit file), and running the software (*.elf file) on the PS. Programming is usually accomplished by downloading the files from the host PC over a JTAG connection, or via Ethernet. In this way, both the PS- and PL-based components of the system are implemented and form part of the test. The GDB debugger is a further alternative (built upon XMD) that facilitates remote debugging. Another alternative is to utilise the in-built Vivado Simulator to recreate the operations of the PL on the host PC [19].

Should the results of testing prompt alteration to the hardware system, the designer can return from SDK to Vivado and make the appropriate changes, before re-exporting the hardware to SDK, updating the BSP, and resuming the software design and testing processes. This software-hardware iteration loop is depicted in Figure 3.2.

It is important to note that not all designs require both an Executable Linkable Format (ELF, *.elf) file and a BIT (*.bit) file to configure the device. The purpose of ELF files is to program the PS, while BIT files are used to program the PL. Therefore, if only one part of the Zynq device is used (PS *or* PL), then only the corresponding file type is needed.

Further information about the GNU compiler, debugging and programming tools can be obtained from the *Embedded System Tools Reference Manual* [17].

3.2.5. System Integration and Testing

System integration is partly accomplished through the development and combined testing of the software and hardware elements of the design. However, there are other factors which need to be taken into consideration too.

System-level debug occurs after the software and hardware components have been individually tested and verified during their separate development and testing stages. Even though both components may be working in isolation, when brought together for the first time, new problems may arise. One method of debugging these system-level problems is through the use of embedded co-debug with hardware/software cross-trigger. This process links the Integrated Logic Analyzer (ILA) hardware debug core in the PL with the Fabric Trace Module (FTM) of the Zynq PS via a pair of trigger in and trigger out signals [20].

When performing co-debug with hardware/software cross-trigger, both the software and hardware development tools are brought together. This allows the user to use breakpoints in the software to trigger hardware data capture and, conversely, hardware breakpoints can halt application debug in the software development environment [20].

Another factor to consider is the method of configuring the device. During development and debugging, the Zynq is typically configured by downloading files from the host PC over a JTAG or Ethernet connection. However, this is generally unsuitable for operation in the field, and it is more usual to boot and configure the Zynq from flash memory [33], [34]. Specific development and validation of the booting process is therefore required in preparation for system deployment.

3.3. SoC Design Teams

SoC designs are often developed by teams rather than individuals. A design team may comprise specialists in system-level, hardware, software, and firmware design, and testing.

As shown in the model of the design flow depicted in Figure 3.2 on page 53, a stage of high-level system design must be completed before hardware or software design can commence. This seeks to define the structure of the system, interfaces, internal and external transactions, constraints, and hardware / software partitioning. There may be system design specialists who lead this process.

Once the high-level design is settled, its individual threads can be progressed. To make best use of resources, hardware engineers and software engineers should be able to progress these aspects of the system concurrently. For example, there may be:

- A hardware design team responsible for architecting and implementing the hardware system, including the design, reuse and integration of IP blocks;
- A firmware design team responsible for designing the boot loader and implementing the drivers necessary to interface software and hardware; and
- A software design team focussing on designing and coding the software elements of the system, and integration with other incorporated elements such as standard operating systems.

Each of these elements will require to be independently verified, and this can involve various techniques including software simulation, RTL simulation, hardware-in-the-loop, the use of virtual platforms, board-based testing, etc. There will be a final integration and testing phase at the end of the project, when all teams will work towards combining their system elements together, and ensuring the desired functionality is achieved. Where necessary there may be further design iterations until the final product is fully tested and meets all requirements.

One of the particular challenges of developing large and complex systems is the definition and management of IP block interfaces in the design, and the integration of these blocks at the system-level. It can be time consuming and error prone to design connections between blocks that use different interfaces, and difficult to successfully integrate them into the system. Vivado and IP Integrator provide a framework to make IP integration faster, easier and less error prone, through support for the industry standard AXI bus interface. This also makes it easier to share and reuse IP blocks between projects, and to acquire new IP from third parties. Vivado features an extensive library of pre-verified IP blocks than can be used in this way, and there is also a wide range of third party IP available. The use of AXI also helps software and hardware teams to progress concurrently, because the AXI standard protocols are well defined.

Another potential difficulty is the need to re-partition the design some way through the development process; for example if the software team discovers that a certain function is more computationally demanding than first envisaged, and would benefit from hardware acceleration. One of the benefits of the Vivado Design Suite is that its Vivado HLS tool can synthesise high-level C code to hardware, meaning that conversion of a software routine to hardware can potentially be undertaken very rapidly based on the existing C code, without the requirement to develop a hardware coprocessor from scratch.

3.4. System-Level IP-Focused Design with Vivado

One of the major factors in modern systems design is the potential for design reuse and rapid development. Time to market is critical in many application areas, and design tools that accelerate the development process, crucially without compromising the robustness of the verification stage or quality of results, bring clear advantages.

The Vivado design flow is based on these principles, and is built on the premise that many system building blocks are exactly that — ready pieces of IP that can be integrated into a project. Unlike older design methods, which typically cater for building systems entirely from the ground up, Vivado focuses instead on exploiting the pre-verified IP available in the Vivado libraries (i.e. cores developed by Xilinx), third party IP developers, or the previous exertions of the designer (and his or her team). To a large extent, the focus of the task has shifted upwards to system integration, rather than low-level hardware design, and the features of the Vivado Design Suite reflect this. Having said that, there is also scope for undertaking custom logic design if the system calls for it.

IP Integrator, used to design the example system shown in Figure 3.4, is the primary embodiment of this IP-focused design method. *IP Integrator* is a feature of the Vivado Design Suite, with which the designer is able to adopt the same 'top-down' approach they would naturally take in conceiving the system hierarchy. IP instances can either be introduced from the existing catalogue where appropriate, or created as black boxes for later population with functional subsystems, and the interfaces between these various elements established. This approach lends itself to rapid development of the hardware system.

An example of IP instantiation is provided in Figure 3.6, which shows a view of the Vivado IP catalogue (note that only a small selection of IPs are visible). Here, a CoOrdinate Rotation DIgital Computer (CORDIC) IP has been selected and dragged into the workspace, or 'canvas'. Following this, the designer would then configure the parameters of the CORDIC design unit, as shown in Figure 3.7, and connect it to other blocks as appropriate to the system being created.

The success of the IP integration stage relies on consistent behaviour and interfacing of IPs. To support this, the Vivado Design Suite includes a related feature, *IP Packager*, that enables IP to be consolidated into standard packages (based on the IP-XACT standard) with the aim of facilitating future design reuse. This is the recommended practice for IP developers and design teams to adopt; in this way, IP designs are made maximally portable

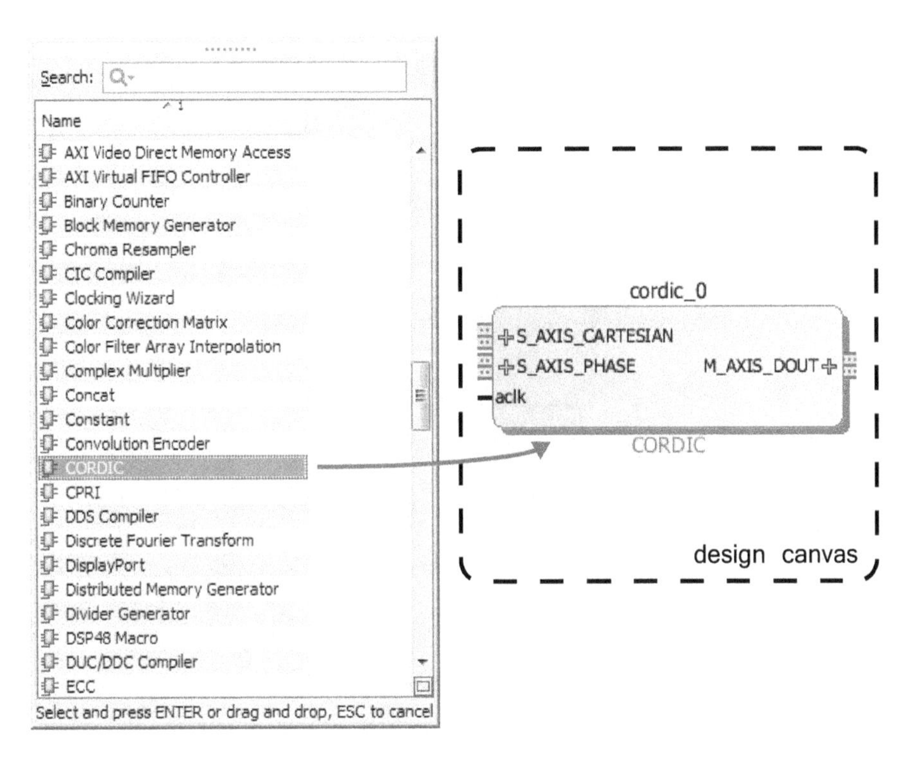

Figure 3.6: A CORDIC IP block is introduced into an IP Integrator system from the catalogue

and reusable. From a design team perspective, building up an in-house repository of easily reusable IP, shared across the organisation, is a compelling method of accelerating product design cycles. IP Packager will be featured in detail in Chapter 18.

There are several tools and design creation techniques that can be used to generate IPs. For example, they may be coded from a HDL such as VHDL or Verilog; generated from a high-level C description by the Vivado HLS tool; or from a System Generator block diagram. These methods will be explored further in Chapter 13.

As a final note on Vivado, it should be highlighted that, although we predominantly feature the GUI aspect of the tools in this book, all of the design tasks may also be undertaken using the industry standard Tool Command Language (TCL) scripting language. This represents a very powerful, repeatable and parameterisable method of driving the design tools.

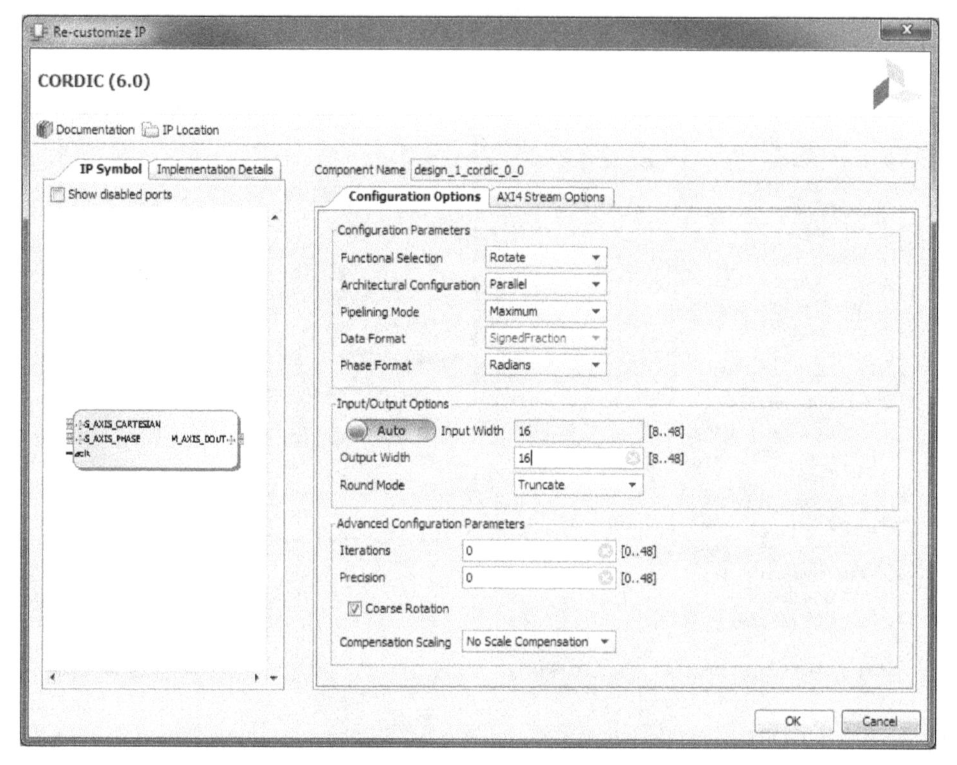

Figure 3.7: Parameterisation of the CORDIC IP block

3.5. The ISE and Vivado Design Suites

Vivado has a considerable number of advantages but, as a relatively new development environment, it may be unfamiliar to some readers. This section is intended to clarify the components within the Vivado Design Suite, and to highlight the differences and similarities with the preceding ISE Design Suite at a practical level (i.e. in addition to the evolution of design philosophy as discussed in Section 3.4). We also provide some brief notes on upgrading existing projects from ISE to Vivado.

3.5.1. Features Comparison

It is useful to summarise Vivado's predecessor, the ISE Design Suite, and to draw parallels between the various components of the two. The intention is to enable those with experience of working with ISE (but not Vivado) to more easily identify the corresponding aspects of the newer tool suite. This information is presented in Table 3.2.

Table 3.2: Correspondence between Vivado and ISE Design Suites

<table>
<tr><th>ISE</th><th>Vivado</th><th>Comments</th></tr>
<tr><td>ISE Project Navigator</td><td>Vivado IDE</td><td rowspan="2">These tools are all used for FPGA and Zynq hardware design. Vivado IDE replaces ISE Project Navigator and PlanAhead in the design flow, and has enhanced functionality and library support. PlanAhead and ISE have similar core functionality, but PlanAhead also has pin and device planning and visualisation facilities.</td></tr>
<tr><td>PlanAhead</td><td>Vivado IDE</td></tr>
<tr><td>Xilinx Synthesis Technology (XST)</td><td>Vivado Synthesis</td><td>Vivado Synthesis is an enhanced synthesis tool for 7 series and subsequent devices.</td></tr>
<tr><td>ISim</td><td>Vivado Simulator</td><td>Vivado Simulator is visually similar to ISim, but it has a new simulation engine with improved performance.</td></tr>
<tr><td>Xpower Analyzer</td><td>Vivado Power Analyzer</td><td>For evaluating the power consumption of designs operating on a target device.</td></tr>
<tr><td>System Generator</td><td>System Generator</td><td>For block-based DSP design. No significant functional changes, but System Generator systems can now be generated to Vivado IP cores.</td></tr>
<tr><td>AutoESL</td><td>Vivado HLS</td><td>Tools for developing IP from high-level C, C++ or System-C descriptions. Vivado HLS represents a rebranded and enhanced version of AutoESL.</td></tr>
<tr><td>Xilinx Platform Studio (XPS)</td><td>IP Integrator</td><td>XPS is used to architect embedded hardware systems using lists, options etc. IP Integrator provides an enhanced graphical environment for undertaking the same task.</td></tr>
<tr><td>Software Development Kit (SDK)</td><td>Software Development Kit (SDK)</td><td>For software systems development. No changes to the purpose or general functionality of this component.</td></tr>
</table>

Table 3.2: Correspondence between Vivado and ISE Design Suites

ISE	Vivado	Comments
ChipScope	Vivado Logic Analyzer	For probing and inspection of real-time signal behaviour on physical devices. The Vivado Logic Analyzer is underpinned by updated hardware cores.
iMPACT	Vivado Device Programmer	Tools for scanning the hardware chain and configuring the identified devices by downloading programming files.

Some of the changes to the software suite are very noticeable, and lead the designer to work in a different way. For example, ISE's Xilinx Platform Studio (XPS) and Vivado's IP Integrator are both environments for designing the hardware aspect of embedded systems, but they have very different user interfaces. XPS uses a series of drop-down lists and text-based configuration options, whereas IP Integrator provides a more graphical interface. Other changes are more subtle; for instance, the synthesis and place & route engines in Vivado constitute improvements over ISE, but these are largely transparent to the user. For more background on this topic, the reader is referred to [9] for an excellent discussion of the origins of the ISE Design Suite, and Xilinx' reasons for reinventing their development tools in the form of Vivado.

It is also worth reiterating that the Vivado flow supports 7 series and Zynq-7000 devices onwards, but is not compatible with older devices (Spartan®, Virtex®-6 and previous FPGAs). Meanwhile, ISE's device support will not extend beyond the 7 series.

3.5.2. Upgrading to Vivado

It is recommended that all new designs are started in the Vivado suite. From version 2013.2 onwards, Vivado has full support for Zynq and, as noted over the last few pages, the IP-centric Vivado design flow is more suitable for systems design, speeding up the design process.

The potential exists to migrate projects created in the ISE flow to Vivado, but not to traverse in the opposite direction. For instance, XPS designs can be updated to IP Integrator, and ISE / PlanAhead projects to Vivado. It is beyond the scope of this book to explain procedures for migration from the ISE flow to Vivado, but comprehensive guidance is available in [24].

One further important difference between ISE and Vivado is that of the constraints file type. In the ISE flow, the UCF file (*.ucf) is used (the acronym standing for User Constraints File), whereas in the Vivado flow, the XDC (*.xdc) file type is adopted (Xilinx Design Constraints). The new file type brings compatibility with the industry standard *Synopsys Design Constraints* [16], augmented with some specific Xilinx constraint types. The syntax for expressing constraints in XDC files differs significantly from UCF, hence established users of ISE should familiarise themselves with the new style. An explicit conversion is required to update UCF files to XDC [24], [25].

Users of System Generator may also notice the major upgrade of the MATLAB / Simulink interface at version 2012b (this is independent of the ISE to Vivado upgrade, but is nonetheless worth being aware of). The improvements are chiefly aesthetic, but there is also a difference in file types: Simulink models previously had the file extension *.mdl, and they are now by default *.slx. Any new models created will automatically use the new file type, which System Generator supports without issue. Any older models created as *.mdl files may still be opened, manipulated and saved, while new models may also be saved as *.mdl for backwards compatibility, if desired.

3.6. Development Boards

At the time of writing. there are a number of development boards available for Zynq, and this section is dedicated to providing an overview of each of these boards. Please bear in mind that others may be released between our time of writing and your time of reading!

Evaluation boards play an important role in the development process, and may be used extensively for incremental testing as the design progresses. They usually have a variety of peripheral interfaces to facilitate prototyping for communications, DSP, video processing, and many other applications, and come supplied with reference designs to demonstrate the use of these facilities.

3.6.1. Zynq-7000 SoC ZC702 Evaluation Kit

This evaluation kit features a board equipped with a Z-7020 Zynq device, which is based on the Artix-7 PL fabric. The kit itself includes a number of components as shown in Figure 3.8; this represents a typical 'evaluation kit' content. The various different parts are numbered and their descriptions provided after the figure, for clarification.

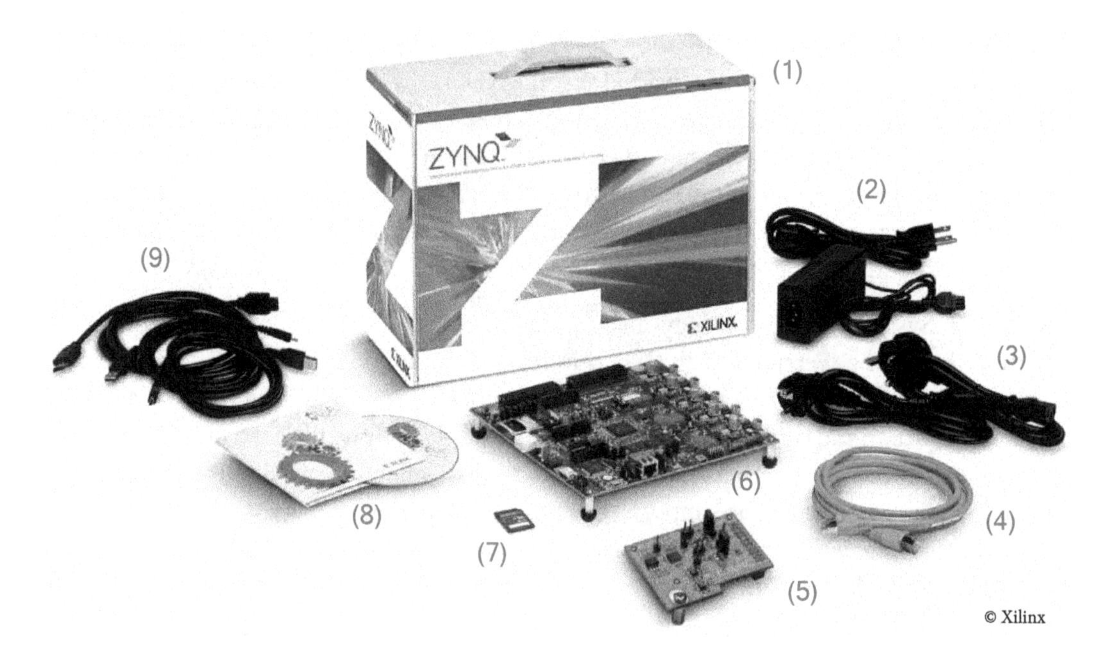

Figure 3.8: Contents of the Zynq-7000 ZC702 Evaluation Kit [37]

The numbered components in Figure 3.8 are as follows:

1. ZC702 Evaluation Kit packaging
2. Power adapter and US power cable
3. EU and UK power cables
4. Ethernet cable
5. AMS 101 Evaluation Board (ADC expansion board)
6. ZC702 Zynq Evaluation Board
7. SD memory card
8. Xilinx Design tools on DVD (device locked)
9. USB cables

Further information about this kit can be found on the Xilinx website [30], [35].

3.6.2. Zynq-7000 SoC Video & Imaging Kit

The Video and Image Processing Kit builds upon the ZC702 Evaluation Kit with some additional equipment to support image and video processing applications. The board, which forms the central element of the kit, is identical to that in the ZC702 Evaluation Kit.

The extra items included in the Video and Imaging Kit are as listed below:

1. Video Expansion Board
2. Image Sensor (video camera) including lenses, cable and tripod
3. HDMI cable
4. HDMI-DVI adapter

Additional information about this kit can be obtained from [32].

3.6.3. Zynq-7000 ZC706 Evaluation Kit

The ZC706 kit features a larger Zynq device than the previous two kits: the Z-7045, which is based on Kintex 7 PL fabric. This is one of the larger Zynq-7000 series devices, and it features GTX transceivers and PCI Express.

The board itself includes an increased memory provision over the ZC702-based boards, a PCI Express connector for use with the PCI Express facilities on the Z-7045 device, and SMA and SFP ('small form-factor pluggable') connectors for use with the embedded GTX transceivers. For more information on the ZC706 kit, please refer to [31].

3.6.4. ZedBoard

The ZedBoard is not just an evaluation kit, but a community as well. We will postpone further discussion on this topic until Chapter 6, which is in fact an entire chapter dedicated to the ZedBoard!

3.6.5. ZYBO

The ZYBO (diminutive of *Zynq Board*) is an ultra-low cost alternative to the ZedBoard featuring the smallest Zynq device, the Z-7010, which is based on the Artix-7 PL fabric. It is aimed at designers looking to get started developing for Zynq but who do not have a requirement for the high density I/O or the FMC connector present in mid-level and above boards. Figure 3.9 demonstrates how the ZYBO manages to include memory, video and audio I/O, Ethernet, various GPIO, six Pmod connectors and more on a compact board, roughly the size of two credit cards place side by side!

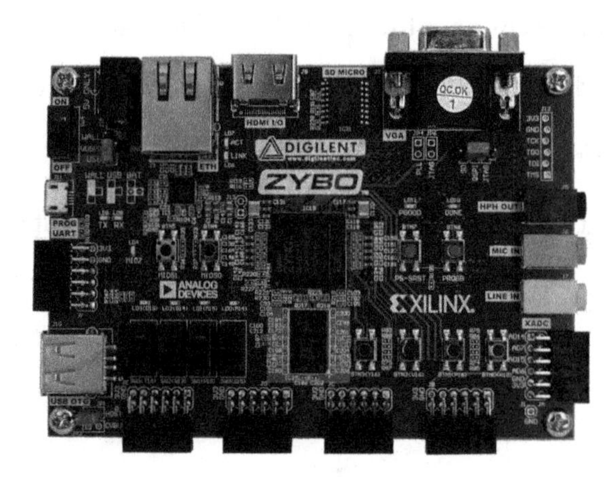

Figure 3.9: The ZYBO Development Board

3.6.6. Third Party Boards

At the time of writing, a number of third party development board are available based on Zynq-7000. These are highlighted below.

OZ745 Zynq SoC Video Development Kit

Produced by OmniTek, this board features a Z-7045 Zynq device, and is targeted at video processing applications, featuring a large number and variety of connectors for video peripherals. More details about this kit are available from the manufacturer [12].

MicroZed Evaluation Kit

The MicroZed is a low-cost development board from Avnet which features the smallest member of the Zynq-7000 series — the Z-7010. The design of the board allows it to operate in one of two modes: as a standalone development board; or as an embedded system-on-module (SOM), whereby it is paired with a compatible carrier card. More details about this evaluation kit is available from the manufacturer [1], or alternatively on the MicroZed community website [8].

The Parallella Board

The Parallella is a credit card sized platform which is built upon the combination of a Xilinx Zynq-7000 device (either the Z-7010 or Z-7020) and an Epiphany multi-core coprocessor by Adapteva [14]. The platform is designed to be affordable, energy efficient and open source. More information on this board is available from the Parallella website [13].

NI myRIO

The NI myRIO teaching platform from National Instruments, is a portable reconfigurable I/O (RIO) device targeted at students who can use it to design control, robotics and mechatronics systems [11]. It features the Zynq Z-7010 device, and is designed to work with the labVIEW system design software. More information on the NI myRIO is available from the manufacturer [10].

3.6.7. Accessories and Expansions

There are some standard connector types that can be used to extend the capabilities of your development board by attaching external modules. In this way, additional functionality can be added — anything from simple input/output devices such as buttons and LEDs, to fully functioned Software Defined Radio (SDR) modules.

The expansion connectors supporting these modules fall into the following categories:

- ***FPGA Mezzanine Connectors (FMCs)*** — A standardised FPGA interface for supporting FMC cards. High data throughput is available, and this type of card is suitable and widely used for data conversion (DAC and ADC), serial connectivity, SDR, and video processing. Examples of currently available FMC cards can be found on the Xilinx website [18].

- ***Pmods*** — This is a simple interface type for adding small peripheral modules (hence the name **Pmod** = **P**eripheral **mod**ule), using either 6-pin or 12-pin interfaces. The name was standardised by, and is a trademark of, *Digilent Inc.*, but other companies such as *Maxim Integrated* also produce Pmods [7]. Typical Pmods are sensors, actuators, data converters, and user I/O devices. There are also some communications transceivers available. Pmods can also facilitate direct wire connections.

- ***XADC Header*** — For connecting a board to interface with the on-chip XADC component for ADC functionality. An example of this type of expansion module is the AMS101 included within the ZC702 Evaluation Kit.

3.6.8. Working with Development Boards

When working with FPGA and Zynq development boards, the user should be mindful of the risk of damage due to Electro-Static Discharge (ESD). This risk can be mitigated by following recommended ESD bench practices, including the use of an ESD mat and wrist straps, grounding of tools, etc. Further guidance for professionals and hobbyists can be found in [3] and [5] respectively.

3.7. Support and Documentation

There are extensive resources available for the Vivado development tools on the Xilinx website, many of which have a Zynq focus. It is also worth highlighting that ZedBoard-specific support materials can also be obtained (more detail on these to follow in Chapter 6).

The Xilinx support site,

http://www.xilinx.com/support.html

includes *User Guides* as the chief source of information, *Tutorials* which guide the reader through hands-on practical examples, and various other documents. There are also some very useful training videos. The above site should be the first port of call for all information on the design flow.

For more specific enquiries, Xilinx Answer Records and Support Forums may also be useful. The support forums provide an opportunity to seek input or guidance from other members of the community, and also to assist others.

http://forums.xilinx.com/

There may also be a set of reference designs accompanying your particular development board.

Finally, it is worth noting that further training resources are available via the Xilinx University Program (XUP) for academics and students [28]. We will return to this topic in Chapter 7, when the opportunities for Zynq and associated tools in education and research will be discussed.

3.8. Chapter Review

This chapter has been concerned with the practicalities of starting to work with Zynq. To do so, you need: (i) design tools, (ii) a development board, and (iii) your imagination! We have covered (i) and (ii) comprehensively, together with the design flow and methodologies for creating Zynq-based systems; (iii) is of course down to you!

Another important aspect of this chapter has been to highlight the philosophy of the Vivado design suite, and its orientation towards system-level design, IP integration, and design reuse. These design principles are relevant to the modern day task of designing SoCs; in particular, the need for rapid systems development.

The next two chapters will actually relate to (iii) to some extent. Having covered the architecture and design processes for Zynq, it is of course important to consider possible applications, and the motivations for choosing a Zynq device over other alternatives.

3.9. References

Note: All URLs last accessed June 2014.

[1] Avnet, "MicroZed Evaluation Kit".
Available: http://www.em.avnet.com/en-us/design/drc/Pages/MicroZed-Evaluation-Kit.aspx

[2] Eclipse website.
Available: http://www.eclipse.org/

[3] ESD Association, "Fundamentals of Electrostatic Discharge, Part 3: Basic ESD Control Procedures and Materials", 2010.
Available: http://www.esda.org/documents/FundamentalsPart3.pdf

[4] T. Feist, "Vivado Design Suite", Xilinx White Paper, WP416, v1.1, June 2012.
Available: http://www.xilinx.com/support/documentation/white_papers/wp416-Vivado-Design-Suite.pdf

[5] C. Harper, "The ESD (Electro-Static Discharge) Guide for the Hobbyist" webpage.
Available: http://www.circuitguy.com/guides/esd/

[6] MathWorks website.
Available: http://www.mathworks.com/index.html

[7] Maxim Integrated, "Pmod-Compatible Plug-In Peripheral Modules" webpage.
Available: http://www.maximintegrated.com/en/design/design-technology/fpga-design-resources/pmod-compatible-plug-in-peripheral-modules.html

[8] MicroZed Community Website.
Available: http://www.microzed.org/

[9] K. Morris, "Kind of a Big Deal: Xilinx Rebuilds Tools - From Scratch", Electronic Engineering Journal (online), May 2012.
Available: http://www.eejournal.com/archives/articles/20120501-bigdeal

[10] National Instruments, "NI myRIO", website.
Available: http://www.ni.com/myrio/

[11] National Instruments, "NI myRIO-1900 User Guide and Specifications", August 2013.
Available: http://www.ni.com/pdf/manuals/376047a.pdf

[12] OmniTek, "OZ745 - Zynq SoC Video Development Kit" product brief.
Available: http://www.omnitek.tv/sites/default/files/OZ745.pdf

[13] Parallella, "Parallella Computer Specifications".
Available: http://www.parallella.org/board/

[14] Parallella, "Parallella Reference Manual", Rev 13.11.25.
Available: http://www.parallella.org/docs/parallella_manual.pdf

[15] R. Sass and A. G. Schmidt, "Partitioning" in *Embedded Systems Design with Platform FPGAs*, Morgan Kaufmann, 2010, pp. 197 - 246.

[16] Synopsys, "Synopsys Design Constraints (SDC)" webpage.
Available: http://www.synopsys.com/Community/Interoperability/Pages/TapinSDC.aspx

[17] Xilinx, Inc., "Embedded System Tools Reference Manual", UG1043, v2014.1, May 2014.
Available: http://www.xilinx.com/support/documentation/sw_manuals/xilinx2014_1/ug1043-embedded-system-tools.pdf

[18] Xilinx, Inc., "FPGA Mezzanine Card (FMC) Standard" webpage,
Available: http://www.xilinx.com/products/boards_kits/fmc.htm

[19] Xilinx, Inc., "Hardware In The Loop (HIL) Simulation for the Zynq-7000 All Programmable SoC", XAPP744, v1.0.2, November 2012.
Available:
http://www.xilinx.com/support/documentation/application_notes/xapp744-HIL-Zynq-7000.pdf

[20] Xilinx, Inc., "Hardware/Software Cross-Trigger for Embedded Design", video.
Available:
http://www.xilinx.com/training/zynq/hardware-software-cross-trigger-for-embedded-design.htm

[21] Xilinx, Inc., "Memory Recommendations: FPGA Memory Recommendations Using the Vivado Design Suite" webpage.
Available: http://www.xilinx.com/design-tools/vivado/memory.htm

[22] Xilinx, Inc., "Standalone (v.4.0)", UG647, April 2014.
Available: http://www.xilinx.com/support/documentation/sw_manuals/xilinx2014_1/oslib_rm.pdf

[23] Xilinx, Inc., "Vivado Design Suite Evaluation and WebPACK" webpage.
Available: http://www.xilinx.com/products/design_tools/vivado/vivado-webpack.htm

[24] Xilinx, Inc., "ISE to Vivado Design Suite Migration Guide", UG911, v2014.1, April 2014.
Available: http://www.xilinx.com/support/documentation/sw_manuals/xilinx2014_1/ug911-vivado-migration.pdf

[25] Xilinx, Inc., "Vivado Design Suite Tutorial: Using Constraints", UG945, v2014.1, April 2014.
Available: http://www.xilinx.com/support/documentation/sw_manuals/xilinx2014_1/ug945-vivado-using-constraints-tutorial.pdf

[26] Xilinx, Inc., "Vivado Design Suite User Guide: Programming and Debugging", UG908, v2014.1, May 2014.
Available: http://www.xilinx.com/support/documentation/sw_manuals/xilinx2014_1/ug908-vivado-programming-debugging.pdf

[27] Xilinx, Inc., "Vivado Design Suite User Guide: Release Notes, Installation and Licensing", UG973, v2014.1, May 2014.
Available: http://www.xilinx.com/support/documentation/sw_manuals/xilinx2014_1/ug973-vivado-release-notes-install-license.pdf

[28] Xilinx University Program webpage.
Available: http://www.xilinx.com/university/index.htm

[29] Xilinx, Inc., "Xilinx Software Development Kit (SDK)" product webpage.
Available: http://www.xilinx.com/tools/sdk.htm

[30] Xilinx, Inc., "Xilinx Zynq-7000 SoC ZC702 Evaluation Kit" webpage,
Available: http://www.xilinx.com/products/boards-and-kits/EK-Z7-ZC702-G.htm

[31] Xilinx, Inc., "Xilinx Zynq-7000 SoC ZC706 Evaluation Kit" webpage,
Available: http://www.xilinx.com/products/boards-and-kits/EK-Z7-ZC706-G.htm

[32] Xilinx, Inc., "Xilinx Zynq-7000 SoC Video and Imaging Kit" webpage,
Available: http://www.xilinx.com/products/boards-and-kits/DK-Z7-VIDEO-G.htm

[33] Xilinx, Inc., "Zynq-7000 All Programmable SoC Software Developers Guide", UG821, v9.0, June 2014.
Available: http://www.xilinx.com/support/documentation/user_guides/ug821-zynq-7000-swdev.pdf

[34] Xilinx, Inc., "Zynq-7000 All Programmable SoC Technical Reference Manual", UG585, v1.6, June 2013.
Available: http://www.xilinx.com/support/documentation/user_guides/ug585-Zynq-7000-TRM.pdf

[35] Xilinx, Inc., "Zynq-7000 EPP ZC702 Evaluation Kit", Product Brief.
Available: http://www.xilinx.com/publications/prod_mktg/zynq-7000-kit-product-brief.pdf

[36] Xilinx, Inc., "Zynq-7000 Platform Software Development using the ARM DS-5 Toolchain", XAPP1185, v2.0, May 2014.
Available: http://www.xilinx.com/support/documentation/application_notes/xapp1185-Zynq-software-development-with-DS-5.pdf

[37] Xilinx, Inc., Zynq ZC702 evaluation kit (image reference).
Available: http://www.xilinx.com/products/boards-and-kits/EK-Z7-ZC702-G-image.htm

4

Device Comparisons (“*Why do I need Zynq?*”)

In the same way that processors or FPGAs can be applied to a multitude of different problems, the same is true of Zynq. There is not one single application for Zynq, but many of them; potential applications include wired and wireless communications, automotive, image and video processing, high performance computing, and numerous others. We will discuss some of these possibilities in more detail later, in Chapter 5.

Before doing so, it is beneficial to consider the characteristics of Zynq as compared to other device alternatives, so that its suitability for these candidate applications can be clearly understood. Amongst the factors to be considered are parallel processing resources, processing capability, bandwidth, latency, and flexibility. As well as these architectural aspects, there are also important practical and commercial considerations to take into account, including Bill of Materials (BOM) costs, physical size, and power consumption.

Over the next few sections, processors and FPGAs will be compared and contrasted with Zynq. Three comparisons will be made: Zynq against an FPGA; Zynq against a processor; and Zynq against the combination of a processor and FPGA together. The last option represents a direct, discrete-component equivalent to Zynq. In each case, it will be observed that Zynq offers a number of possible advantages.

Towards the end of the chapter, we will introduce the specific support within the Vivado suite for rapidly porting functionality from software into hardware implementation. It will be seen that this 'high-level synthesis' approach can help to quickly achieve a favourable partitioning between hardware and software elements, thus exploiting the Zynq architecture.

4.1. Device Selection Criteria

Before proceeding to compare Zynq with other device alternatives, it is useful to briefly review the parameters against which these options will be assessed. These can be grouped into five themed categories as given in Table 4.1.

Table 4.1: Factors involved in selecting a device (or devices)

Categories	Factors
Device capabilities	Processor performance Logic performance Memory performance Support for high-speed arithmetic Support for I/O and communications Security features and support for secure booting Bandwidth between processor and logic Latency between processor and logic
Commercial factors	Bill of materials Developments costs (see below) Integration (and implied costs) Longevity of device availability and technical support Quality and reliability Ease (cost) of providing field upgrades
Design and development	Time to market Fast, convenient and reliable design flows Integrated verification Integration with other development tools Support for team-based design flows Support for design reuse Support for industry standard design formats Support for desired design entry methods Documentation and technical support

Table 4.1: Factors involved in selecting a device (or devices)

Categories	**Factors**
Physical device characteristics	Physical size Power consumption Ease of integration and resulting PCB complexity Connectivity Ruggedness Radiation sensitivity Supported temperature range
Flexibility	Scalability *(larger and smaller devices available and can be adopted with little or no rework)* Portability *(designs in standard forms that can be ported from/to other platforms)* Re-programmability *(ability to change functionality in the field or even dynamically at run time)* Ease of partitioning *(ability to partition functionality between hardware and software)* Expansibility *(ease of integrating new functionality)*

The technical and commercial requirements of any particular design task will dictate the relative priorities of these factors. In other words, the most important criteria for Project X in Company A (who might be designing a ground-based military radar system) may not match those of Project Y in Company B (who are developing a low cost embedded sensor node for managing green buildings). We will discuss applications and related considerations further in Chapter 5.

Notwithstanding all of the above, one factor that is usually near the top of the list, along with technical performance criteria, is cost. That is: (i) the BOM costs; and (ii) the development costs involved in developing products based on the target device(s). As will be discussed during the remainder of this chapter, the capabilities of Zynq are such that it can potentially replace two devices, which constitutes a cost saving. There are also significant improvements in productivity possible due to the new Vivado Design Suite and its associated flows, leading to a saving in development time and cost compared to previous methods.

4.2. Comparison A: Zynq versus FPGA

An important point to reiterate is that the programmable logic of Zynq is equivalent to that of an FPGA. The PL of the smaller Zynq devices corresponds to the fabric of Artix-7 FPGAs, while the larger ones are equivalent to Kintex-7. Our discussions over the next few pages will focus on the differing possibilities for implementing embedded processors on these devices.

In embedded applications, it is usual to require one or more processors to orchestrate the system, support software, and co-ordinate exchanges with peripheral components. FPGAs have been commonly used with soft processors for more than a decade, and as these devices are deployed for ever more sophisticated applications, the need for processor-based systems grows.

As mentioned in Chapter 2, the Zynq architecture includes a hard processor, but it is also possible to build soft processors using programmable logic. Current standard FPGAs are composed of programmable logic without a hard processor facility, and therefore it is reasonable to explore the option of soft processors in greater detail. This will help us to understand why Zynq may be considered 'better' for certain applications than an all-FPGA solution including a soft processor.

In essence, the question here is, "*What can Zynq do for me that an FPGA with a soft processor can't?*".

In order to make appropriate comparisons and hence answer this question, we must first define the standard FPGA soft processors that are available. There is one primary type, the MicroBlaze, which is a 32-bit soft processor with extensive support within the Xilinx toolflow, and there are also some further options.

4.2.1. MicroBlaze Processor

MicroBlaze is the principal soft processor type and is supported within both the Xilinx ISE and Vivado design flows, including the most recent releases. Multiple MicroBlaze instances can be deployed on a single device, if desired. There is no implied licensing cost of using MicroBlaze processors in system designs, commercially or otherwise.

Configuration

One of the benefits of using a soft-processor is that its configuration is flexible. The MicroBlaze has a number of different architectural options which can be included or excluded from the processor implementation depending on the requirements of the target application. For example, the FPU can be excluded if the system does not call for floating point computation, thus reducing the footprint of the processor implementation on the FPGA (i.e. the amount of resources it requires). In a more general sense, the configuration of the MicroBlaze can be customised to optimise for operating frequency, performance, or area; alternatively, it can be specified such that a suitable balance between these three metrics is achieved. This is done very easily in Vivado using a configuration wizard.

MicroBlaze resource utilisation varies with configuration, starting at approximately 900 LUTs, 700 FFs, and 2 Block RAMs for the 'minimum area' option, and rising to about 3800 LUTs, 3200 FFs, 6 DSP48E1s and 21 Block RAMs for the 'maximum performance' configuration. Sample floorplans of these, targeting the Zynq XC7Z020, are shown in Figure 4.1.

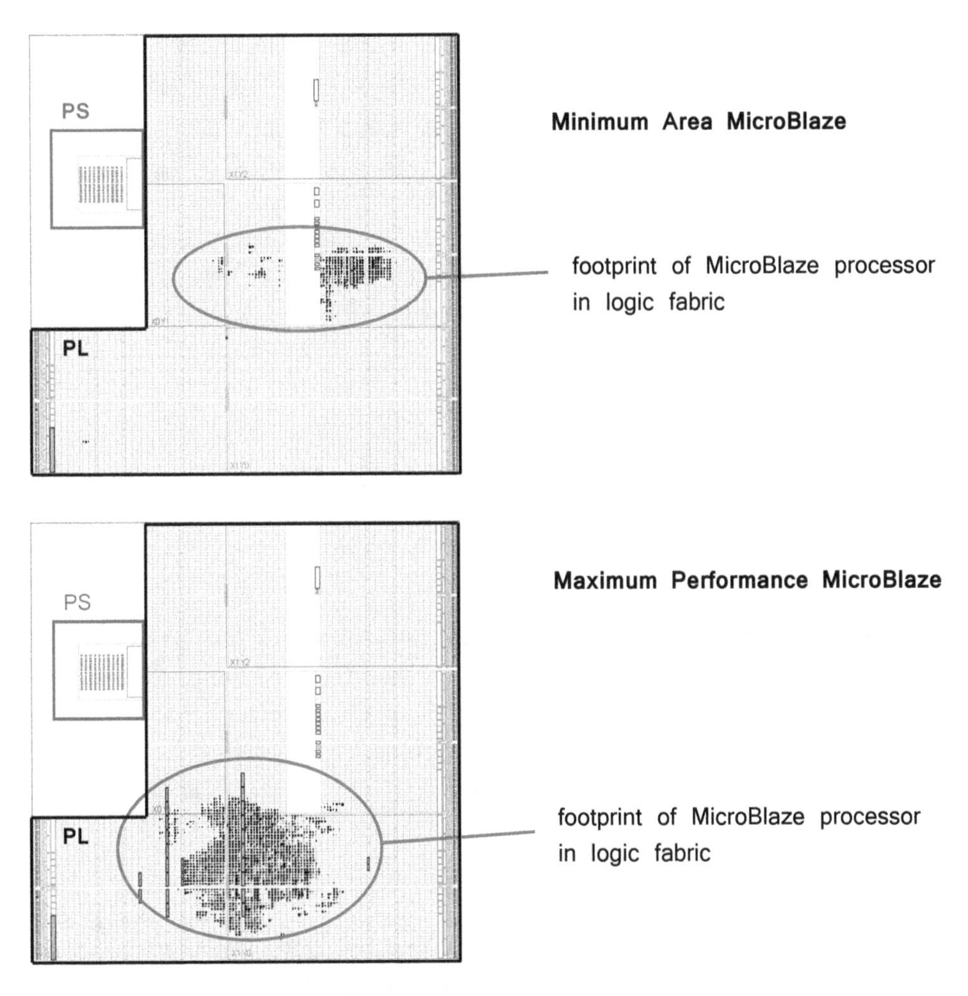

Figure 4.1: Floorplans of 'minimum area' (top) and 'maximum performance' (bottom) MicroBlaze soft processor implementations

As mentioned earlier, the number of implemented MicroBlaze instances is also configurable, and this adds another dimension of flexibility. For example, the closest MicroBlaze equivalent to the (dual-core) ARM Cortex-A9 processor would comprise two MicroBlaze instances.

Processing Performance

The maximum frequency attainable by MicroBlaze is dependent on its configuration, which is customisable, and also other factors such as placement and routing on the PL. To provide a rough indication, a typical MicroBlaze configuration might achieve about 70% of the maximum frequency of the PL, which equates to, at most, two or three hundred MHz — this compares to the ARM processor's maximum operating frequency of 800MHz to 1GHz.

Processor performance is normally evaluated using a benchmark. In order to quantify the performances of the ARM Cortex-A9 and MicroBlaze, and thus compare them, two widely used benchmarks can be used:

- ***DMIPs (Dhrystone Millions of Instructions Per second) [19]*** — The quantity DMIPs expresses the number of operations achieved per second undertaken by the processor when running the *Dhrystone* standard test application. Dhrystone is a synthetic application (i.e. it is not representative of real work), specifically designed to exercise the processor with a representative set of processor operations.

- ***CoreMark score [6]*** — CoreMark establishes a simple numerical 'score' for processor performance, which can be directly compared with the scores of other processors. The CoreMark application serves the same purpose as Dhrystone, but its content is tailored to execute a set of operations truer to typical embedded processor usage.

DMIPs and CoreMark are measured (as opposed to calculated) metrics for quantifying processing capability. Although they are both derived from running specific, freely available test applications on the processor being evaluated, there are some fundamental differences between them. For a range of reasons, CoreMark is generally considered to be a more robust and realistic benchmark than the older Dhrystone method, and indeed ARM recommends the use of CoreMark [1], [5].

According to Xilinx literature [8], the three MicroBlaze configurations listed in Table 4.2 can achieve no more than 260DMIPs on Zynq in speed grade -3, whereas the dual-core ARM is projected to reach 5000DMIPs (2500DMIPs per core), assuming a PS clock frequency of 1GHz [17]. This indicates that the ARM processor offers approximately a 20 times greater processing performance than a single MicroBlaze core. DMIPs figures should however be treated with a little caution, and these figures considered 'best case' results.

Table 4.2: Maximum MicroBlaze and ARM Cortex-A9 performance on Zynq (DMIPs)

Processor Type	**Configuration**	**Processing Performance (DMIPs)[a]**
MicroBlaze	Area optimised (3-stage pipeline)	196[b]
	Performance optimised with branch optimisations (5-stage pipeline)	228[b]
	Performance optimised without branch optimisations (5-stage pipeline)	259[b]
ARM Cortex-A9	1GHz; both cores combined; 2500 per core	5000[c]

a. All figures are based on speed grade -3 (the fastest grade) and should be considered best case values.
b. Statistic from [8].
c. Statistic from [17].

CoreMark figures can also be obtained to compare the Zynq ARM processor with MicroBlaze. At the time of writing, however, the latest available MicroBlaze score arises from a Virtex-5 FPGA implementation, rather than Zynq or a 7 series FPGA, and for a single MicroBlaze core only [6]. These are given in Table 4.3 for operating frequencies as stated (figures of CoreMark/MHz are also available). The figures show a large difference between the capability of the ARM Cortex-A9 processor on Zynq, and that of MicroBlaze.

Table 4.3: Maximum MicroBlaze and ARM Cortex-A9 performance on Zynq (CoreMark)

Processor Type	**Configuration**	**Processing Performance (CoreMark)[a]**
MicroBlaze	125MHz; 5-stage pipeline (Virtex-5)	238
ARM Cortex-A9	1GHz; both cores combined	5927
ARM Cortex-A9	800MHz[b]; both cores combined	4737

a. Statistics from [6].
b. This figure does not relate to the fastest available speed grade.

Other Features and Factors

There are several important differences between the MicroBlaze and ARM Cortex-A9 processors. Among them: the MicroBlaze is a single core processor compared to ARM's dual-core; the ARM has a richer instruction set than the MicroBlaze; the MicroBlaze FPU implements only single precision floating point, whereas the ARM also supports double precision; and the cache configuration of the MicroBlaze provides a single level cache, whereas the ARM has a two-level cache with greater capacity. These architectural and functional differences help to account for the difference in performance between the two processor types.

Taking all of these factors into account, it is obvious that the ARM Cortex-A9 processor is better equipped than the MicroBlaze. Even so, MicroBlaze is still a very suitable choice for many applications. In the specific context of Zynq, a MicroBlaze can act as a useful 'subordinate' processor to the ARM. For example, a MicroBlaze could be used to control a subset of the PL system functionality. Given its status as a soft resource, it is also true that multiple MicroBlaze processors can be implemented in the PL if desired, and if the available logic permits. This, together with their inherent configurability, renders them a very flexible processing resource.

To make a simple and direct comparison between Zynq's ARM processor and a MicroBlaze processor implemented on an FPGA, it is clear that the ARM has a significant advantage in terms of the processing facilities and performance available. Therefore, it can be seen that Zynq brings a clear advantage to the implementation of processor-intensive applications: it offers a level of performance unattainable by a standard FPGA.

4.2.2. MicroBlaze MicroController System

In 2012, a lightweight version of the MicroBlaze was introduced: the MicroBlaze Micro Controller System (MCS) [7]. It was designed for controller applications, and has a fixed architecture including an area-optimised MicroBlaze processor, combined with data and program memory and a standard set of peripherals. Although the basic architecture is predefined, there are some lower-level configuration options which affect the ultimate footprint of the MCS implementation on an FPGA. A rough cost is 550 - 700 LUTs and 300 - 600 FFs, or more if debug features are included [7]. Timing performance varies depending on the target device and the other elements of the implemented system.

As is the case with the MicroBlaze Processor, the MCS is compatible with the Zynq PL. An MCS instance may form part of a Zynq-based SoC design, typically implementing low-level control functionality under the supervision of applications running on the more capable ARM.

4.2.3. PicoBlaze

The PicoBlaze is a microcontroller rather than a processor (i.e. it comprises other facilities besides the processing element, and supports a limited but useful set of operations). However, it is worth including PicoBlaze here for completeness, and to establish how it differs from the similarly named *MicroBlaze*. This 8-bit soft microcontroller IP has a very small footprint (a few tens of slices plus program memory) and is capable of implementing finite state machines and other simple control functionality [10]. The designs for PicoBlaze can be obtained as a download directly from the Xilinx website, and the fileset includes VHDL and Verilog for the core PicoBlaze controller, together with optional functionality such as UART and SPI interfacing.

Given that it is an 8-bit controller, PicoBlaze functionality is limited, and incomparable to that of a Zynq ARM processor. However, a PicoBlaze instance can run at over 200MHz in Kintex-7 logic fabric, in most cases as fast as the logic it may be controlling [4]. Therefore, the PicoBlaze can be viewed as a valuable resource of a different kind. It is possible that this compact controller may play a useful role within a Zynq or MicroBlaze based embedded system, controlling lower-level functionality.

4.2.4. ARM Cortex-M1

ARM offers a 'soft-core' microcontroller, the ARM Cortex-M1, which is optimised for FPGA implementation. Therefore in Zynq, this core would be implemented in the PL section of the device to complement the processing undertaken by the ARM Cortex-A9. Like the MicroBlaze, the configuration of the Cortex-M1 can be specified according to user requirements, meaning that the logic resources required to implement the core may be minimised.

4.2.5. Other Processor Types

There are some other FPGA embedded processors which it is useful to be aware of, and these can be categorised as *soft* and *hard* processors. We have already discussed in detail the Zynq ARM processor (a hard processor) and the MicroBlaze (a soft processor), and here we briefly review other processors falling into these categories.

Soft Processors

MicroBlaze is the most prevalent soft processor in Xilinx FPGA and SoC designs, due to both the integrated and extensive support provided for it, and its excellent implementation and performance characteristics. However, it is not the only soft processor available, and third party processor IP is available as an alternative, or to cater for niche applications.

Example third party processors include ***LEON4*** and ***OpenRISC***. As an example, product information for LEON4 states its performance as 1.7DMIPs/MHz or 2.1 CoreMark/MHz, and that it can reach up to 125MHz on a Virtex-5 device, while the area requirement is given as 4,000 LUTs [3]. OpenRISC is a collaborative open source project hosted by OpenCores; performance and area statistics are not readily available. In both cases, the processor cores are not exclusively for use on FPGAs, but can also be targeted at ASIC implementation. This is also true of the OpenSparc project. ***OpenSparc*** is an open-source, 64-bit Reduced Instriction Set Computer (RISC) processor developed by Sun MicroSystems, one particular version of which specifically targets FPGA implementation. There have been two major releases of OpenSparc to date: T1 (in 2006), and T2 (in 2008). The 64-bit architecture clearly differentiates OpenSparc from other soft processors, which are generally 32-bits, and although of interest to the research community, it has not been widely adopted commercially. This is possibly a reflection that the more mature 32-bit processors satisfy the majority of current FPGA-based embedded processing requirements.

Hard Processors

The single hard processor to be discussed is the IBM ***PowerPC***®, which was included as a hard processor in the Virtex-II Pro (released in the 2002 [11]) and subsequently in a subset of Virtex-4 and Virtex-5 FPGAs [12], [13]. Each of these FPGAs includes either one or two PowerPC (PPC) units.

Similarly to the ARM processor in the Zynq, the PowerPC hard processor offers a considerable performance advantage compared to a MicroBlaze implemented in the logic fabric of the same device. Taking the example of the most advanced PowerPC-equipped FPGA available, the PowerPCs in the Virtex-5 can each achieve up to 1,100 DMIPs (i.e. 2,200 DMIPs in total using one of the larger, two-unit devices), whereas MicroBlaze performance is around 240 DMIPs [14], [15]. Comparing these figures with those in Table 4.2 for the Zynq, it is evident that the Zynq ARM processor possesses more than double the processing capability of the PowerPC on the Virtex-5. It is not possible to make a direct comparison using the CoreMark benchmark at the time of writing, due to the lack of published performance figures.

The older, PowerPC-based embedded solutions established a clear motivation for devices combining hard processors alongside FPGA general purpose logic, and therefore can be considered near-direct predecessors of the Zynq All-Programmable SoC. These belong to older devices which, although still in production at the time of writing, are unsupported by the latest design tools and processes. They are therefore not recommended for new designs; Zynq is preferred. Zynq represents a new generation technology and offers significant improvements in hard processor performance, power consumption, programmable logic fabric, device integration, and tool support.

4.2.6. Summary Comments

In this section we have compared the Zynq processor against the alternative of an FPGA upon which a soft core processor is implemented. It has been noted that there are various types of soft processing resources available, including cores provided by Xilinx and by third parties. It is also possible, albeit with significant design effort, to design your own soft processor in-house.

By far the most prominent type of soft processor is Xilinx's *MicroBlaze*, which has customisable functionality and can be configured to optimise its processing performance, operating frequency, or area (or a combination thereof). MicroBlaze is integrated into Vivado and extensive support is available. It has been noted that MicroBlaze has a number of useful attributes, but that it cannot provide processing performance close to that of Zynq's ARM Cortex-A9 processor. MicroBlaze is however a very flexible and useful resource in its own right, and these two different processor types should be viewed as complementary. Figure 4.2 provides a brief graphical summary of the architectural comparison made in this section, where a MicroBlaze has been utilised as a representative soft processor implemented on an FPGA.

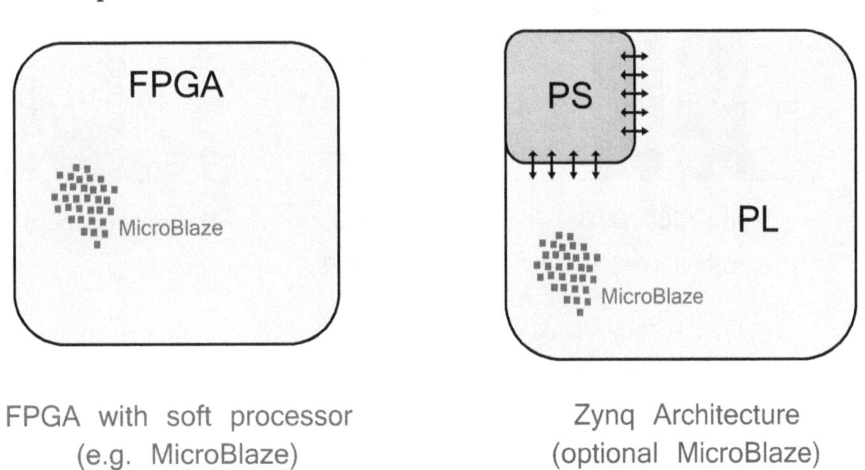

Figure 4.2: Comparison between an FPGA with soft processor (MicroBlaze), and a Zynq device with hard processor (ARM) and optional soft processor (MicroBlaze).

Other FPGA based embedded processor options have also been surveyed, and it has been shown that the processing facilities of Zynq are markedly enhanced compared to previous FPGA hard processor technologies, namely those based on the PowerPC. Figure 4.3 provides a graphical comparison of the considered processor types, based on published performance figures (given in term of DMIPs).

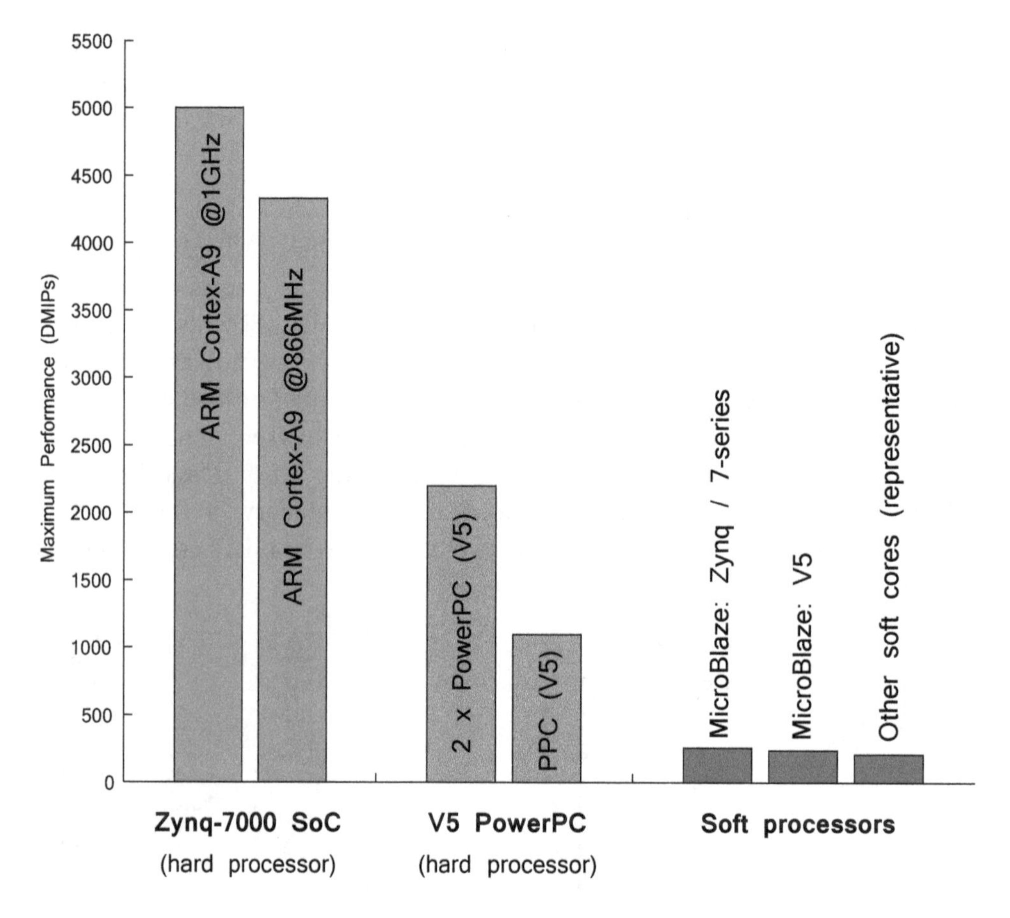

Figure 4.3: Performance comparison of hard and soft processor options (indicative only - extrapolated and based on best case)

Notably, the left and central groups represent hard processors, while the group on the right corresponds to a selection of soft processors. There is clearly a significant performance gap between the two. The Zynq-based ARM processor (leftmost group) operates at up to 866MHz (in the Artix-based devices) or up to 1GHz (in the Kintex-based devices), leading to two different levels of performance. These both provide a marked improvement compared to the PowerPC hard processors most recently incorporated in Virtex-5 FPGAs, which Zynq effectively supersedes.

In the Zynq, the high performance ARM Cortex-A9 hard processors can, if desired, be augmented with MicroBlaze or other soft processors to effectively form two layers of processing. This is effectively the architecture depicted on the right hand side of Figure 4.2.

4.3. Comparison B: Zynq versus Standard Processor

Those considering a standard processor for their next application may wish to evaluate the Zynq as an alternative. In this case, the question is slightly different: "*What might be gained by combining programmable logic with my processor?*".

To a large extent, the answer depends on the application area, and the type of processing that the processor is required to perform. Having chosen a processor, for instance a General Purpose Processor (GPP) or a Digital Signal Processor (DSP), it must be assumed that there is a need to support software routines, or an operating system with applications. It might be that the software implements computationally intensive operations which result in a bottleneck, and would benefit from hardware acceleration using the additional PL available in the Zynq. The benefit of PL is that it can be fully tailored and optimised for a particular task, whereas processors are more general purpose.

Programmable logic offers an ideal resource for implementing algorithms inherently parallel in nature; for example in signal and image processing, where mathematical operations are performed on a large number of samples or pixels simultaneously. Although there are specific resources available in some processors to cater for this type of application (such as the NEON engine in the ARM Cortex-A9), their performance cannot necessarily match that of optimised, task-specific hardware processing blocks implemented in FPGA-type programmable logic.

A wide variety of processors are available, and as mentioned in Section 4.2.1, their performances can be evaluated and compare using a standard benchmark. It is particularly convenient that the website of the Embedded Microprocessor Benchmark Consortium (EEMBC) provides a database of submitted CoreMark scores [6]. Through this, it may be confirmed that Zynq compares favourably with other implementations of the ARM Cortex-A9 architecture.

4.3.1. Processor Operation

The resources of a processor are fixed, and usually limited to one, two or four processing cores (more in some cases), which are required to operate at a specific clock frequency. The cost of implementing a desired software implementation is measured in terms of *clock (execution) cycles*, which will of course require some specific amount of time to execute at the desired clock frequency; the more complex the required processing, the longer the execution time will be. The efficiency of the implemented algorithm is also important, such that it does not include redundant operations.

To give an example, let's assume that we have a single-core processor operating at 1GHz. If a particular routine requires 1,825 execution cycles, then it will take 1.825μ s to complete, assuming that the processor is dedicated to this one particular routine. In the case that subsequent modifications are made, adding a further 500 cycles of computation, then the duration of the routine will extend to 2.325μ s. These two simple examples are illustrated in Figure 4.4. Augmenting the functionality of a routine via the addition of operations naturally increases the execution time, and this may be acceptable (within reason) for non time critical applications.

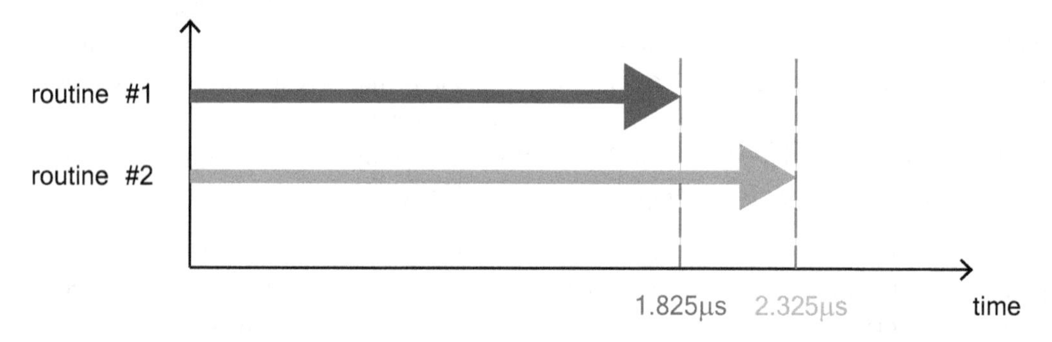

Figure 4.4: Visualisation of processing time for two variations of a software routine

Non Real-Time Operation

If we consider the behaviour of a generic processor, it has a finite number of timeslots (clock cycles) that are occupied — or not — by particular operations scheduled onto them. Some of these operations may take a single cycle; others several cycles. More specifically, the occupation of processor cycles can be represented in terms of program functions or tasks. It might be that the processor supports a number of different tasks that repeat regularly, or occur on an ad hoc basis, and which are scheduled onto the processor according to their relative priorities. This model of operation is depicted in Figure 4.5. Notice that, as the processor is a serial resource, it accommodates only one task during any one timeslot; this reflects the serial nature of processor operation. Of course, it is also true that processors are increasingly 'multi-core', meaning that they have two or four processing cores (for instance), each of which can process tasks in serial. Here we restrict the example to a single core for clarity.

As the processor becomes 'busy', the level of occupation of timeslots increases, and therefore its performance in terms of executing software routines may become slower. This is analogous to driving on a trunk road: when it is congested with traffic, the travelling time is likely to be longer. It is also true that the timeliness of completing particular tasks is variable as a result of the sharing of the processor resources between different tasks. In certain applications, this lack of determinism and ability to guarantee timing cannot be tolerated, and real-time processing is required.

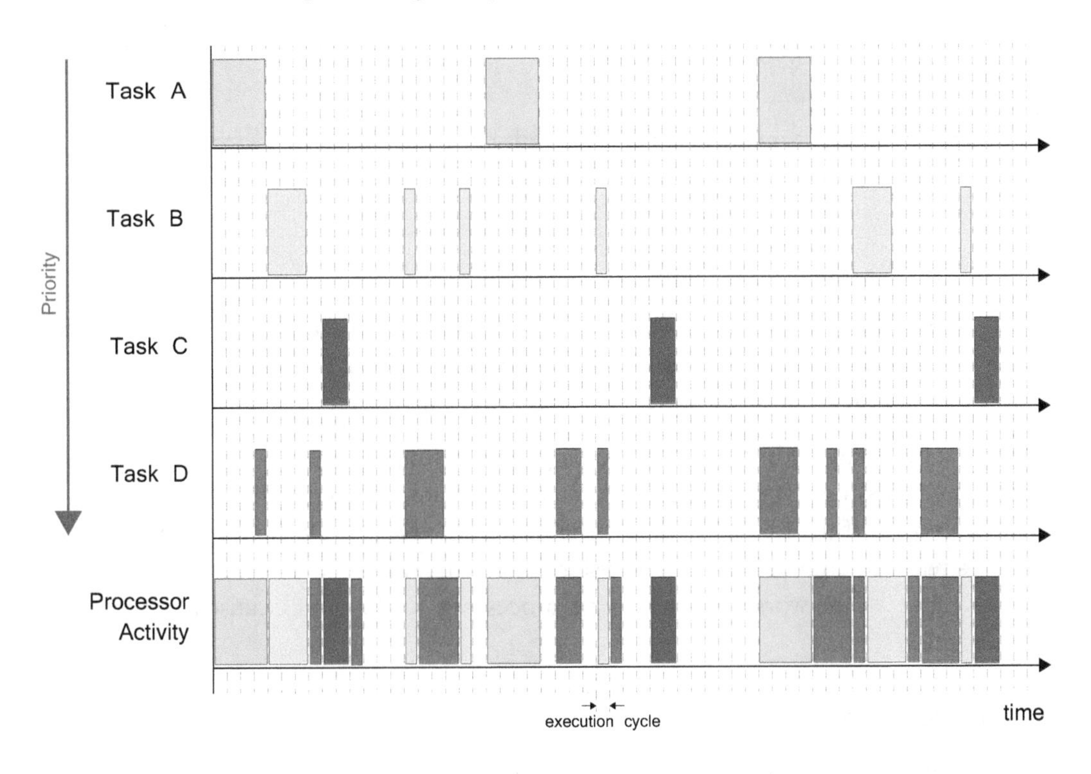

Figure 4.5: Processor operation: execution cycles

Real-Time Operation

When aiming for real-time operation, it should be considered that there is a 'budget' of execution cycles available, within which the desired software applications or algorithms must be executed. For example, a real-time video processing application demands a processing resource that can keep pace with data arriving at the desired frame rate and resolution; otherwise, the processor will still be busy with the previous frame(s) as new data arrives. This scenario is depicted in Figure 4.6 — notice that there are effectively processing deadlines by which all processing relating to each individual frame must be completed.

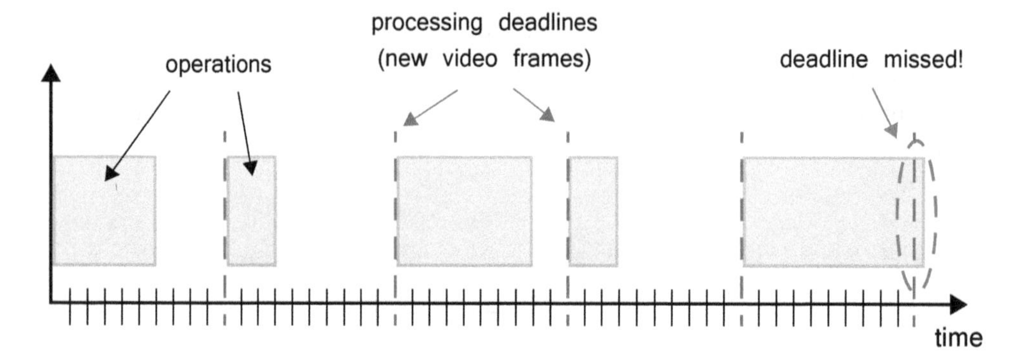

Figure 4.6: Real-time processing

Real-time systems can be classified as *soft* real-time or *hard* real-time. If *soft* real-time, the system performance is likely to degrade, usually temporarily, as a result of a missed processing deadline; if *hard* real-time, the system may fail completely.

In the above example of a video processing application, the consequence of skipping a deadline and missing a frame is likely to cause a brief loss of quality, however the system will soon recover if the following processing deadlines are met. On the other hand, if the deadlines relate to an industrial safety system, then this would be considered a hard real-time system as, in the worst case, missing a processing deadline may cause a catastrophic failure of the system.

4.3.2. Execution Profiling

When considering the activity of the processor, it is often useful to consider its activity with respect to the various tasks it must support. For instance, if the processor spends 80% of its execution cycles on a particular task, this should be investigated to establish whether any parallelism can be identified. If so, then partitioning this functionality into hardware can result in a significant speed-up overall.

For example, consider that we have a software routine which performs FIR filtering of some data, implemented in software on a processor. It may be observed that the processor spends a large proportion of its execution cycles performing the filter arithmetic to generate the results. As the computation of FIR filtering is highly parallel (i.e. multiple operations may take place simultaneously), this an ideal candidate for hardware acceleration. The NEON unit of the ARM Cortex-A9 processor is a suitable resource due to its

SIMD capabilities, although it has a limited number of processing 'lanes', as reviewed in Chapter 2, which may limit its performance. Better still, FPGA logic fabric permits a high-speed, fully parallel version of the filter to be implemented. This 'off-loading' of operations from the processor into hardware has the potential to greatly reduce the overall execution time, as illustrated in Figure 4.7. It also frees up the processor, such that it can undertake other operations using the processing cycles vacated by the accelerated function.

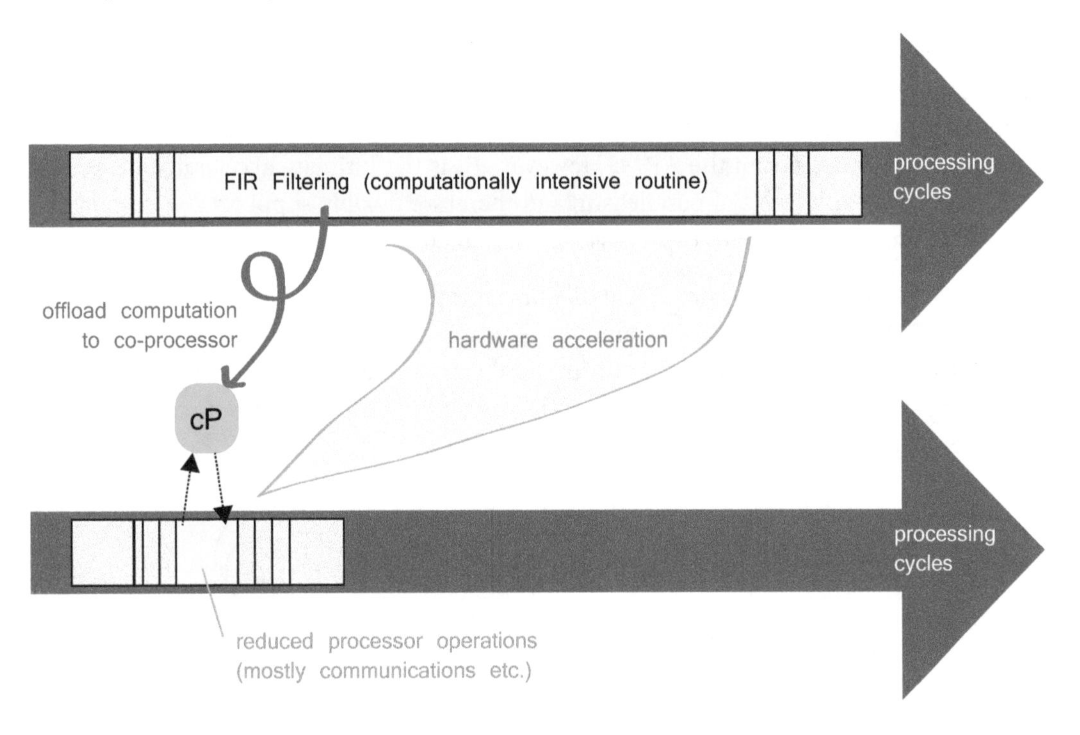

Figure 4.7: Processor activity before (top) and after (bottom) hardware acceleration of FIR filtering

One aspect to be aware of when partitioning functionality from software into hardware is the implied overhead of communicating between the two parts of the system. The time taken to transfer data and instructions between software and hardware constitutes an additional latency that offsets the processing speed-up; if the overhead is too large, the benefit of hardware acceleration is lost. As will be explained in the next section, an implication of using Zynq, as compared to a discrete processor and FPGA, is that the communication overhead is low; this is due to the tight coupling between the PS and PL components of the device.

4.3.3. Summary Comments

In this section, we have reviewed the general operating principles of processors, and considered the need for real-time processing in some applications. It has also been demonstrated that, while certain intensive routines *can* be implemented on a processor, they may represent a significant burden on its resources, potentially causing the overall performance of the processor to suffer.

The possibility of off-loading such routines onto a co-processor was put forward as a solution to this problem, and can be particularly effective when the nature of the off-loaded computation is parallel, and hence well-suited to the architecture of acceleration engines like the NEON processor in the ARM. However, PL is the ultimate accelerator because it can support arbitrary levels of parallelism, and therefore flexibly supports a wide range of algorithms, and even multiple co-processors simultaneously.

Figure 4.8 shows a simple, conceptual diagram comparing a processor in isolation, and a Zynq device. Adding even a modest amount of programmable logic (equivalent to one of the smaller Zynq devices) enables significant hardware acceleration, and therefore can be a compelling alternative to a processor alone, freeing up the processor's resources to support other aspects of system performance.

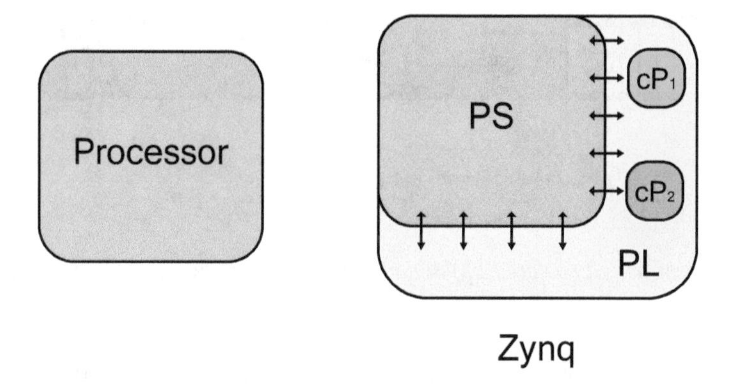

Figure 4.8: Comparison between a processor alone, and a small Zynq device

4.4. Comparison C: Zynq versus a Discrete FPGA-Processor Combination

The last architecture to compare with Zynq is that of a processor-FPGA combination, i.e. where these two elements exist as physically separate components. The motivation for this type of system is usually that the system needs to support both computationally intensive data-flow type processing (ideally suited to the FPGA), and sophisticated software algorithms or applications (ideally suited to a dedicated processor). It may be that the software element of the system extends beyond the scope of MicroBlaze implementation, meaning that both an FPGA and a processor are deemed to be required.

The Zynq gives the option of a single-chip replacement for this configuration, and the two possibilities are depicted in Figure 4.9.

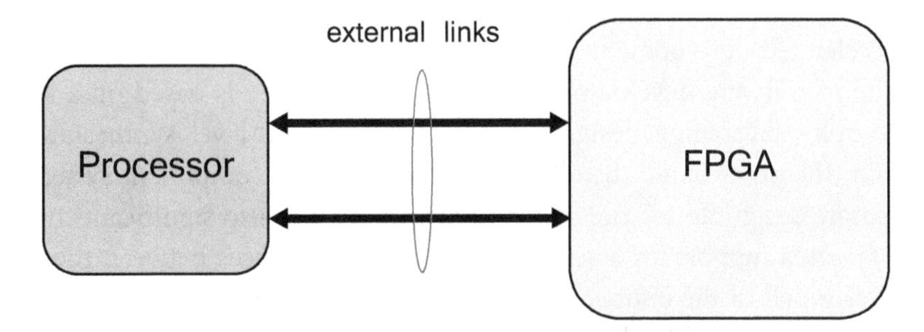

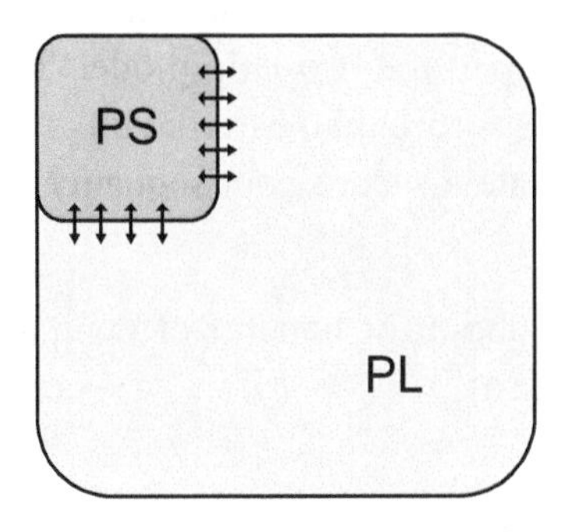

Figure 4.9: Comparison between discrete-component and Zynq-based combinations of processing system and programmable logic

The Zynq solution is advantageous on a number of fronts. Firstly, with reference to one of the most important parameters identified in Section 4.1, it should be recognised that opting for a single device has the potential to reduce the BOM. The board-level system hardware architecture is simplified due to the reduced number of components, which further contributes to cost savings, and potentially also improves reliability.

The use of a Zynq device also enables the physical size of the system to be reduced, while power consumption may be significantly lower, in terms of both (i) device operation and (ii) inter-chip communication. The external links of a discrete component system expend more power than the comparatively more local connections between the PS and PL in the Zynq; this contributes to the overall power saving associated with the Zynq solution. Other power savings are achieved due to the smaller physical size of the Zynq device, its 28nm device architecture, and its tight memory integration.

From a design perspective, Zynq also provides the potential for productivity gains, leading to accelerated development times. This can be attributed to the integrated design flow and suite of software development tools for Zynq, which is based on a system-level design philosophy, leveraging design reuse and rapid, high-level synthesis of C based algorithms. On the other hand, the alternative for a discrete component system might be two very separate design flows, tool sets, and processes. It is also significant that the Zynq design flow features support for a set of standard AXI interfaces between the PL and PS, which removes much of the effort of designing and implementing appropriate interfaces. The extensive availability of third party AXI-compatible IP is a further advantage.

Finally, the overhead of communicating between the processor and FPGA over external connections is avoided; as mentioned previously, this may be the cause of bandwidth constraints and increased latency in the two-chip model. The internal connections in the Zynq device are inherently more secure than external links and, in fact, additional security features are integrated to facilitate a secure boot sequence, and to defeat tampering [16], [18].

Extensive further discussion about the benefits of Zynq as compared to a two element, discrete solution is available in [16].

4.5. Exploiting the Zynq Architecture and Design Flow

One particularly powerful aspect of the Vivado design flow is its tool for high-level synthesis, *Vivado HLS*, which allows hardware (destined for implementation on the PL) to be generated from a C-based software description.

HLS will be discussed in detail in Chapters 14 and 15, but for the purposes of the current discussion, it is useful to briefly summarise the motivation for its use. The HLS design method permits rapid creation of designs, as a result of describing functionality at a higher level of abstraction than the traditional RTL-level (as used by HDL and related design entry methods). Using HLS, the designer can influence the synthesis of C code to hardware by applying specific directives and constraints that relate to aspects of the generated hardware.

The use of HLS is particularly compelling in the context of Zynq system development, because its architecture comprises both PS and PL. It means that functional parts of the system can very easily be ported from software destined for execution on the ARM, to hardware for implementation on the PL, simply by retargeting C code with only minor modifications.

Changing the realisation of system elements represents a different hardware/software partitioning, which may achieve performance or implementation benefits. For example, in Figure 4.10, the functional element F_4 has been moved from software to hardware implementation (potentially via the use of HLS), and the element F_1 has been shifted from a hardware implementation to a software routine. The adapted system architecture may be found to facilitate increased data throughput, for example.

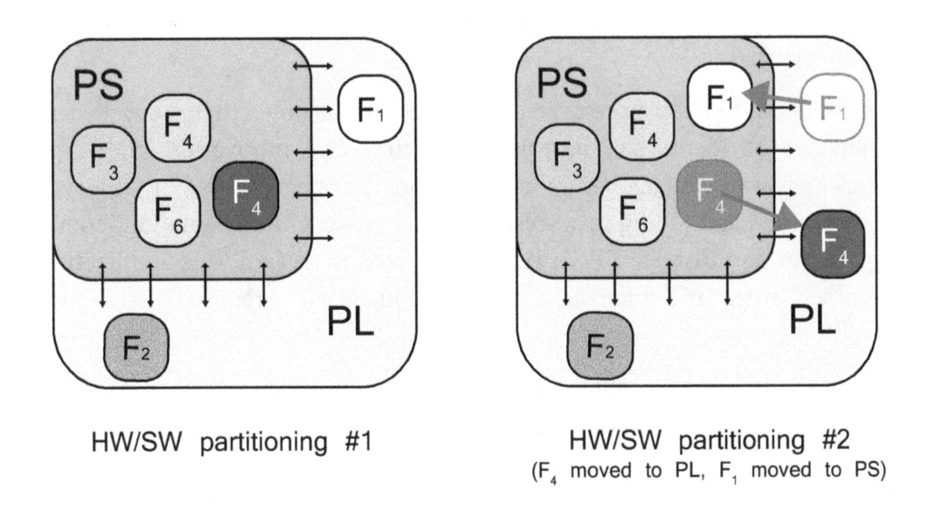

Figure 4.10: Two examples of hardware/software partitioning based on Zynq

When developing complex systems, the implication that software/hardware partitioning can be very readily accomplished with the support of HLS methods is of significant benefit. As required, the design team can investigate different partitionings before committing to a final system architecture, with a reduced time overhead in doing so.

4.6. Chapter Review

This chapter began by reviewing factors influencing the choice of a target device (or devices), from the perspective of a system developer. A number of such factors were identified, including: the architecture, configuration and performance of the device; commercial considerations; the supporting design flow and development tools; physical characteristics; and flexibility. With these in mind, the used of Zynq has been motivated through comparisons with other system implementation options, namely using an FPGA or processor in isolation, and an FPGA-processor combination as discrete components.

A processing resource is a necessity in most embedded applications, and the support for processors in FPGAs has been reviewed, covering both soft and hard processors. The performance of current and historical FPGA-based embedded processors has been compared with Zynq's ARM processor, and it has been shown that the ARM outperforms all available alternatives.

The general operation of processors was reviewed. It was noted that, if parallelism within the software can be identified, there is great scope and motivation to retarget these functions to hardware as significant acceleration can be achieved in this way. Such hardware/software partitioning can result in a system implementation combining a discrete processor and discrete FPGA but, as has been discussed, this architecture has its own difficulties, many of which arise from the interfacing required between the two physical chips.

Given all of these factors, the fact that Zynq provides both high performance processing and FPGA-type programmable logic fabric in an integrated device is of great interest. This represents the best of both worlds, and gives the designer the flexibility to partition the system between software and hardware elements according to requirements. One of the facilities which helps with this process is *Vivado HLS*, a tool which converts C or C++ algorithms into hardware descriptions suitable for implementation in Zynq's PL. This means, for example, that computationally intensive sections of software can be quickly retargeted for hardware acceleration. HLS will be the subject of a 'spotlight' chapter later in the book (Chapter 14), followed by a specific review of the Vivado HLS development tool (Chapter 15).

Next, we will consider some of the applications for which the Zynq is particularly well-suited, based on the observations made in this chapter. Application areas include wireless communications and software defined radio, smart networks, automotive and industrial, image processing, robotics, and many others.

4.7. References

Note: All URLs last accessed June 2014.

[1] ARM, "CoreMark Benchmarking for ARM Cortex Processors", Application Note 350, Issue A, 2013. Available: http://infocenter.arm.com/help/topic/com.arm.doc.dai0350a/DAI0350A_coremark_benchmarking.pdf

[2] ARM, "Cortex-M1 Processor" webpage. Available: http://www.arm.com/products/processors/cortex-m/cortex-m1.php

[3] Aeroflex Gaisler, "LEON4 32-bit Processor Core", product information sheet, January 2010. Available: http://www.gaisler.com/doc/LEON4_32-bit_processor_core.pdf

[4] K. Chapman, "PicoBlaze for 7 Series FPGAs", KCPSM6 Release 8, March 2014. Available (as part of a download) from: http://www.xilinx.com/ipcenter/processor_central/picoblaze/member/

[5] EEMBC website, "CoreMark FAQ" webpage. Available: http://www.eembc.org/coremark/faq.php

[6] EEMBC website, "CoreMark Scores" webpage. Available: http://www.eembc.org/coremark/index.php

[7] Xilinx, Inc., "LogiCore IP MicroBlaze Micro Controller System", Product Specification, DS865, v1.1, April 2012. Available: http://www.xilinx.com/support/documentation/sw_manuals/xilinx14_1/ds865_microblaze_mcs.pdf

[8] Xilinx, Inc., "MicroBlaze Soft Processor Core" webpage. Available: http://www.xilinx.com/tools/microblaze.htm

[9] Xilinx, Inc., "PicoBlaze 8-bit Embedded Microcontroller User Guide", UG129, June 2011. Available: http://www.xilinx.com/support/documentation/ip_documentation/ug129.pdf

[10] Xilinx, Inc., "PicoBlaze Soft Processor" webpage. Available: http://www.xilinx.com/products/intellectual-property/picoblaze.htm

[11] Xilinx, Inc., "Virtex-II Pro and Virtex-II Pro X Platform FPGAs: Complete Data Sheet", DS083 (v5.0), June 2011. Available: http://www.xilinx.com/support/documentation/data_sheets/ds083.pdf

[12] Xilinx, Inc., "Virtex-4 Family Overview", Product Specification, DS112, v3.1, August 2010. Available: http://www.xilinx.com/support/documentation/data_sheets/ds112.pdf

[13] Xilinx, Inc., "Virtex-5 Family Overview", Product Specification, DS100, v5.0, February 2009. Available: http://www.xilinx.com/support/documentation/data_sheets/ds100.pdf

[14] Xilinx, Inc., "Virtex-5 FPGAs: The Ultimate System Integration Platform", Virtex-5 Family Brochure, 2008. Available: http://www.xilinx.com/publications/prod_mktg/Virtex_family_brochure.pdf

[15] Xilinx, Inc., "Xilinx Extends Platform FPGA Performance with Award Winning MicroBlaze Soft Processor", Xilinx Press Release #0695, 9th October 2006.
Available: http://www.xilinx.com/prs_rls/2006/embedded/0695microblaze5.htm

[16] Xilinx, Inc., "Zynq-7000 All Programmable SoC", Xilinx Backgrounder, 2013.
Available:
http://www.xilinx.com/publications/prod_mktg/zynq-7000-generation-ahead-backgrounder.pdf

[17] Xilinx, Inc., "Zynq-7000 All Programmable SoC Overview", Preliminary Product Specification, DS190, v1.5, September 2013.
Available: http://www.xilinx.com/support/documentation/data_sheets/ds190-Zynq-7000-Overview.pdf

[18] Xilinx, Inc., "Zynq-7000 Technical Reference Manual", UG585, version 1.7, February 2014.
Available: http://www.xilinx.com/support/documentation/user_guides/ug585-Zynq-7000-TRM.pdf

[19] R. York, "Benchmarking in context: Dhrystone", ARM White Paper, March 2002.

5

Applications and Opportunities (*"What can I do with it?"*)

The application areas for Zynq have much in common with those of FPGA and certain processor-based devices, and therefore it is useful to survey the general landscape of applicable products and systems. As will be shown over the next few pages, examples include: automotive, military, aerospace, image processing, wired and wireless communications, medicine, industrial control, and many others.

Having compared Zynq with three principal alternatives in Chapter 4, it has been established that Zynq provides a distinct set of features and characteristics compared to other available solutions. It is therefore useful to build on these observations, and to explore and consider applications for which Zynq is particularly well-suited. The Zynq architecture can be leveraged to meet the demands of applications with significant requirements in terms of both high performance computation, and sequential, processor intensive functionality. With this in mind, we will focus on three of them as case studies: Software Defined Radio (SDR), Smart Systems and Networks, and Image and Video Processing.

Another important avenue to explore is that of the Zynq ecosystem, and the opportunities which surround Zynq for developers of, for example, IP blocks, operating systems, and other software solutions.

5.1. An Overview of Applications

In reviewing applications for Zynq, FPGAs, and related devices, a few important areas can be identified. The possibilities are extensive, and this is just a representative selection.

5.1.1. Automotive

Cars nowadays contain a significant amount of electronics, ranging from engine management, to the control of functions such as windows, mirrors and lighting, to navigation and infotainment systems. Advanced Driver Assistance Systems (ADAS) refers specifically to the collection of systems provided in cars for driver safety and convenience, which can include: lane departure warning systems; road sign recognition (e.g. to warn the driver when entering a lower speed limit); parking assistance; head-up displays; and even monitoring of driver awareness. Examples are shown in Figure 5.1.

© Xilinx

© Xilinx

Figure 5.1: Automotive systems (left: head-up display; right: lane and road sign recognition)

FPGAs, and now Zynq devices, can be used to realise these automotive systems [10], [50]. The processing capabilities of Zynq make it particularly well suited to such systems, while the opportunity to reduce the number of component devices is advantageous in a market sensitive to cost and power, and which often has physical space constraints.

5.1.2. Communications

FPGAs are established platforms for undertaking the computationally intensive processing required in both wireless communications and wired, packet-based systems. This field is diverse, and includes transceivers for terrestrial and satellite transmissions, mobile backhaul infrastructure, wired network equipment, radar, sonar, Global Positioning System (GPS), and many other communication systems. Figure 5.2 shows a small selection of examples.

In wireless communications, the radio spectrum is under increasing pressure, and meanwhile the number of wireless systems and standards continues to expand. The concept of *flexible radio* indicates the potential to better utilise the radio spectrum, and consolidate radio equipment into a single device that is capable of dynamically changing its operation. Zynq is an ideal flexible radio platform — more to follow on this in Section 5.3. In wired communications, a similar level of flexibility is sought via the use of 'softly defined networks', which have the ability to upgrade functionality under software control [39].

© Xilinx

© Xilinx

© Xilinx

Figure 5.2: Communications systems (left: wireless basestation; middle: satellite groundstation; right: wired network switches

5.1.3. Defence and Aerospace

Defence systems include a wide variety of communications, image processing, aviation, navigation, and transport systems, as well as weapons related technology. Defence electronics usually require an increased level of robustness compared to civil applications, with extended temperature ranges and security features [58].

An area gaining interest is the concept of the 'networked battlefield', wherein military personnel and equipment are inter-connected with aircraft, satellites, signals intelligence equipment and other defence systems [38]. The aim of the networked battlefield is to gather, filter and share information, and hence optimise the effectiveness of miliary operations.

Civil aerospace applications include navigation and on-board flight systems, satellite and ground communications, and radar systems.

5.1.4. Robotics, Control and Instrumentation

Industrial and scientific processes, ranging from manufacturing and utilities, to high energy physics experiments, require precise control and instrumentation. Figure 5.3 shows illustrative examples of a plant control room, electricity generation with wind turbines, and apparatus for high energy physics experiments at CERN [4].

Figure 5.3: Control and instrumentation systems: (left: industrial control room; middle: wind turbines; right: high energy physics experiments

FPGAs and Zynq devices are very suitable platforms because they can perform fast, real-time processing and handle multiple sensor inputs and actuator outputs simultaneously, leveraging the capabilities of the PL. Zynq adds further potential for systems integration and operational flexibility. For instance, the performance of a control loop could be monitored, and its configuration altered if required, using software control. The PS can also support a real-time operating system and/or GUI, if required.

Motor control algorithms are particularly important in some branches of industry. For example, a survey of US manufacturing industry estimated that about 50% of the electrical energy consumed in industry could be attributed to 'machine drives', i.e. motors, pumps and fans [48]. Zynq is well suited to motor control, due to the high bandwidth link between the PS and PL, enabling a tight feedback loop, and leveraging the DAC sampling capabilities offered by the AMS block [26].

5.1.5. Image and Video Processing

Image and video processing encompasses a number of different applications including cameras for consumer and professional use, video compression and storage systems, broadcast equipment, display technology, industrial process monitoring, security and surveillance, and many others.

The processing capabilities of Zynq are particularly valuable to 'embedded vision' applications, which require both deterministic processing of large amounts of pixel data, and software algorithms for extracting information from images (suited to the PL and PS, respectively) [8]. This theme will be explored further as a case study in Section 5.5.

5.1.6. Medical

An important issue in medical diagnosis is "seeing" inside the body, and this requires medical imaging equipment such as Computer Tomography (CT) scanners, ultrasound and Magnetic Resonance Imagers (MRIs). Enhancing and displaying the images obtained via this equipment requires sophisticated image processing algorithms to be performed, often on large data sets. As for other image processing applications, the Zynq mix of PL and PS capabilities supports both high-speed parallel processing, and software-based algorithms.

Further applications in medicine include the control of instruments in robot-assisted surgery, and real-time surgical imaging, such as endoscopic equipment, patient monitoring equipment and home health technologies [24].

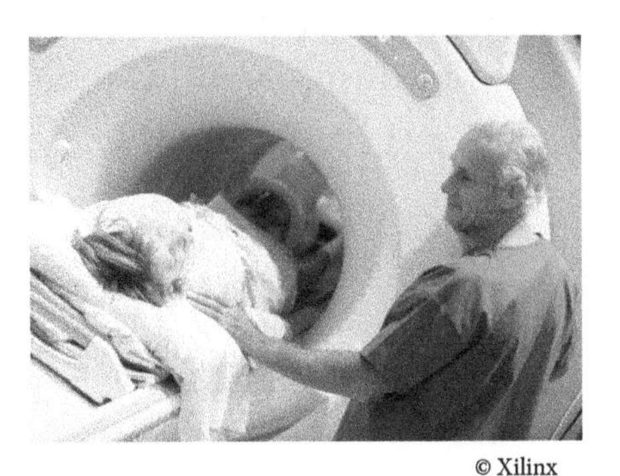

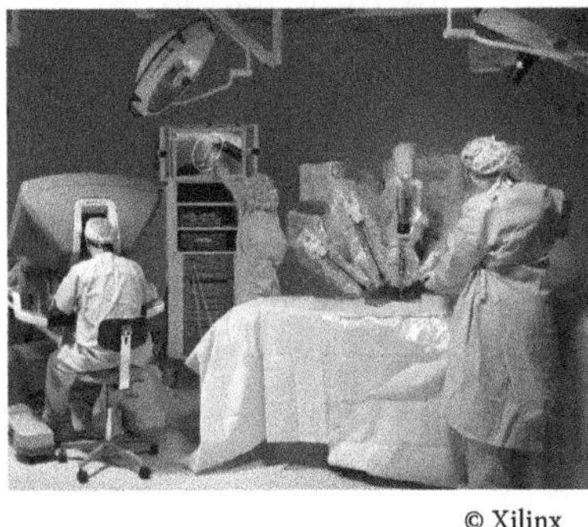

Figure 5.4: Medical applications: (left: MRI scanning; right: robot assisted surgery)

5.1.7. High Performance Computing (HPC)

The umbrella term of 'high performance computing' covers applications with requirements for fast processing of large datasets, which can typically be accelerated with dedicated hardware processing [44]. HPC includes, but is not limited to, such diverse applications as financial modelling [54], analytics for oil and gas exploration, processing of data from scientific experiments, radio astronomy [15], and analysis of captured radar signals. Also included are the infrastructure for data centres and cloud computing facilities that often underpin HPC applications.

5.1.8. Others and Future Applications

The preceding pages have provided an overview of some of the key application areas for FPGAs and Zynq. There are, however, many more current and potential applications than can be covered in detail here ranging from audio signal processing to Zynq-based drones!

A good way to keep up with the latest developments is to read the Xilinx ***Xcell Journal***, which is published quarterly and contains a variety of interesting articles on Xilinx technology and applications. Many of these focus on Zynq, or are relevant to Zynq. The Xcell Journal can be accessed online at the web address:

http://www.xilinx.com/about/xcell-publications/xcell-journal.html

Additionally, the ***Xcell Daily Blog*** often highlights the latest Zynq-relevant design news, tutorials, and application stories. The Xcell Daily Blog is available at:

http://forums.xilinx.com/t5/Xcell-Daily-Blog/bg-p/Xcell

Some recent application examples featured in the Daily Blog include *Phenox*, a Zynq-based drone that is being developed by researchers in Tokyo, via a KickStarter project [35], the open-source instrumentation platform, *Red Pitaya* [36], and the collaborative open source project, *AXIOM*, which is creating an 'open cine camera', based around the ZedBoard Zynq development board [1].

5.2. When Can Zynq Really Help...?

As the summary of application areas in Section 5.1 has illustrated, Zynq is suitable for a wide variety of applications, and there are many more possibilities than can be presented here.

Many of these applications are suited to both FPGAs and Zynq, so it reasonable to pose the question "When can Zynq really help?". In answering the question, we can identify two particular types of processing which may be required in a target application:

- High-speed, parallel, deterministic processing; and
- Sequential, dynamic, unpredictable processing.

These two processing styles are ideally suited to the PL and PS of Zynq, respectively. Zynq provides the greatest benefit to applications requiring *both* of these processing styles, particularly when implementation would otherwise require the use of two discrete processing chips (e.g. an FPGA for high-speed vector processing, and a processor running an operating system). Moreover, where the application involves close cooperation between these two processing elements, the consolidated Zynq architecture can permit power savings and design simplifications. The low-latency and high-bandwidth link between the PS and PL is an advantage, and is particularly significant in systems with fast real-time processing requirements and feedback loops.

It is likely that different characteristics of Zynq will appeal most to different markets. For instance, some application areas will be particularly cost sensitive — especially in high volume products with low price points — and will benefit most from the reduced bill of materials achievable via using a single Zynq device, as compared to two separate chips. In other cases, the most valuable aspects might be the time-to-market acceleration achieved via Zynq's standards-based and IP-centric design flow, the reduction in power consumption, or the low latency links between processing and logic resources.

As demonstrated in Chapter 4, Zynq offers enhanced embedded systems performance compared to an FPGA alone, and therefore some products that traditionally incorporated FPGAs will naturally transition to Zynq in future development iterations, when new features and performance improvements are integrated.

Over the next three sections, we will present a small selection of case studies, focussing on three important and progressive areas: communications, smart networks, and image and video processing. In each case, the particular suitability of Zynq for each application is demonstrated.

5.3. Communications: Software Defined Radio (SDR)

The first application we will consider is that of flexible radio. The rate of change in wireless communications is so rapid that systems capable of adapting their operation would be hugely valuable. Zynq is a platform capable of realising this flexibility.

5.3.1. Trends in Wireless Communications

One of the significant trends over recent years has been the rapid expansion in demand for wireless connectivity and, in tandem, the number of radio standards in use. For instance, your smartphone is likely to have separate radios for the cellular network, supporting 2G, 3G and now 4G, and also WiFi (with its several variations), Bluetooth and GPS reception. Each of these will require at least one discrete integrated circuit. Notebook computers, tablets, e-readers, televisions and even cars are also equipped with at least some of this functionality. Furthermore, away from the sphere of the consumer, multi-standard radios are needed in transportation, emergency services, military and many other sectors.

In all of the above markets, it would make sense to consolidate the radio transceivers for these various standards into a single, programmable device. Doing so would provide a number of useful benefits:

- Reducing the required hardware from several chips to potentially just one, with associated benefits for the cost, size, and power consumption of radio terminals.
- Improving geographical portability, i.e. allowing radio functionality to be adjusted in response to location, where different modes of operation apply.
- Facilitating ready integration of support for new standards, without any need to replace the hardware.

SDR is a technology that can enable these advantages to be achieved.

5.3.2. Introducing Software Defined Radio (SDR)

The concept of a *Software Defined Radio* — a radio which can be reconfigured while in operation — is not a new idea, and has been around in some form since the mid-1990s [29]. The term can refer to different aspects of reconfiguration, and thus means different things to different people; however, here we regard it as the ability to support multiple radio standards on a single device, as controlled by software. Thus, SDR represents a mechanism for achieving consolidation of radio functionality as described above.

SDR has been predominantly associated with military applications to date. In particular, the US military started its first major SDR programme, Joint Tactical Radio System (JTRS), in 1998 (since evolved into the Joint Tactical Networking Centre (JTNC) [22]), with the basic aim of reducing the amount of radio equipment soldiers were required to carry while on operations. The resulting military SDR radios have typically been physically large and expensive to produce, and these characteristics are incompatible with the requirements of mass-market civil, commercial and consumer communications. The availability of flexible, capable hardware platforms at acceptable cost levels has stimulated change, and the applicability of SDR to non-military applications is now gaining interest. Meanwhile, military SDR continues to advance, with a significant goal being convergence to a common standard for SDR, in order to support and enhance communications between allies [62].

5.3.3. SDR Implementation and Enabling Technologies

The use of reconfigurable processing platforms underpins the development of modern SDR. While early SDRs achieved switchable functionality in part through component redundancy, now SDR has vastly enhanced applicability and possibilities, because adaptable radios can exploit the capabilities of devices like FPGAs and Zynq. To understand why this is, we must first consider the definition of SDR (according to the IEEE and SDR Forum, [55]). These organisations jointly define SDR as:

"Radio in which some or all of the physical layer functions are software defined."

This definition confirms that the software defined aspects reside in the physical layer (PHY), i.e. the part of the radio directly adjacent to the Radio Frequency (RF) circuitry and air interface. The PHY is computationally intensive, implementing high-speed filtering and other arithmetic-based DSP algorithms, and exchanging data with the DAC and ADC. As some parts of the PHY layer have a very high degree of computational complexity, software can only be used to define the behaviour of these elements — it is unsuitable for implementing the processing itself. The less complex computation, such as modulation and coding, can be performed in either software or hardware. SDR therefore calls for the close integration of a processor running software to control PHY functionality, and a high-speed, computationally parallel resource to undertake PHY processing. FPGA-type programmable logic is ideal for implementing the PHY, especially given its capabilities for dynamic reconfiguration (to be discussed further in Section 5.6), while an ARM processor provides a suitable platform for SDR software. The combination of these two key components of the SDR architecture into a single device, such as Zynq, can therefore be considered a perfectly suited solution, and in fact there are already SDR products on the market based on Zynq [60]. As reviewed in Chapter 4, Zynq offers the benefits of a reduced bill of materials, lower power consumption, and tighter integration compared to a two-chip configuration. All of these characteristics are highly compatible with the aims of SDR.

As reviewed in [23], wireless standards can vary in several aspects of PHY processing, including packet format, bit rate, error correction coding, modulation scheme, pulse shape and carrier frequency. The reader is referred to [17] for an excellent tutorial on communications architectures, SDR, and the motivations for developing these more adaptable and interoperable radios. Achieving an architecture capable of supporting several standards is an interesting problem, especially when the bit rate and derived parameters are not conveniently related. The facilities of modern FPGA and Zynq devices can, however, enable such systems to be achieved.

Figure 5.5 provides a simple example of an SDR, in which the Intermediate Frequency (IF) carrier frequency of a transmitter can be programmed by setting a register via software. The register applies a step size input to a Numerically Controlled Oscillator (NCO), thereby controlling its frequency of oscillation.

As mentioned above, many other aspects of a radio can also be reconfigured, for example the modulation scheme could be dynamically selected from a pre-defined set of schemes, and the pulse shape could be changed by reprogramming filter coefficients. Where the change of wireless standard implies a fundamentally different structure of computation (for instance changing from a Code Division Multiple Access (CDMA)-based

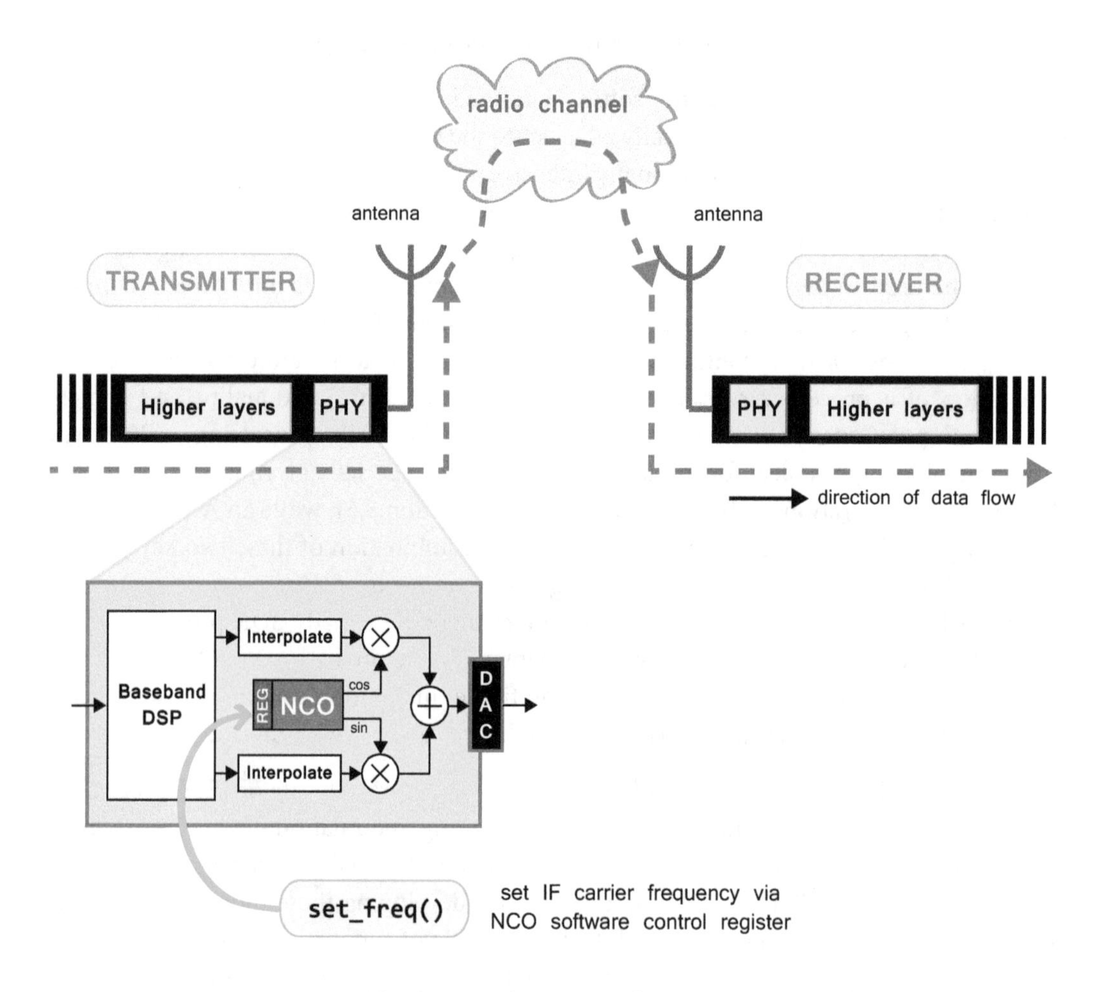

Figure 5.5: Example of SDR with IF carrier frequency programmability

standard to an Orthogonal Frequency Division Multiplexing (OFDM)-based standard), switching the functionality is best accomplished by reconfiguring part of the FPGA hardware. This can be achieved using Dynamic Partial Reconfiguration (DPR) techniques, which will be covered in Section 5.6.

5.3.4. Cognitive Radio

Many see the natural evolution of SDR to be *Cognitive Radio*, in which the adaptability of the radio is exploited to enhance use of the RF spectrum [23], [41] [53]. At present, much of the spectrum is licensed and yet under-used, even while frequency bands supporting certain classes of use are under extreme pressure. *Dynamic spectrum access*, i.e. employing radio equipment with the capability to sense local conditions and respond accordingly, is an attractive proposition, as this could ultimately enable better spectrum utilisation. The implementation of cognitive radio does however need to address the various risks it inherently poses, not least to protect licensed users of the spectrum.

Research into cognitive radio is an active area, in both academia and industry [20], [32], [52]. Developing better ways to access and manage the RF spectrum could have huge commercial and societal impacts, but the complexity of the problem is also significant. Like SDR, realisations of cognitive radio solutions will need to couple high-speed, parallel, and reconfigurable hardware with a processor capable of supporting complex software algorithms, and therefore Zynq is clearly a strong candidate for research and development in this area. For instance, researchers at Trinity College, University of Dublin have developed the *Iris* software framework for cognitive radio [45], and consider its operation on Zynq in [49]. At the time of writing, the first commercial cognitive radios are beginning to emerge [56], and this could prove to be a very exciting area of technological innovation.

5.4. Smart Systems and Smart Networks

Staying with the communications theme, the term 'smart' is becoming near ubiquitous, the idea being that adding intelligent systems around us can enhance our lives, meanwhile achieving cost savings and environmental benefits. The realisation of smart systems implies smart networks to underpin them, and we consider both over the next few pages.

5.4.1. What is a Smart System?

The term *smart* actually finds its way into a number of application areas, including smart grids, smart buildings, smart homes, smart transport, smart cities, smart agriculture, and others. A pertinent question would be, what makes these specifically *smart* systems?

Actually there is no single established definition of *smart*. It is however interesting to note the definition applied by the Organisation for Economic Co-operation and Development (OECD) [33], which will serve our discussion in this chapter:

"An application or service is able to learn from previous situations and to communicate the results of these situations to other devices and users. These devices and users can then change their behaviour to best fit the situation. This means that information about situations needs to be generated, transmitted, processed, correlated, interpreted, adapted, displayed in a meaningful manner and acted upon."

The role of a smart system can therefore be interpreted as optimising *something* in response to the circumstances it observes. But what might that something be? This question is addressed in the next section, in which a selection of smart systems are described.

5.4.2. Examples of Smart Systems

A number of smart systems types were highlighted in the introductory part of this section — these are not the only smart systems around, but they are arguably some of the most prominent at the time of writing. It is useful to understand what these example systems do, and their respective benefits.

- The term ***smart grid*** refers to the augmentation of the electrical power distribution network (the 'grid') with sensors, telecommunications networks, and automation, such that it can be more effectively managed. The primary potential benefits of smart grids are: (i) to increase energy efficiency, which helps the environment and enables cost savings; (ii) to improve the network's reliability and response to local faults; and (iii) to automate the collection of data for metering purposes.

- ***Smart agriculture*** refers to the enhancement of crop, land, and livestock management through the use of monitoring and automated systems. This might include the adjustment of greenhouses to the optimum temperature and humidity, the control of outdoor irrigation and drainage systems in response to local weather conditions, and the monitoring of livestock health, with updates and alerts provided to the farmer [9].

- ***Smart transport*** refers to the integration and dynamic management of transport networks, infrastructure, signalling and passenger information. Smart transport systems can encompass both public and private transport, including freight as well as passengers. The aims of smart transport can include management of traffic flow (minimisation of congestion), reduction of emissions, and the considerable economic benefits of faster and more reliable connections [40]. Taken in the urban context, smart transport is considered an important aspect of the smart city [6].

- A ***smart building*** uses a network of sensors and actuators to adjust aspects of its operation in response to detected conditions and the current population, e.g. lighting, heating and ventilation systems. A smart building uses less energy, and is therefore 'greener' and cheaper to run, while also providing a more pleasant environment for its inhabitants [42]. The term *smart buildings* tends to refer to non-residential buildings such as government and private offices, schools, airports, shops, and factories.

- ***Smart homes*** use similar principles to smart buildings, bringing cost and environmental benefits to the domestic setting. In addition to managing the use of heating, lighting and electricity usage, the smart home may also include aspects of security and entertainment in the home. An illustration of a smart home is provided in Figure 5.6.

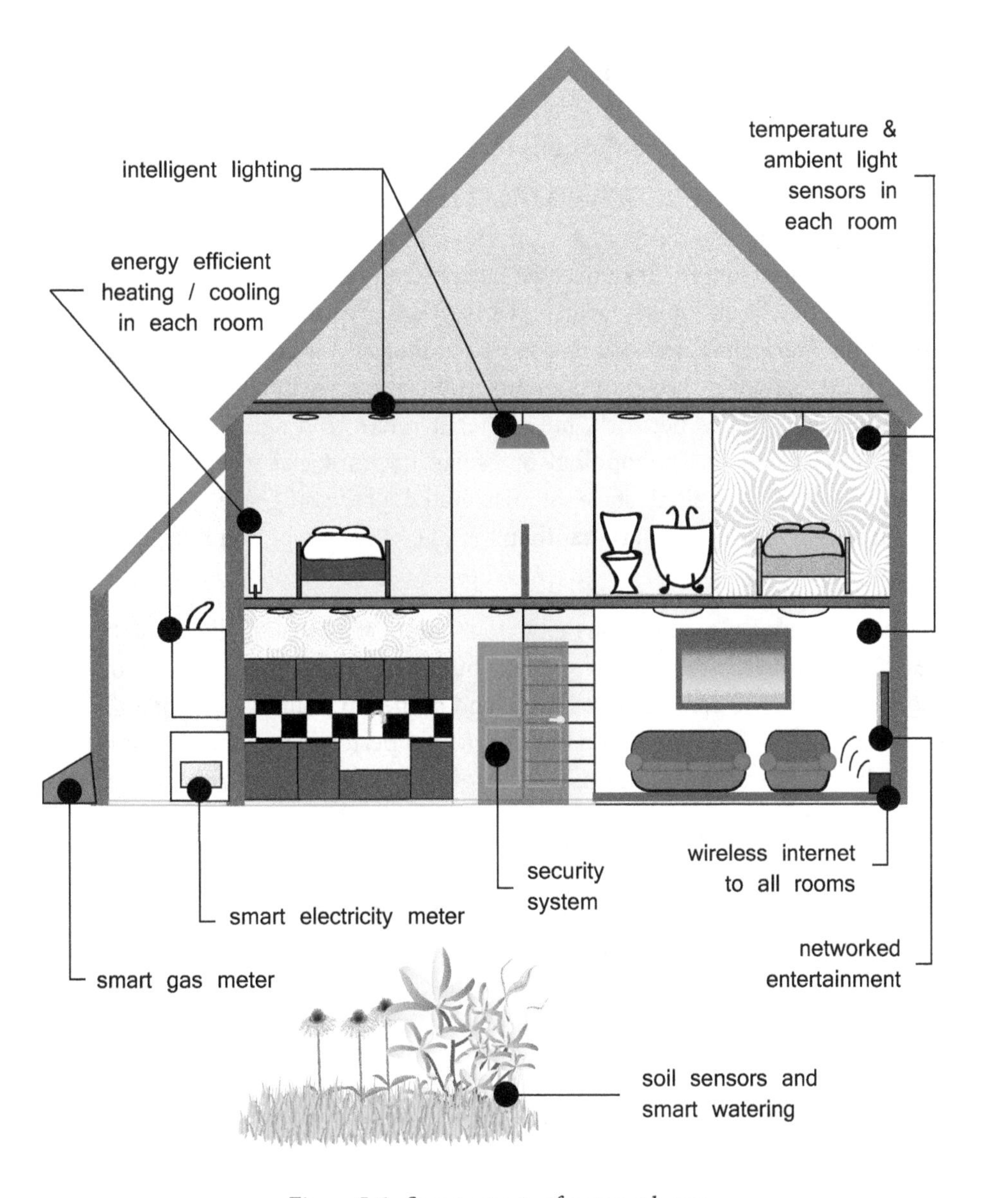

Figure 5.6: Some aspects of a smart home

- ***Smart cities*** aim to integrate and intelligently manage the various systems of the city, for the purposes of sustainability, economic growth, and citizens' quality of life. This may include transport, buildings, utilities, etc., but also internet provision in public areas, monitoring and maintenance of infrastructure, delivery of council services, security, and disaster management [43], [51].

Of course, these are just a few examples of smart systems. There are others, and the portfolio of 'smart' is likely to expand in the future.

5.4.3. Smart Networks: Communications for Smart Systems

It is clear that all of these smart systems will require communications infrastructure to underpin them. Given the variety of scenarios described, these 'smart networks' may however be very different. For instance, we have mentioned home and building networks (very local, and mostly indoors), farm networks (larger area, outdoor rural setting) and urban networks (very large area, a mixture of outdoor and indoor, dense, and dynamic). The term 'smart network' therefore does not prescribe a particular technology, or even specify a wired or wireless network, but rather it refers to a network that is adaptable, expandable, and intelligent. The topology of a smart network is also flexible. Smart systems could entail at least an initial stage of distributed processing (i.e. processing on the individual nodes of the network), rather than for all raw data to be communicated back to a central point.

Of all the cited examples, the smart city would seem the most challenging, but also the most enticing prospect. Cities are complex, but they are also economic powerhouses, centres of education, culture and innovation, and homes to millions of people. The impact of improving the economic, social and environmental performance of a city could be huge, and this is evident from the aspirations of cities already engaged in smart city projects [16].

Smartness would be embedded into various aspects of city life, but there could be no one-time retrofit to turn a city into a smart city, without thought to further development. Smart networks will evolve, and therefore their infrastructure must be capable of evolving. They should also be adaptable in response to particular events, for example emergencies or damage to the network itself, which could require the routing of network traffic to be dramatically altered. The smart networks underpinning smart systems must therefore support intelligence, dynamism, and expansibility.

With these factors in mind, the relationship between smart systems, smart networks, and Zynq can be understood. The resources that Zynq provides enables the embedding of intelligence (e.g. software algorithms running on the PS), support for sensing, processing and analysing data (input/output facilities on the PL), communications interfaces and networking support [57]. Zynq is reconfigurable and integrated, with various sizes of device available to cater for different complexities of smart networks and systems.

5.4.4. Related Concepts

Smart cities and smart networks are closely related to some other current themes in technology, including *Machine-to-Machine* (M2M) communications, which refers to networked nodes that function independently (without human intervention); and the *Internet of Things* (IoT), a collective term for the networks formed by these nodes.

The information generated by smart networks, especially where there is a large amount of such information, can be termed *Big Data*. Big Data implies sets of data that are very large and/or complex, and the effective analysis of Big Data may require specialist algorithms and methods, but the intended outcome is to enable better decision-making than would otherwise be possible [5].

Perhaps the most widely familiar of these terms, the *Cloud* is a computing resource which resides remotely, and can be accessed from anywhere. Typically cloud computing resources are very flexible and scalable, and consequently there is great interest in processing big data in the cloud. Conceivably, a smart city could produce Big Data from its smart network of intelligent, Zynq-based nodes, for analysis in the cloud and subsequent feedback to the city control room, enabling better decisions to be made both by the smart system itself, and the humans overseeing its operation.

5.5. Image and Video Processing, and Computer Vision

The area of image and video processing is diverse, and features in consumer and commercial products, as well as finding use in medicine, industry, defence, and security, amongst many other areas.

Image processing is concerned with single, or 'still' images, whereas video processing refers to a temporal series of images ('frames'), usually at a specific frame rate. Computer vision (or 'machine vision' or 'embedded vision') systems add an aspect of intelligence, such that meaning can be extracted from images or video data. As appropriate, decisions may then be taken based on this information.

5.5.1. Image and Video Processing

There are a number of techniques particular to image and video processing, which can generally be classified as: (i) enhancing or otherwise altering images; (ii) extracting information from them; or (iii) compressing image and video data. Many books have been written on these subjects, and there is insufficient scope to present detailed review here, other than to mention a few representative examples. The reader is referred to [13] for theoretical background and examples, and [3] for an implementation perspective.

Image processing is characterised by its huge parallelism: individual images, or equivalently video frames, are composed of two dimensional arrays of pixels, and there may be more than 1,000 pixels in both the X and Y axes. For instance, Full High Definition (HD) video is usually defined as 1920 x 1080 pixels, equivalent to 2,073,600 pixels in total. Consider also that each pixel comprises three channels in order to represent colour! With 8 bit channels for red, green and blue data, this corresponds to 24 bits per pixel, and 49,766,400 bits per single HD image. Images from a digital camera might be considerably bigger. Bearing in mind these large dimensions, any processing involving individual pixels therefore requires a huge number of computations to be performed in parallel, a task which is ideally suited to an FPGA, or the PL part of the Zynq device. Of course, it is desirable to reduce the amount of data to be processed, and some image processing can be performed at a higher level of abstraction, operating on less data to perform more sophisticated algorithms, such as might be undertaken in software. We will return to this topic in Section 5.5.3 and later in Section 5.5.4.

Much of the focus in video processing is concerned with compression. Single images are often compressed in the spatial domain to reduce their file size, but there is a more acute need for compression in video applications where, with frame rates of up to 100 frames per second (or higher in some specialist applications), a much larger amount of data is involved, and there may also be real-time processing constraints. The temporal domain does however provide a further opportunity for compression, as consecutive video frames normally exhibit a high degree of commonality. Algorithms can be used to send full, fresh frames only a fraction of the time, and in between, to code the difference between frames. Video compression standards are well established, and continue to evolve and improve [21], [30].

5.5.2. Computer Vision

Computer vision systems can act upon still images or video, and are able to extract meaningful information from the content of images. An example might be recognising shapes, colours, or the sizes of objects within an image. This can have a number of practical applications, ranging from manufacturing, to surveillance, traffic management, food processing, medicine, biometrics and even space exploration.

To provide an illustrative example, it might be desirable to identify individual people within Closed Circuit Television (CCTV) footage, such that the number of people present in the captured area can be determined. Furthermore, a computer vision system might be able to detect events such as fights or disturbances. Commercial examples of Zynq-enabled computer vision are now beginning to emerge, such as visual scene understanding [46]. Of course, as mentioned above, there are also more routine uses of computer vision, such as classifying the quality of fruit and vegetables based on size, shape, and surface defects! [47].

5.5.3. Levels of Abstraction

Image processing (as a whole, i.e. to include video processing and computer vision) can be segmented into three levels of abstraction, which are characterised by the amount of image data being processed, and the amount of knowledge available regarding the content of the image. As will be discussed in Section 5.5.4, this selection of abstraction levels, and the types of processing implied, make Zynq a very suitable platform for image and video processing, and computer vision in particular.

The abstraction levels are shown in Figure 5.7 and clarified below (in order from bottom to top):

- ***Pixels*** — Pixel-level processing represents the lowest level of abstraction, with the highest amount of data to process, and very little knowledge about the content and meaning of the image. Pixel-level processing can include adjustments to the colour balance or contrast of an image, or neighbourhood filtering (such as smoothing

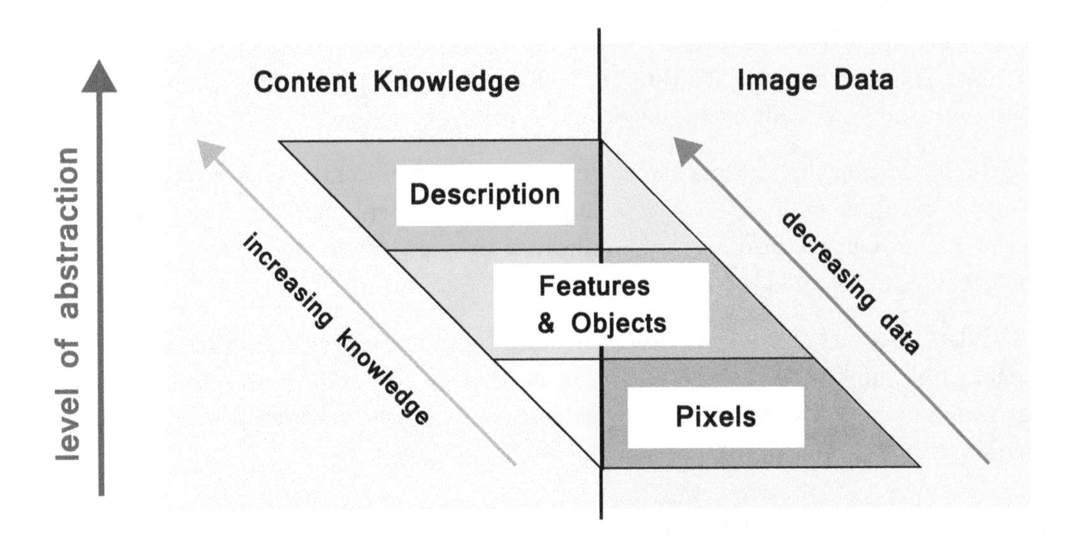

Figure 5.7: Abstraction in image processing

operations to reduce noise, or Sobel filtering to highlight edges [3], [13]). Usually these pixel-level tasks, of which there may be several stages, precede other operations, and hence they are referred to as 'pre-processing'.

- ***Features and Objects*** — As part of the process of extracting information from an image, features and objects are detected. This can include the identification of lines, curves, shapes and regions. Various techniques can be employed to help accomplish the transition from pixels to features and objects, including the Hough transform, colour identification, thresholding, and morphology [3], [13].

- ***Description*** — Having identified relevant features or objects in the image, the final stage is to attain knowledge of what the image represents, i.e. a description of its content. At this stage, the information is not an image, but a textual or numeric description. Achieving a description may involve extracting parameters of the identified features and interpreting them, or classifying the image based on a comparison with pre-determined criteria, or a training set [3], [13].

Computer vision requires features and objects to be identified, as described above, such that meaning can then be extracted from the image. This can involve a complex set of operations and software algorithms. For instance, it might be necessary to analyse the detected lines within an image, to determine whether it contains a vehicle, and possibly even to classify the type of vehicle as a motorcycle, car, bus, etc. This stage would represent a transition to the highest level of abstraction shown in Figure 5.7, wherein a description of the image content is achieved. For example, in Figure 5.8, the image shows cars at a junction. The number of cars can be detected and recorded using computer vision algorithms, and potentially even categorised according to the direction of travel.

Video processing applications may additionally call for object tracking, e.g. if a person of interest is identified in CCTV footage, that person's movements can be tracked over time as they move through the field of view. Similarly, if the view of the junction was provided as video, then vehicles could be tracked.

The data obtained from image processing and computer vision algorithms may be used as data within higher-level applications. For example, in the vehicle recognition scenario, the computer vision system could provide statistics about vehicles passing through a junction, to be used for traffic management and urban planning.

5.5.4. Implementation of Image Processing Systems

Considering the implementation of image processing systems in general, it is significant that different types of processing are required to operate on different types and volumes of

Figure 5.8: Computer vision can be used to detect cars at a junction

data. Noting again Figure 5.7, the starting point may be a very large amount of pixel data on which simple, repetitive operations are performed, while at higher levels of abstraction, computer vision techniques operate on smaller sets of data (perhaps representing lines or shapes), but using more complex algorithms.

For this reason, Zynq is a highly optimised platform for image processing. The PL is well suited to fast, parallel operations like those required for pixel-level image processing. Computer vision functionality can be implemented in software for execution on the Zynq PS, and integrated with higher-level software applications as required. The transition between the two, via the detection of features and objects within the image, might be accomplished using the PL with appropriate interfacing to the PS, or by leveraging the SIMD facilities of the NEON processor. Extensive support for NEON is available in third party image and video processing products [28].

In addition to the device architecture, we should also consider the role of Xilinx and third party development tools in enabling the design of image processing systems for Zynq. The following are worthy of particular note:

- ***Xilinx IP blocks*** — A number of IP blocks are available in IP Integrator for image and video processing applications, including video memory, image enhancement, and colour adjustment functionality.

- ***OpenCV*** — Open Computer Vision (OpenCV) is an open source project providing a set of C/C++ libraries for image and video processing [34]. The facilities of OpenCV can be used to develop software algorithms for running on the PS.

- ***Vivado HLS Video Libraries*** — Vivado HLS includes specific support for image and video processing, via a library of functions synthesisable to HDL. These can replace selected OpenCV functions, and therefore functionality can be easily partitioned into hardware if desired [31].

- ***MATLAB / Simulink*** — Extensive facilities are available for image and video processing, and computer vision in MATLAB and Simulink [27]. In addition to providing relevant functions and a development environment, developed algorithms can be converted to C/C++ code for implementation on Zynq.

5.5.5. Computer Vision on Zynq Example: Road Sign Recognition

Computer vision has several applications in automotive systems and traffic management, including driver safety and assistance systems, intelligent transport systems, traffic enforcement, traffic flow analysis, automatic number plate recognition, and more [7], [14].

An interesting application was addressed in [37], built upon the Zynq platform. The problem was to identify road signs based on captured images, with the aim of informing driver assistance systems (or even autonomous vehicles) about the environment surrounding the vehicle, and any restrictions in force. The system comprises three parts: pre-processing in the Zynq PL (colour adjustments etc.), further processing on the PS (morphology, edge and shape detection), and finally classification of signs by comparing them against a database (also in the PS). The authors made the point that the system was designed in a total of only six weeks, due to the convenience of AXI, the support provided by the Xilinx tools, and functionality available via the OpenCV library.

5.6. Dynamic System-on-Chip

All of the applications mentioned in this chapter either demand, or could benefit from, a platform that is flexible, in terms of the functionality implemented in the PL. As will be discussed in this section, the technique of DPR provides scope for a part (or parts) of the PL to be completely reconfigured during run time.

5.6.1. Run Time System Flexibility

Zynq All Programmable SoCs provide flexibility at design time, as a result of the two different parts of the device, the support of development tools and processes, and the use of AXI interfacing. There is an additional degree of convenience due to the Vivado HLS tools, and the potential to implement C-described functionality on either the PS or PL.

At run time, software on the PS can control functionality residing in the PL by passing commands, and other parameters such as filter coefficients, via software registers or shared memories. This constitutes run time flexibility.

While software control of hardware affords a good degree of scope to alter functionality during run time, there are some circumstances where the desired flexibility extends beyond setting parameters, and more fundamental changes are required to the component(s) implemented in hardware. For example, it might be desirable to realise an SDR that can support multiple communications standards with different architectures, only one of which is actively used at any given time. In fact, any application which requires different hardware components, but does not need them all simultaneously, could benefit from time-multiplexing of the relevant functionality.

5.6.2. Dynamic Partial Reconfiguration (DPR)

The technique of DPR involves designating a region (or regions) of the PL for reconfiguration during run time. These areas are referred to as Reconfigurable Partitions (RPs), and their functionality can be completely altered while the rest of the PL continues to operate. Importantly, reconfiguration of an RP is achieved without affecting any other part of the PL.

There may be multiple RPs on a Zynq or FPGA device, and each RP has a set of Reconfigurable Modules (RMs) associated with it. Here we will focus on a single RP for the benefit of clarity.

An RP may have any arbitrary number of corresponding RMs available, but only one of them occupies the RP at any given time, and hence, functionality is time-multiplexed onto a specific part of the PL. This concept is demonstrated in Figure 5.9.

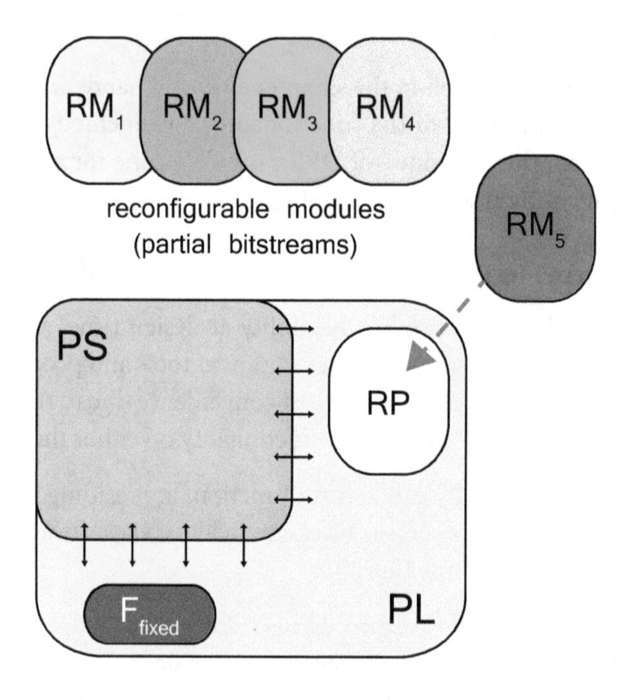

Figure 5.9: The use of dynamic partial reconfiguration to time multiplex modules in the PL

From the designer's perspective, the RP must be sized to accommodate the largest of the RMs deployed on it, because this ensures that sufficient PL resources are available for all of the designs to be supported. The use of DPR also implies that a set of partial bitstreams are created in advance (one for each of the RMs) and appropriately stored, such that these files can be downloaded onto the PL to configure the RP during run time. In operation, the software application hosted on the PS usually coordinates the execution of DPR.

5.6.3. DPR Application Examples

DPR has clear relevance to two of the applications featured in this chapter in particular.

Let us consider first the example of SDR. In this case, we suppose that the PHY of a wireless communications transceiver is implemented on the Zynq, with software control. The SDR may need to accommodate fundamentally different hardware structures depending on the set of wireless standards to be supported. *Without* DPR, this would necessitate implementation of all required architectures in parallel, implying considerable resource cost; *with* DPR, the required functionality may be consolidated to a single architecture comprising: (i) components common to all standards; and (ii) RPs with associated RMs, for all sections whose hardware structures vary between standards [18].

This concept for a flexible SDR transmitter architecture is shown in Figure 5.10, where it is assumed that the SDR comprises four functional blocks: coding, modulation, a transform, and a digital upconverter. We assume that the last of these components is common to all variations on the architecture, while there are different RMs for each of the others. The appropriate coding RM, modulation RM, and transform RM are chosen at run time and coordinated by the PS. (Please note that this is a simplified architecture for the purposes of discussion, and that the digital upconverter is assumed common for illustration only — it need not be.)

It is worth bearing in mind that DPR complements rather than replaces the other available methods for flexibly defining functionality; DPR is appropriate only where the underlying hardware structure needs to be changed. Some of the other components of the radio are better realised without DPR, but simply using software control, such as the NCO shown in Figure 5.5 on page 110.

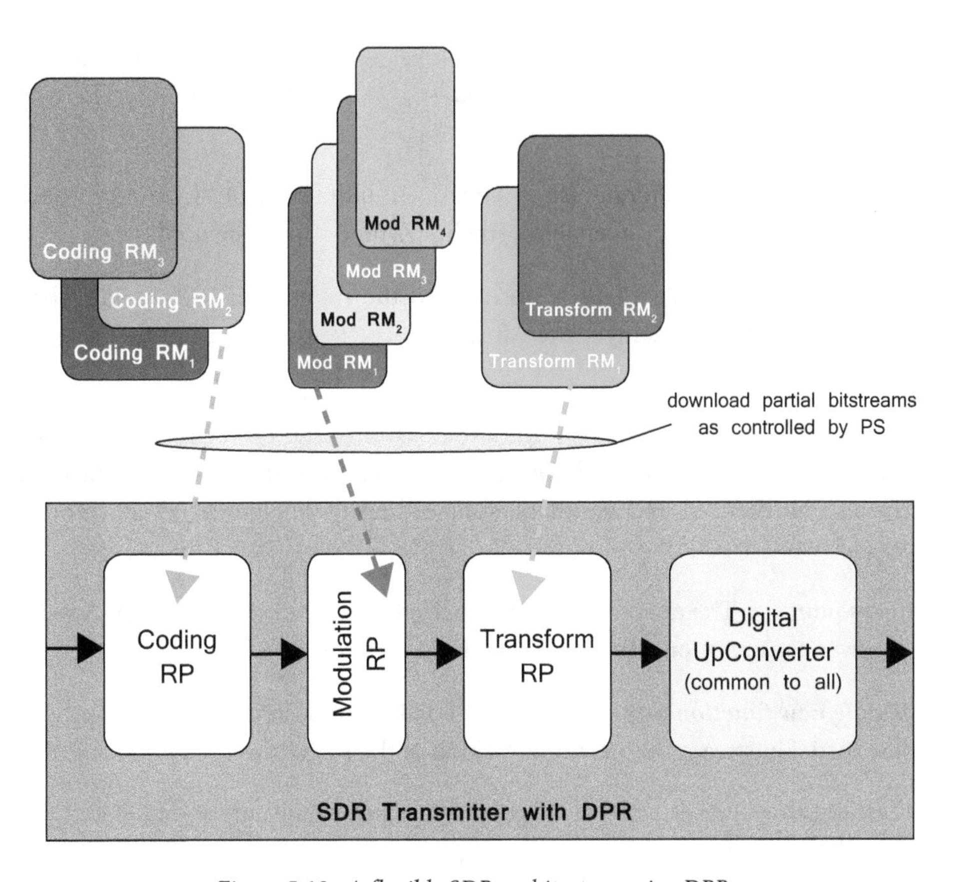

Figure 5.10: A flexible SDR architecture using DPR

Secondly, considering that image processing algorithms often comprise several different stages of processing, DPR offers the potential to make a dynamic selection of filters. For instance, in [25], DPR is used to offer a selection between a Sobel edge detection filter, and a Sepia filter for altering the colours in the image. There is an RM for Sobel and an RM for Sepia, and the chosen filter type is downloaded as a partial bitstream to the RP.

Image processing applications without tight real-time processing requirements can also exploit DPR to iterate through the stages of an image processing algorithm serially, perhaps even using a single RP. A design of this type was demonstrated in [11] for the FPGA implementation of a biometric fingerprint recognition system, and the scheme would be equally applicable to Zynq. Similarly, [10] mentions an example in the automotive field, where processing could switch from lane departure warning to rear-view camera depending on the car's direction of travel.

5.6.4. Benefits of DPR

The run time flexibility afforded by DPR is a great benefit, to the extent that it might be easy to forget about some of the other advantages of using this technique. These can be summarised as follows:

- ***Potential to use a smaller device*** — The time multiplexing of PL-based components made possible via DPR can enable a smaller Zynq device to be used.
- ***Feature richness*** — As there is no limit to the number of RMs, aside from that implied by available storage, then it is possible to provide a greater selection of functionality.
- ***Lower power consumption*** — If the use of DPR enables the use of a smaller Zynq device, this implies lower static power consumption. Dynamic power consumption can also be reduced as a result of not needing to implement all PL modules in parallel.
- ***Cost savings*** — The use of a smaller device enables a direct cost saving. A reduction in power consumption also reduces operational costs.
- ***Adding new functionality*** — New functionality can be added simply by providing a new partial bitstream file, i.e. no reworking of the PL hardware design is needed.

For these reasons, DPR is a useful technique, and potentially applicable in a number of different application areas.

5.7. Further Opportunities: the Zynq 'EcoSystem'

The discussion in this chapter so far has focussed on applications of Zynq, and implicitly that these areas represent opportunities for developing products based on Zynq. There is also another type of opportunity — arguably a less obvious one — relating to software and hardware support products for Zynq. These add value to the core product and associated development tools, and thus enhance its possibilities. Many third party developers are already part of this activity.

During the remainder of this section, we describe the current shape of the Zynq 'ecosystem', and review some of the opportunities for developers to contribute to, and benefit from, the ecosystem [61].

5.7.1. What is the Ecosystem?

The notion of a technology ecosystem will be familiar to many through their experiences of consumer electronics. Consider smartphones in particular. The phone manufacturer provides the phone and basic operating system, and it may host an 'app store' or similar, where users can obtain additional applications to run on their phone, should they desire. The apps in the app store are almost exclusively developed by companies other than the phone manufacturer, and there may be several thousands of different apps available, catering both for mainstream interests like organising music or checking the weather forecast, to more specialised tasks such as tracking the path of your cycle rides. There is also an array of hardware accessories available for smartphones, much of which is from third parties, including cases and screen protectors, portable speakers, chargers, car kits and more.

The situation with Zynq is somewhat similar. Xilinx manufactures the silicon devices, a small set of development boards, and provides the core set of development tools, but there are also opportunities for other companies to develop software tools, applications and hardware-based products around Zynq. The Zynq ecosystem includes software development environments, specialist software libraries, IP development tools, packaged IP, OSs and middleware, software applications for deployment on Zynq, virtual platforms, hardware development boards, add-on modules, and other accessories. Over time, and as applications of Zynq consolidate and diversify, its ecosystem will become a richer resource. Additionally, ARM processors have an ecosystem of their own, and this can be exploited too [2].

A graphical representation of the Zynq ecosystem is provided in Figure 5.11, showing a representative (but not comprehensive) selection of the software, hardware and supporting resources available. The ecosystem includes Xilinx Alliance Program partners and other third party developers of Zynq-based or Zynq-relevant products and services [59], [61].

5.7.2. What is the Opportunity?

Xilinx encourages third parties to develop and market products and services around Zynq. The market for these offerings is primarily the community of companies developing Zynq-based products, which is growing steadily; however, given that the device is comprised of two standard parts (an ARM processor and FPGA fabric), there is potential for portability to these platforms too. For example, the opportunity for IP developers is to create IP blocks valuable to users of Zynq, and these would also be applicable for general FPGA usage. The Zynq ecosystem therefore represents a business opportunity for technology companies to become involved in.

Aside from commercial ventures, there is also scope for open source projects and free-to-use software. In fact, several components of the ecosystem as depicted in Figure 5.11 are open source, for instance the widely used OpenCV libraries for computer vision [34], and the FreeRTOS real-time operating system [12].

From the opposite perspective, i.e. that of a product developer, the ecosystem provides an opportunity to obtain important resources for creating systems based on Zynq. It represents an opportunity to leverage third party IP, software components, development tools etc., enabling developers to accelerate their design and verification processes, and therefore to reach the market more quickly than would otherwise be possible.

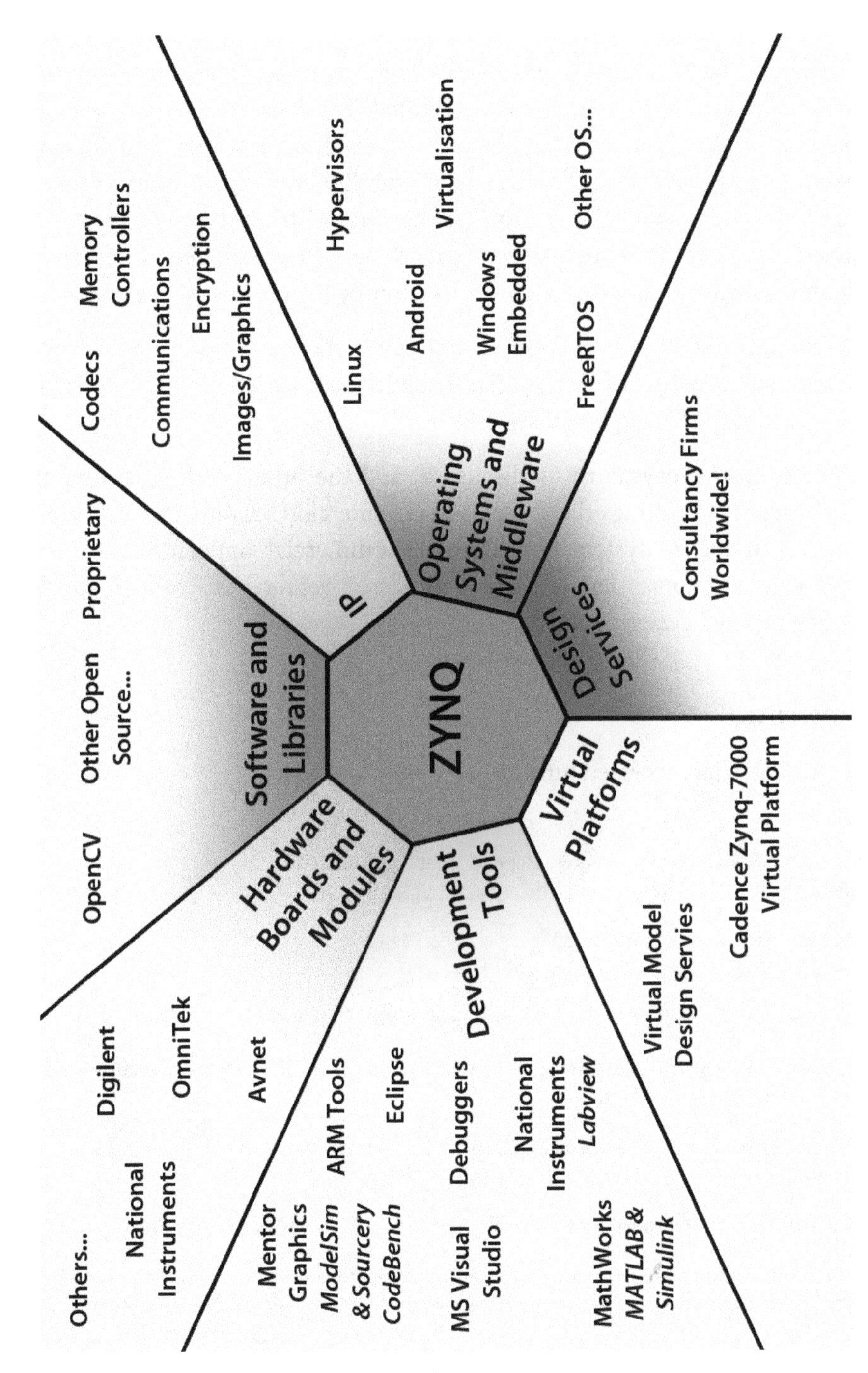

Figure 5.11: The Zynq ecosystem

5.8. Chapter Review

This chapter has demonstrated Zynq's wide-ranging applicability by reviewing a selection of applications for which it is well suited. Three particular areas were reviewed in detail as case studies, namely SDR, smart systems and networks, and image and video processing. In each case, it was shown that flexibility and expansibility is important, and therefore that Zynq is an ideal platform. The need for different types of processing was also highlighted: in applications such as computer vision, there is a need for both high-speed, parallel processing, and 'smarter' algorithms running in software on a processor.

The technique of DPR was introduced, and it was shown that DPR can offer a number of benefits, not just in terms of operational flexibility, but also the cost, power consumption and upgrade-ability of Zynq based systems.

Lastly, the Zynq ecosystem was discussed, and the broad and expanding portfolio of third party product offerings (both free and commercial) was reviewed. It was noted that participating in the ecosystem constitutes a commercial opportunity for providers of resources, while for consumers of these resources, it represents a valuable mechanism for accelerating product design cycles.

5.9. References

Note: All URLs last accessed June 2014.

[1] Apertus.org, "AXIOM Alpha Progress Report" webpage, May 2013.
Available: https://www.apertus.org/article-axiom-alpha-update

[2] ARM Ltd., "ARM Community" webpage.
Available: http://www.arm.com/community/

[3] D. Bailey, *Design for Embedded Image Processing on FPGAs,* Wiley-IEEE Press, 2011.

[4] T. Barnaby, "FPGAs Help Measure Trajectory of Particles in CERN's Proton Synchrotron", *Xcell Journal,* Fourth Quarter 2009, pp. 22 - 26.

[5] BBC News, "How New York is Releasing its 'Big Data' to the Public", 12th October 2013.
Available: http://www.bbc.co.uk/news/technology-24505860

[6] A. Bohandi, "Smart Transport", *Engineering & Technology,* Vol. 7, Issues 6, July 2012, pp. 70 - 73.

[7] N. Buch, S. A. Velastin and J. Orwell, "A Review of Computer Vision Techniques for the Analysis of Urban Traffic", *IEEE Transactions on Intelligent Transportation Systems,* Vol. 12, No. 3, September 2011, pp. 920 - 939.

[8] B. Dipert, J. Alvarez and M. Touriguian, "Embedded Vision: FPGAs' Next Notable Technology Opportunity", *Xcell Journal*, First Quarter 2012, pp. 14 - 19.
Available: http://www.xilinx.com/publications/archives/xcell/Xcell78.pdf

[9] A. Dembosky, "Silicon Valley links with Salinas Valley to make Farming 'Smart'", *Financial Times (ft.com)*, 28th June 2013.
Available: http://www.ft.com/cms/s/0/55656cf2-dff4-11e2-bf9d-00144feab7de.html#axzz2iIdk82Be

[10] F. Fons and M. Fons, "FPGA-based Automotive ECU Design Addresses AUTOSAR and ISO 262622 Standards", *Xcell Journal*, First Quarter 2012, pp. 20 - 31.
Available: http://www.xilinx.com/publications/archives/xcell/Xcell78.pdf

[11] F. Fons and M. Fons, "Making Biometrics the Killer App for FPGA Dynamic Partial Reconfiguration", *Xcell Journal*, Third Quarter 2010, pp. 24 - 31.
Available: http://www.xilinx.com/publications/archives/xcell/Xcell72.pdf

[12] FreeRTOS website.
Available: http://www.freertos.org/

[13] R. C. Gonzalez and R. E. Woods, *Digital Image Processing*, 3rd Edition, Pearson, 1998.

[14] A. Guan, S. H. Bayless, and R. Neelakantan, "Connected Vehicle Insights: Trends in Computer Vision", *Intelligent Transportation Society of America, Technology Scan Series 2011-2012.*
Available: http://www.itsa.org/knowledgecenter/technologyscan/computer-vision-report

[15] G. Hampson et al, "Xilinx FPGAs Beam Up Next-Gen Radio Astronomy", Xcell Journal, Second Quarter 2011, pp. 30 - 35.
Available: http://www.xilinx.com/publications/archives/xcell/Xcell75.pdf

[16] G. P. Hancke, B. de Carvalho e Silva, and G. P. Hancke Jr., "The Role of Advanced Sensing in Smart Cities", Sensors 2013, pp. 393 - 425.
Available: http://www.mdpi.com/1424-8220/13/1/393

[17] f. harris and W. Lowdermilk, "Software Defined Radio: part 22 in a series of tutorials on instrumentation and measurement", *IEEE Instrumentation and Measurement Magazine*, Vol. 13, Issue 1, February 2010, pp. 23 - 32.

[18] K. He, L. Crockett and R. Stewart, "Dynamic Reconfiguration Technologies Based on FPGA in Software Defined Radio System", *Journal of Signal Processing Systems,* Vol. 69, Issue 1, October 2012, pp. 75 - 85.

[19] T. Hill, "Motor Drives Migrate to Zynq SoC with Help from MATLAB", Xcell Journal, Second Quarter 2014, pp. 32 - 37.
Available: http://www.xilinx.com/publications/archives/xcell/Xcell87.pdf

[20] IMEC, "Wireless Communication: Cognitive Radio" webpage.
Available: http://www.imec-nl.nl/nl_en/research/green-radios/cognitive-radio.html

[21] International Telecommunications Union, "H265: High Efficiency Video Coding", Recommendation H.265 (04/13), April 2013.

[22] "Joint Tactical Networking Centre" webpage.
Available: http://www.jtnc.mil/Pages/Home.aspx

[23] F. K. Jondral, "Software-Defined Radio - Basics and Evolution to Cognitive Radio", *EURASIP Journal on Wireless Communications and Networking*, Vo. 2005, Issue 3, August 2005, pp. 275-283.

[24] K. Khan, "FPGAs Help Drive Innovation in Complex Medical Systems", *Medical Electronics Design* website, April 2012.
Available: http://www.medicalelectronicsdesign.com/article/fpgas-help-drive-innovation-complex-medical-systems

[25] C. Kohn, "Partial Reconfiguration of a Hardware Accelerator on Zynq-7000 All Programmable SoC Devices", Xilinx Application Note, XAPP1159, v1.0, January 2013.
Available: http://www.xilinx.com/support/documentation/application_notes/xapp1159-partial-reconfig-hw-accelerator-zynq-7000.pdf

[26] Y. Lin, "Using Xilinx Devices to Solve Challenges in Industrial Applications", *Xilinx White Paper,* WP410, v2.0, October 2012.

[27] MathWorks, Inc., "Image and Video Processing" webpage.
Available: http://www.mathworks.com/image-video-processing/

[28] Mike Mitchell, "Zynq for Video Applications", Xilinx presentation, 2012.
Available: http://www.xilinx.com/Attachment/52101/video_opps_rv3.pdf

[29] Joe Mitola, "The Software Radio Architecture", *IEEE Communications Magazine*, May 1995, pp. 26 - 38.

[30] Moving Pictures Experts Group website, "MPEG-4" webpage.
Available: http://mpeg.chiariglione.org/standards/mpeg-4

[31] S. Neuendorffer, T. Li and D. Wang, "Accelerating OpenCV Applications with Zynq-7000 All Programmable SoC using Vivado HLS Video Libraries", Xilinx Application Note, XAPP1167, v2.0, August 2013.
Available: http://www.xilinx.com/support/documentation/application_notes/xapp1167.pdf

[32] Nokia Research Center, "NRC Presents Cognitive Radio" webpage.
Available: https://research.nokia.com/page/9401

[33] OECD, "Building Blocks for Smart Networks", *OECD Digital Economy Papers*, No. 215, OECD Publishing, 2013.
Available: http://dx.doi.org/10.1787/5k4dkhvnzv35-en

[34] OpenCV website,
Available: http://opencv.org/

[35] Phenox Lab, "Phenox" webpage.
Available: http://phenoxlab.com/?page_id=296

[36] Red Pitaya website.
Available: http://redpitaya.com/

[37] M. Russell and S. Fischaber, "OpenCV Based Road Sign Recognition on Zynq", *Proceedings of the 11th International Conference on Industrial Informatics*, Bochum, Germany, July 2013, pp. 596 - 601.

[38] M. Santarini, "Xilinx FPGAs to Power Next-Generation Networked Battlefield", *Xcell Journal*, Fourth Quarter 2009, pp. 8 - 14.
Available: http://www.xilinx.com/publications/archives/xcell/Xcell69.pdf

[39] M. Santarini, "Xilinx's New SDNet Environment Enables 'Softly' Defined Networks", *Xcell Journal*, Second Quarter 2014, pp. 8 - 13.
Available: http://www.xilinx.com/publications/archives/xcell/Xcell87.pdf

[40] J. Shankleman, "Public Transport Gets Smart", *The Guardian*, 9th January, 2013.
Available: http://www.theguardian.com/public-leaders-network/2013/jan/09/centro-public-transport-travel-systems

[41] A. Shukla et al, "Cognitive Radio Technology: A Study for Ofcom - Volume 1", QinetiQ Ltd. consultancy report, February 2007.
Available: http://stakeholders.ofcom.org.uk/binaries/research/technology-research/cograd_main.pdf

[42] D. Snoonian, "Smart Buildings", *IEEE Spectrum*, August 2003, pp. 18 - 23.

[43] E. Strickland, "Cisco Bets on South Korean Smart City", *IEEE Spectrum*, August 2011, pp. 11 - 12.

[44] P. Sundararajan, "High Performance Computing Using FPGAs", *Xilinx White Paper*, WP375, v1.0, September 2010.

[45] P. D. Sutton et al, "Iris: An Architecture for Cognitive Radio Networking Testbeds", *IEEE Communications Magazine*, September 2010, pp. 114 - 122.

[46] Teradeep webpage:
http://www.teradeep.com/

[47] J.A. Throop, D. J. Aneshansley, W. C. Anger, and D. L. Peterson, "Quality evaluation of apples based on surface defects: development of an automated inspection system", *Postharvest Biology and Technology*, Vol. 36, Issue 3, June 2005, pp. 281-290.

[48] U. S. Energy Information Administration, "Electricity Use by Machine Drives Varies Significantly by Manufacturing Industry", October 2013,
Available: http://www.eia.gov/todayinenergy/detail.cfm?id=13431

[49] J. van de Belt, P. D. Sutton, and L. E. Doyle, "Accelerating Software Radio: Iris on the Zynq SoC", *Proceedings of the IFIP/IEEE 21st International Conference on Very Large Scale Integration (VLSI-SoC)*, October 2013, pp. 294 - 295.

[50] G. Velez et al, "A Reconfigurable Embedded Vision System for Advanced Driver Assistance", *Journal of Real Time Image Processing*, Springer, March 2014.

[51] N. Walravens and P. Ballon, "Platform Business Models for Smart Cities: From Control and Value to Governance and Public Value", *IEEE Communications Magazine*, June 2013, pp. 72 - 79.

[52] B. Wang and K. J. R. Liu, "Advances in Cognitive Radio Networks: A Survey", *IEEE Journal of Selected Topics in Signal Processing*, Vol. 5, No. 1, February 2011, pp. 5 - 23.

[53] J. Wang, M. Ghosh, and K. Challapali, "Emerging Cognitive Radio Applications: A Survey", IEEE Communications Magazine, March 2011, pp. 74 - 81.

[54] S. Weston et al, "FPGAs Speed the Computation of Complex Credit Derivatives", Xcell Journal, First Quarter 2011, pp. 18 - 25.
Available: http://www.xilinx.com/publications/archives/xcell/Xcell74.pdf

[55] Wireless Innovation Forum, "What is Software Defined Radio?" webpage,
Available: http://www.wirelessinnovation.org/introduction_to_sdr

[56] xG Technology, "xG Technology Ships World's First Comprehensive Cognitive Radio System to Walnut Hill Telephone Company", press release, 16th October, 2013.
Available: http://www.xgtechnology.com/2013-Press-Releases/xg-technology-ships-worlds-first-comprehensive-cognitive-radio-system-to-walnut-hill-telephone-company.html

[57] Xilinx, Inc., "A Generation Ahead: Smarter Networks", Backgrounder, 2013.
Available: http://www.xilinx.com/publications/prod_mktg/smarter-networks-backgrounder.pdf

[58] Xilinx, Inc., "Defense Grade Zynq-7000Q AP SoCs" webpage.
Available: http://www.xilinx.com/products/silicon-devices/soc/zynq-7000q.html

[59] Xilinx, Inc., "Xilinx Alliance Program" webpage.
Available: http://www.xilinx.com/alliance/index.htm

[60] Xilinx Inc., "Xilinx's Zynq-7000 All Programmable SoCs Enable Mobilicom's Advanced Peer-to-Peer Software-Defined Radios", press release, 16th July 2013.
Available: http://press.xilinx.com/2013-07-16-Xilinxs-Zynq-7000-All-Programmable-SoCs-Enable-Mobilicoms-Advanced-Peer-to-Peer-Software-Defined-Radios

[61] Xilinx, Inc., "Zynq 7000 AP SoC Ecosystem" webpage.
Available: http://www.xilinx.com/products/silicon-devices/soc/zynq-7000/ecosystem/index.htm

[62] C. Zammariello and A. Lorelli, "Towards SDR Standardisation for Military Applications", *European Defence Agency news article*, January 2012.
Available: http://www.eda.europa.eu/info-hub/news/12-01-11/Towards_SDR_standardisation_for_military_applications

6

The ZedBoard

Now that we have addressed some of the fundamental questions relating to the Zynq platform, it is appropriate to consider one of the most popular development boards for working with Zynq — the *ZedBoard*. To explain the name, 'Zed' stands for **Z**ynq **E**valuation and **D**evelopment. The ZedBoard is just one of the development and evaluation boards available at the time of writing. It is priced at a suitable level for students and enthusiasts, and it forms the focus of an online community of ZedBoard users.

In this chapter, the ZedBoard architecture and features will be introduced, together with some basic information about getting started, and the support integrated into the Xilinx design tools for ZedBoard. Currently available resources for creating ZedBoard designs, including tutorials, videos and support, will also be covered, with a special mention for the ZedBoard online community.

6.1. Introducing Zed

The ZedBoard is a low-cost, community-based board which features a XC7Z020 Zynq device. It is a joint venture between Xilinx, Avnet (the distributor), and Digilent (the board manufacturer).

Although suitable as a development platform for industry, the ZedBoard is also targeted at students, academics, and hobbyist users, with specific materials to cater for new Zynq users, and addressing the learning curve of beginners. The community aspect derives from a website run by Avnet that is dedicated to supporting users (ZedBoard.org), which will be covered in detail in Section 6.7.

6.2. ZedBoard System Architecture

As mentioned in the introductory comments, the ZedBoard features a ZC7Z020 Zynq device. This is one of the smaller devices in the Zynq-7000 range, and it is based on the Artix-7 logic fabric, with a capacity of 13,300 logic slices, 220 DSP48E1s, and 140 BlockRAMs. The device also contains an XADC hard IP block, although it does not feature high-speed transceivers or PCIExpress blocks.

There are a number of peripheral interfaces on the ZedBoard:

- GPIO: in total, 9 x LEDs, 8 x switches, 7 x push buttons
- Audio codec (Analog Devices ADAU1761, supporting line in, line out, microphone (in), and headphone (out))
- Video (HDMI)
- Video (VGA)
- Organic Light Emitting Diode (OLED) display
- Pmod interfaces (x 5)
- Ethernet
- USB-OTG (peripherals)
- USB-JTAG (programming)
- USB-UART (communication)
- SD card slot (located on the underside of the board)
- FMC interface
- XADC header
- Xilinx JTAG header

Additionally, the Zynq device interfaces to a 256Mbit flash memory and 512MB DDR3 memory, both of which are found on the board. There are two oscillator clock sources, one at 100MHz, and the other at 33.3333MHz.

The features residing on the front of the ZedBoard are highlighted in Figure 6.1.

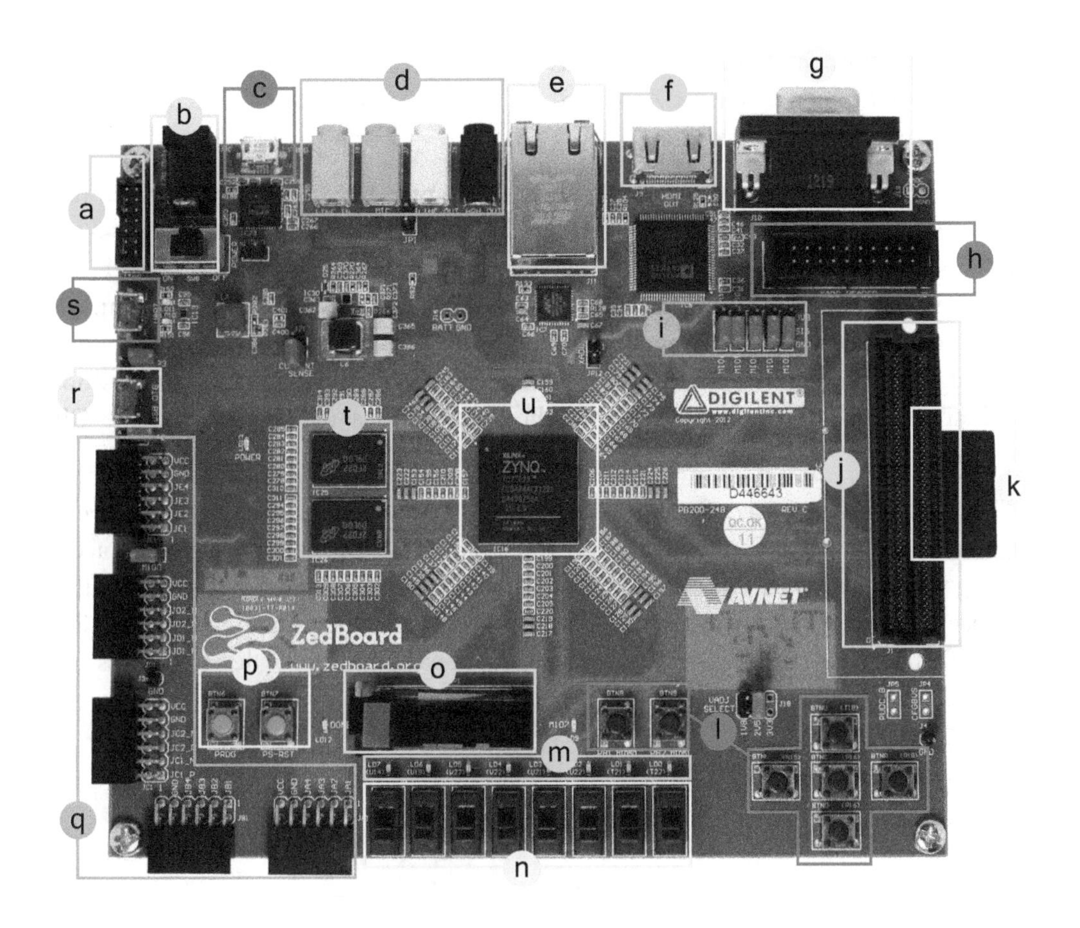

- a Xilinx JTAG connector
- b Power input and switch
- c USB-JTAG (programming)
- d Audio ports
- e Ethernet port
- f HDMI port (output)
- g VGA port
- h XADC header port
- i Configuration jumpers
- j FMC connector
- k SD card (underside)
- l User push buttons
- m LEDs
- n Switches
- o OLED display
- p Prog & reset push buttons
- q 5 x Pmod connector ports
- r USB-OTG peripheral port
- s USB-UART port
- t DDR3 memory
- u Zynq device (+ heatsink)

Figure 6.1: ZedBoard layout and interfaces (front)

6.3. The Design Flow for ZedBoard

The design flow for the ZedBoard is as described in Chapter 3, with the additional benefit of specific support for the ZedBoard within the tool flow, and this can be exploited to streamline the design process.

When creating a project in Vivado, the user is prompted to specify the intended target for the design. At this point, the target *part* could be specified as the **xc7z020clg484-1** (i.e. the Zynq device on the ZedBoard), but notably there is also an option to select a target *board*. Here, choosing the **ZedBoard Zynq Evaluation and Development Kit** means that, in addition to correctly targeting the appropriate Zynq device, the design tools have knowledge of the specific facilities and peripheral connections available on the ZedBoard (for example, the numbers of LEDs, switches and buttons which comprise the GPIO interface).

This selection stage is confirmed in Figure 6.2. At the time of writing, there are two revisions of the ZedBoard ('c' and 'd'), and you should consult the documentation accompanying your board to check which applies. A number of other board targets are available. Likewise if using one of those (e.g. the ZC702 or ZC706 Zynq evaluation boards), this can be specified to ensure specific support for the target board during the design process.

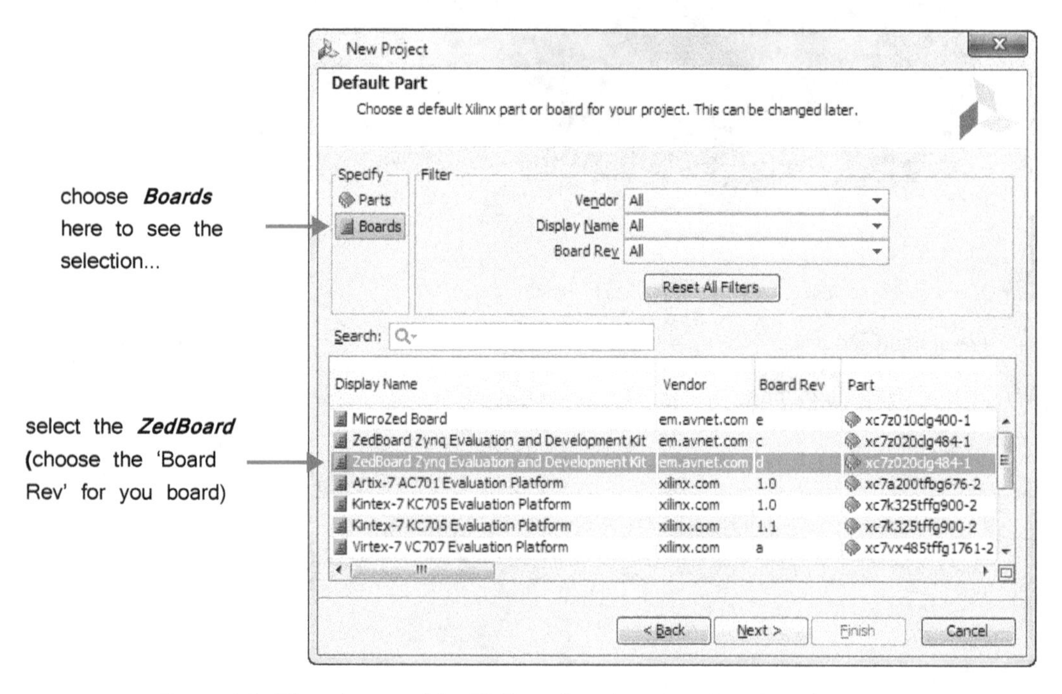

Figure 6.2: The selection of the ZedBoard as the target for a Vivado project

6.4. Getting Started with the ZedBoard

This section provides a brief review of the ZedBoard kit, and the initial steps required to complete the hardware setup of the board.

6.4.1. What's in the Box?

When you open the box for the first time, you will find a number of items inside, in addition to the ZedBoard itself. These are helpfully demonstrated in a video produced by Avnet and posted on the ZedBoard.org website [3]. The same video points out the different features of the ZedBoard, as were reviewed in Figure 6.1, and confirms the steps required to run the out-of-the-box design.

To confirm the contents of the ZedBoard box, you should find the following items inside:

- The ZedBoard, packaged in an anti-static bag
- A DVD of software design tools
- A power adapter with attachments for US and EU socket types
- A USB-A *to* micro-USB-B cable
- A micro-USB-B *to* USB-A female adapter cable
- A 4GB SD card
- A leaflet with getting started information [7]

It is worth bearing in mind that those not living in an area where US or EU sockets are used will additionally need to source a conversion adapter for their local socket type. For instance, those based in the UK require a US-to-UK or EU-to-UK adapter.

A common requirement, especially for early-stage tutorials that you might follow, is to interact with the board via the UART interface. This will require a second USB-A *to* micro-USB-B connection (note that there is only one such cable provided, so a second may need to be sourced). Depending on your intended usage of the ZedBoard's peripherals, cables for audio, video, ethernet, etc. might also be needed.

6.4.2. Hardware Setup

The ZedBoard must be connected to a power supply and, in the default configuration, to the host PC for JTAG programming over USB (i.e. the USB-A *to* micro-USB-B connection). An additional connection should be made if intending to use the UART to facilitate simple board-PC communication using the Terminal application. The locations of these two connectors on the ZedBoard are shown in Figure 6.3. Note that there is also a third micro-USB port which is reserved for connecting USB peripherals, if used (this can be seen at the bottom of the photograph).

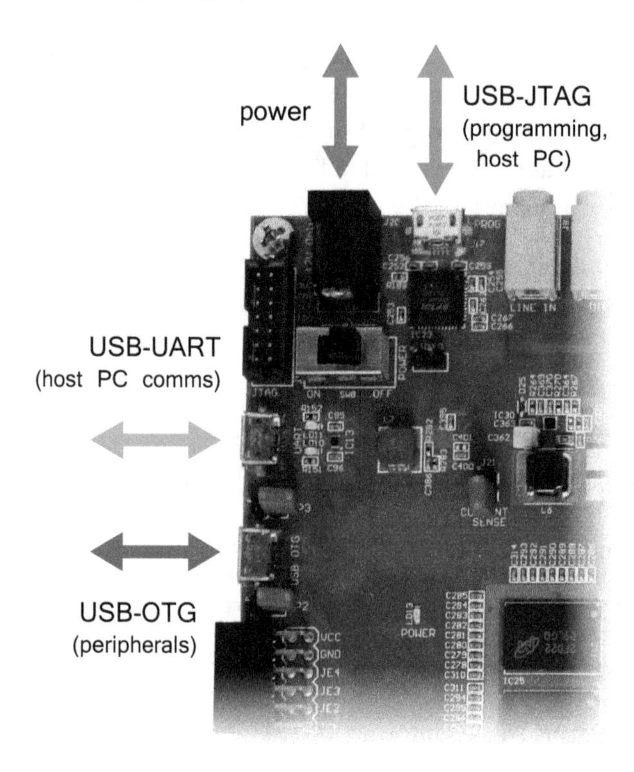

Figure 6.3: Top left corner of ZedBoard, clarifying power and USB connectors

It is important to note that the correct device drivers must be installed in order to use the USB-UART connection. These can be sourced from Cypress, and full details of the installation process can be found in [2], including links to download the relevant driver files.

6.4.3. Programming the ZedBoard

The ZedBoard can be programmed in four different ways. These are:

- ***USB-JTAG*** — This is the default and most straightforward method of programming the ZedBoard, given that it can be done directly over the USB-micro-USB cable supplied in the ZedBoard kit.

- ***Traditional JTAG*** — A Xilinx JTAG connector is available on the board and may be used in place of the USB-JTAG connection, if desired. This will require a different type of cable than is included in the ZedBoard kit: a *Xilinx Platform USB* cable [11], or a *Digilent USB-JTAG programming cable* [10].

- ***Quad-SPI flash memory*** — The flash memory on the board is non-volatile and therefore can be used to store configuration data which persists when the board is powered off. Using this method removes the requirement for a wired connection to program the Zynq device.

- ***SD card*** — There is an SD slot on the underside of the ZedBoard. This facility can be used to program the Zynq with files stored on the SD card, thus requiring no wired connections for programming. This method is featured in the ZedBoard *Getting Started Guide* [6].

Notably, the JTAG methods are ideal for the development phase, when it is convenient to establish a USB / JTAG connection between the development PC and the ZedBoard. The other two methods are more portable, and ultimately more representative of the type of programming method which would be used in the field. It is also significant that JTAG is a non-secure mechanism, while the other methods are secure; this is another factor in favour of the flash and SD-based configuration methods when outside the lab environment. Booting will be discussed in detail in Chapter 24, in the context of Linux-based systems.

The ZedBoard user specifies the method of booting / programming via a set of jumper pins, which are positioned above the *Digilent* logo, as clarified in Figure 6.4. Of the five jumpers in the group, the central three are used to define the source for programming the board (JTAG, flash memory, or SD card), the jumper on the far right controls the JTAG mode, and the jumper on the far left determines whether the internal PLL is used.

These last two options require further clarification. The JTAG mode refers to the method of debugging using JTAG; this can either be *cascaded*, in which case a single JTAG connection is used to interface to the debug access ports in both the PS and PL, or *independent*, wherein the debug access ports in the PS and PL are accessed separately, requiring one cable for each. The PLL mode determines whether the process of configuring the device includes a phase of waiting for the PLL to lock, before starting the boot process. The alternative is to bypass the PLL, but in this case, booting takes longer.

Extensive further information about configuration options and processes is available in the *Zynq-7000 Technical Reference Manual* [13].

Table 6.1 specifies permitted jumper settings, and selected examples are also given in Figure 6.5.

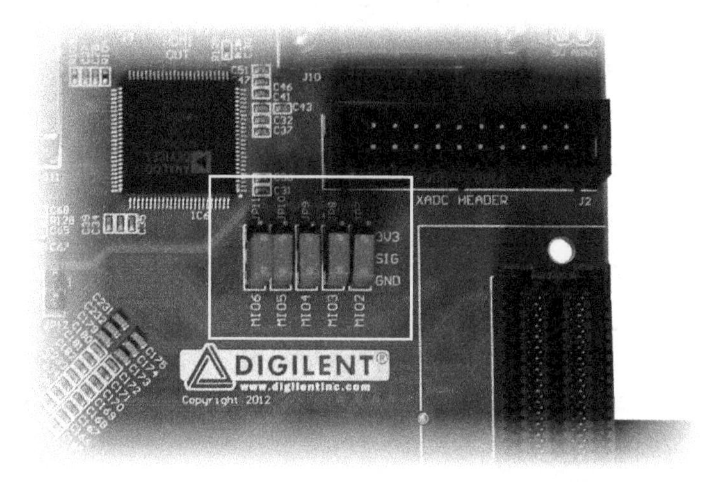

Figure 6.4: The set of jumpers determining configuration options

Table 6.1: Configuration jumper settings on the ZedBoard [8]

	MIO[6]	**MIO[5]**	**MIO[4]**	**MIO[3]**	**MIO[2]**
In Xilinx Technical Reference Manual...	Boot_ Mode[4]	Boot_ Mode[0]	Boot_ Mode[2]	Boot_ Mode[1]	Boot_ Mode[3]
			JTAG Mode		
Cascaded JTAG[a]	-	-	-	-	0
Independent JTAG	-	-	-	-	1
			Boot Device		
JTAG	-	0	0	0	-
Quad-SPI (flash)	-	1	0	0	-
SD Card[a]	-	1	1	0	-
			PLL Mode		
PLL Used[a]	0	-	-	-	-
PLL Bypassed	1	-	-	-	-

a. Denotes default setting.

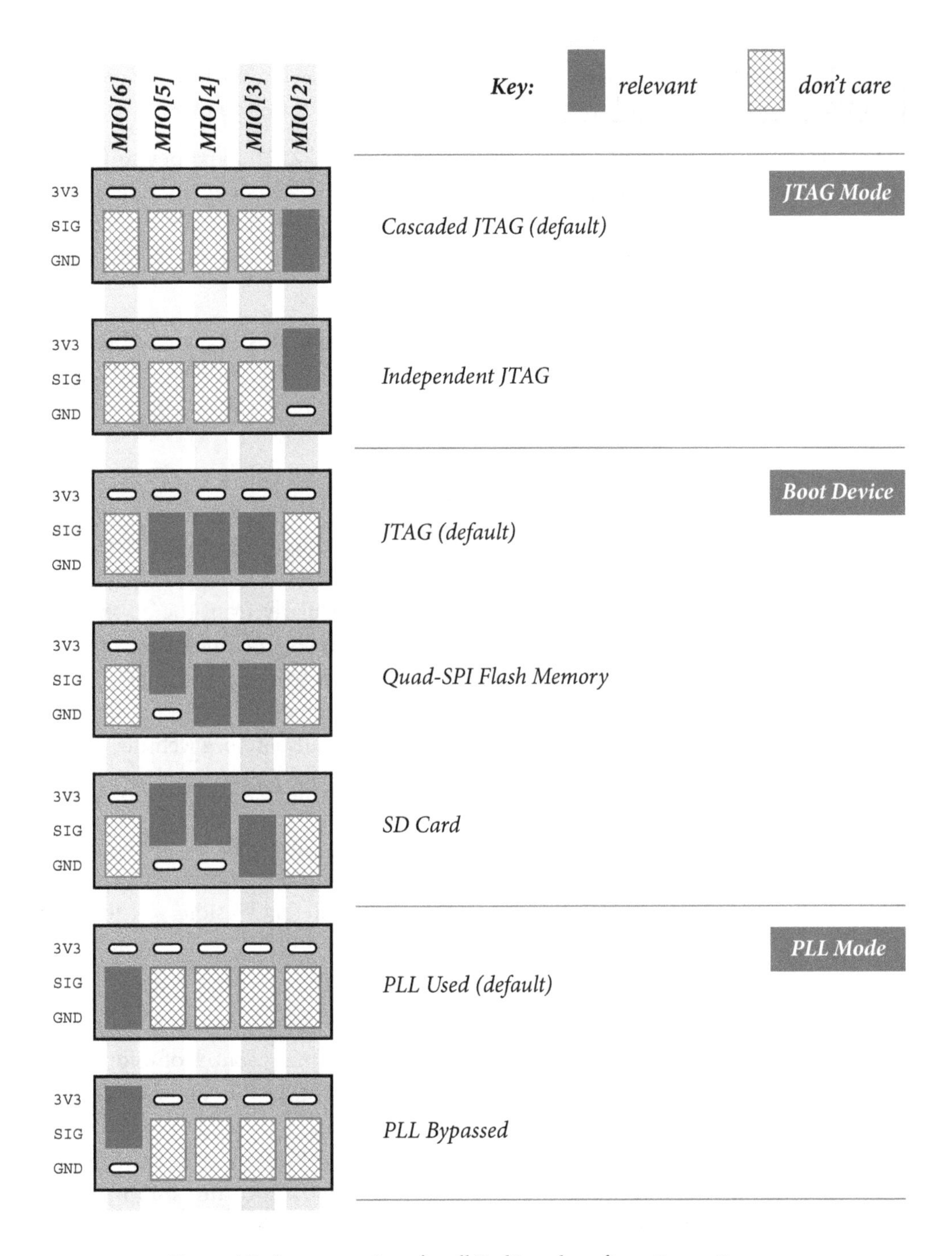

Figure 6.5: Jumper settings for all ZedBoard configuration options

6.5. MicroZed

The MicroZed is a Zynq-based development board following on from the success of ZedBoard. It is a smaller board with fewer peripherals, and cheaper to purchase.

The default configuration of the MicroZed features a ZC7Z010 Zynq device (a smaller one than on the ZedBoard). The MicroZed can be used either as a standalone development board, or placed into a carrier card as a module. As compared to standalone operation, using the MicroZed as a module hosted by a carrier board has the advantage that additional input / output functionality becomes available.

Although we are concerned primarily with the ZedBoard in this chapter, MicroZed and its associated carrier card are both featured and supported on the www.ZedBoard.org website.

6.6. Documentation, Tutorials and Support

Whether a new user, or one with prior experience of Zynq, it is probable that you will need to refer to documentation for the ZedBoard at some stage. It is useful to highlight the various sources of information and key documents, particularly from the perspective of those with little prior knowledge in the area. This includes documentation about the ZedBoard itself, as well as demonstrations and tutorials on working with the board, and other sources of support. Apart from where specifically noted, all of the resources cited below are available from the ***Documentation*** pages of the ZedBoard.org website:

http://www.zedboard.org/support/documentation

Although it is considered useful to review the materials and support available on the ZedBoard website, please bear in mind that these web resources are inevitably dynamic in nature, and may change over time. In particular, it is expected that versions will change and new materials will appear.

6.6.1. Documentation about the ZedBoard

In terms of documentation about the board itself, the primary source of information is the ZedBoard ***Hardware User's Guide*** [8], which reviews all of the functional aspects of the ZedBoard, including clock and reset sources, power circuitry, configuration methods, jumper settings, on-board memory, peripheral interfaces, pin voltages, and expansion headers. *ZedBoard Errata* documents provide necessary updates to this information [4], [5]. The *Hardware User's Guide* is a very useful reference, especially when developing designs featuring the peripherals on the board.

Master constraints files are available, which provides the superset of input/output connection assignments. Constraints are provided as both a UCF (*.ucf) file, for use in the Xilinx ISE flow, and as an XDC (*.xdc) for the Vivado flow.

Another important document relates to the USB-UART connection, which requires particular driver files to be downloaded and installed on the host PC. Guidance on this process may be found in the ***Cypress CY7C64225 USB-to-UART Setup Guide*** [2].

Further board-related documentation available on www.ZedBoard.org includes schematics, mechanical drawings, and the bill of materials. All of these documents are located under the ***Support / Documentation*** section of the website [14].

6.6.2. Demonstrations and Tutorials

A number of demonstrations and tutorials are available to assist with getting started. To highlight a few of these, the following may be particularly useful:

- The ***Getting Started Instructions*** card [7] — a short guide included in the kit to confirm that the ZedBoard is operating correctly;
- The ***Overview of the ZedBoard Kit*** video [3] — confirmation of the contents of the ZedBoard kit, with a demonstration of how to connect up the board and get started;
- The ***Getting Started Guide*** [6] — featuring a number of demonstrations of the ZedBoard's capabilities;
- The ***ZedBoard: Zynq-7000 AP SoC Concepts, Tools, and Techniques (CTT)*** guide [12] — a useful set of hands on tutorials introducing both the PL and PS; and

Some guides may have accompanying sets of files, and may be specific to a particular version of the Xilinx tools, so it is worth checking version compatibility before starting.

6.6.3. Online Courseware

There are several coursewares available on the ZedBoard.org website, all of which are freely available to anyone who registers on the site. Additionally, members of the academic community (academics, researchers and students) can obtain further resources via Xilinx University Program. Academic materials will be covered in Chapter 7.

The primary training material on the ZedBoard.org website is the SpeedWay Training produced by Avnet. At the time of writing, there are four courses available: ***Introduction to Zynq***, ***Implementing Linux on the Zynq-7000 SoC***, ***Software Defined Radio on Zynq***, and ***Debugging ARM Processor Systems***.

Another source of training, of an informal style, are the blogs posted by ZynqGeek, a member of the ZedBoard.org community. These are generally focussed on particular procedures (sometimes stimulated by posts in ZedBoard.org's forum feature) and cover such topics as a simple "Hello World" example, how to create and interface with a custom peripheral in PL, and how to create a Linux kernel. The community aspect of ZedBoard.org will be discussed further in Section 6.7.

6.6.4. Other ZedBoard Resources and Support

Often it is instructive to investigate the designs of others, and the various ***Reference Designs*** posted on ZedBoard.org provide a useful resource. These are found under ***Support / Reference Designs***, and constitute a set of standard projects developed and maintained by Avnet and other ZedBoard partners. The projects include a Quadrature Phase Shift Keying (QPSK) digital up and down converter demo, a Zynq-based video design based on HDMI, and running desktop Linux on the Zynq. Some of these require additional modules such as Pmods or FMC cards, which are not part of the basic ZedBoard kit.

6.7. ZedBoard.org Community

The community aspect of the ZedBoard, which is hosted by the www.ZedBoard.org website, differentiates it from other development boards, and creates an environment where ZedBoard users can access resources, exchange ideas, and help each other to troubleshoot problems. It is useful to briefly point out the highlights of this website from the community perspective; certain other aspects, particularly documentation, have already been covered earlier in the chapter.

6.7.1. Community Projects

Augmenting the set of reference designs, there is also a page for community projects, which are designs contributed by members of the ZedBoard.org website. These projects can either be completed designs that the developer(s) would like to share, or incomplete systems, regarding which they would like to seek help, or encourage community participation. Any member of the community can request to post a community project, which involves submitting a short application for subsequent review and approval.

The topics of projects contributed to date include outputting to the OLED display, creating an electrical motor controller, and using the AMS101 add-on module for mixed signal applications. The projects cover a range of subjects and start with simple systems. Some of these projects may require additional modules, but the majority require only the integrated facilities of the ZedBoard.

6.7.2. Blogs

Another useful part of the ZedBoard.org community are the links to blogs by Zynq enthusiasts. At the time of writing, there are two particularly active blogs, *ZynqGeek* and *Zynq from Scratch*. The individual blog posts are usually based on practical issues, and take an enthusiastic and accessible whirl through some aspect of Zynq-based design, whether fundamentals, hardware design, or software development. Many of these follow a helpful, tutorial style. Blog posts are often stimulated by others' questions or comments, which adds to the community feel.

6.7.3. Support Forums

With over 3,000 ZedBoards sold to the general community in the first year after its release (in July 2012), and a similar number to the academic market [9], the platform has a strong and growing user base. This translates into a healthy ZedBoard community, which is apparent from the forums. These provide an opportunity for ZedBoard users to ask or respond to questions, providing an encouraging environment for beginners (in particular) to seek help. Forum facilities are provided in both English and Chinese languages.

ZedBoard users can also post questions and responses on the appropriate Zynq-related forums on the Xilinx website, for access to a wider community of potential respondents:

http://forums.xilinx.com/

6.8. Chapter Review

This chapter has introduced the ZedBoard, a low-cost evaluation and development board featuring the Zynq XC7Z020 device. The important features of the ZedBoard have been reviewed, and in particular the various physical interfaces of the board have been identified and their purposes explained. We have also seen that the design tools include specific support for the ZedBoard (as well as other development boards) which can help to expedite the design process.

The ZedBoard is partly aimed at encouraging use of Zynq within the academic and enthusiast communities, and as such there are a variety of resources available which cater especially for those starting out in Zynq design. Over the preceding pages, we have confirmed the contents of the ZedBoard kit, explained how it can be set up, and summarised the key sources of information for getting started with the ZedBoard. Lastly, we have discussed the community aspect of the ZedBoard, noting in particular the community projects, as well as the forums, which are a 'live' resource for discussing technical problems and exchanging ideas with others working on the same platform.

ZedBoard caters for a variety of users, including the academic community in particular. In the next chapter, we will continue this theme by focussing on the use of both Zynq and ZedBoard in education, research, and training.

6.9. References

NOTE: All URLs last accessed June 2014.

[1] Analog Devices, "Analog Devices FMC - Communications Board: Analog Connectivity with Xilinx Zynq", 2012.
Available:
http://www.analog.com/static/imported-files/overviews/FMC-Communications_Product_Highlight.pdf

[2] Avnet, "Cypress CY7C64225 USB-to_UART Setup Guide", version 1.3, January 2014.

[3] Avnet, "Overview of the ZedBoard Kit", video.
Available: http://www.zedboard.org/videos/overview-zedboard-kit

[4] Avnet, "ZedBoard Rev C.1 Errata", revision 1.4, January 2014.

[5] Avnet, "ZedBoard Rev D.2 Errata", revision 1.1, January 2014.

[6] Avnet, "ZedBoard Getting Started Guide", version 7.0, January 2014.

[7] Avnet, "ZedBoard Getting Started Instructions".

[8] Avnet, "ZedBoard (Zynq Evaluation and Development) Hardware User's Guide", version 2.2, January 2014.

[9] BusinessWire website, "Avnet Electronics Marketing Celebrates One Year of ZedBoard", 23rd July, 2013.
Available: http://www.businesswire.com/news/home/20130723005579/en/Avnet-Electronics-Marketing-Celebrates-Year-ZedBoard

[10] Digilent, Inc., "XUP USB-JTAG Programming Cable" webpage.
Available: http://www.digilentinc.com/Products/Detail.cfm?NavPath=2,395,716&Prod=XUP-USB-JTAG

[11] Xilinx, Inc., "Platform Cable USB II" webpage.
Available: http://www.xilinx.com/products/boards-and-kits/HW-USB-II-G.htm

[12] Xilinx, Inc., "ZedBoard: Zynq-7000 AP SoC Concepts, Tools, and Techniques", Vivado 13.2, July 2013.
Available: http://www.zedboard.org/support/design (requires login)

[13] Xilinx, Inc., "Zynq-7000 Technical Reference Manual", UG585, version 1.7, February 2014.
Available: http://www.xilinx.com/support/documentation/user_guides/ug585-Zynq-7000-TRM.pdf

[14] ZedBoard.org "Documentation" webpage.
Available: http://www.zedboard.org/support/documentation

7

Education, Research and Training

This chapter will consider the possibilities and opportunities for Zynq in academic research and teaching, as well as the training available to industry and the wider community. We set this within the context of a growing interest in creating networked embedded systems, and the engineering skills that will be required to meet current and emerging technology challenges in this and other areas of SoC interest.

As part of our review of education, we will highlight the opportunities for conveying a variety of aspects of SoC design through Zynq, and the potential to use Zynq as a common platform for several related academic subjects. The extensibility of Zynq and related academic development boards are shown to provide a very suitable platform for both formal taught classes, and project based learning. Some examples of each will be provided.

In research terms, Zynq has widespread applicability, and therefore the potential for academic research (particularly applications-led research), is wide and varied. We highlight some papers already published, and mention further topics which are expected to be fertile areas of interest and investigation.

The Xilinx University Program (XUP) will be introduced, and its role as a support mechanism for universities in respect of both teaching and research described. Finally, we end the chapter by reviewing sources of training particularly applicable for the industrial and hobbyist communities.

7.1. Technology Trends and SoC Education

One of the major technology trends over recent years has been the dramatic increase in internet connectivity. Whereas 10 years ago the primary source of internet connection was via a PC, consumers now expect wireless data on their smartphones and tablets, not only when at home or in the office, but also while mobile. Many households also make use of TV-on-demand services, via internet-enabled televisions, set-top boxes and games consoles, which allows them to watch their favourite programmes at any convenient time. As a combined result of these and other technologies, aggregate internet traffic is on a steep incline, and in fact global IP traffic is expected to surpass 1.0 zettabytes per year in 2015 [4].

As well as the growth in traffic, we are also witnessing an overall increase in the number of internet-connected devices. In fact, as of approximately 2008/2009, the number of such devices worldwide increased beyond that of the global population! [8]

The proliferation of connected devices is supported by Internet Protocol version 6 (IPv6), which supersedes IPv4 and removes the pressure on internet address space [3]. Partly due to this increased scope for connectedness, M2M has started to emerge, since the early 2010's, as a significant source of internet traffic. As described in Chapter 5, M2M refers to a class of devices that can be connected to the internet to produce, consume, or otherwise share data without the direct intervention of humans. M2M relates closely to the concept of the IoT, a term which neatly encapsulates the need for autonomous machines to communicate and network with each other independently [2].

These M2M terminals can also be considered embedded systems, i.e. systems which are not general purpose computers, but which are designed and optimised to undertake a specific function. For instance, M2M might include sensors and actuator nodes distributed through an industrial process to monitor and adjust manufacturing conditions, with the aim of increasing productivity, minimising waste, etc. As discussed in Chapter 5, application areas of IoT include *smart cities*, a loose concept covering several aspects of city life; *smart grids*, electrical power distribution networks that can dynamically sense and adapt, in order to maximise power delivery performance and self-protect against faults; and the *smart home*, wherein sensors and actuators manage the light levels, temperature, security and entertainment systems in an environmentally and economically sensitive way.

The growth of M2M-style internet traffic is projected to continue over the next few years, and even to outstrip the growth in more conventional traffic types [4]. A 2013 report projected that there will be 212 billion devices connected to the internet by 2020, of which around 30 billion will be autonomous [14]. The implication is that more internet-connected embedded systems will be required for the realisation of smart grids, smart cities, smart buildings, smart health (and several other smarts!), asset tracking, 'big data', the industrial internet, and many applications yet to emerge. The potential opportunities for innovation, commercial success and societal benefits in this area are huge. The importance of the SoC to these applications is clear, given the demands for complete autonomous systems to be implemented at modest cost, with low power consumption, and often in physically compact form.

This aspect of technology advance is significant when contemplating the evolution of curricula in electronics and communications. It will be important for engineering institutions to educate young engineers with the skills and mindset to develop the networked embedded systems required for the next-generation internet.

Of course, the IoT is not the only factor driving SoC. As reviewed in Chapter 5, applications are many and varied, including advanced communications and radar systems, broadcast technology, machine vision, automotive and aerospace, and considerably more. Collectively, these technology areas will stimulate new commercial opportunities, and new problems for the engineers of the future to solve.

7.2. University Teaching with Zynq

The majority of this section focuses on the development of courses in electronics and computing, and identifies educational themes which may be demonstrated, supported or enhanced through the use of Zynq All Programmable SoC and its associated design flow. Before doing so, it is useful to first review the use of Xilinx development tools and boards in the academic environment, from a general perspective.

7.2.1. Teaching with Xilinx Tools and Boards

The Vivado Design Suite can be used in university teaching to equip students with industry-relevant experience of building SoC designs using high-level design tools. The constraints of time-limited lab sessions may be different to industrial environments, but the increases in productivity that apply to commercial development can be similarly leveraged in academia, thus enabling students to achieve more during their limited lab times. In the past, students undertaking FPGA design using VHDL or Verilog could not be expected to develop a large system or complex IP block, due to the limitations of their timetables, and the adoption of a low-level, HDL-based design methodology.

With Zynq and the system-level design methods possible with Vivado, it is feasible for students to build more sophisticated SoC designs, containing both custom hardware and software, within the time constraints of a university module or project.

From a 'hands-on' practical perspective, Zynq academic development boards feature a selection of on-board peripherals, providing everything that is needed for embedded systems development. These range from simple I/O, to Terminal communication, audio, video, and data interfaces. Boards can also be further extended with add-on modules. Students engagement is typically enhanced by activities which interact with the real world using facilities like these. One of the implicit benefits of adopting Zynq is that the development boards can actually be used as a common platform for several different teaching areas, spanning electronic engineering to computer science, allowing university departments to rationalise existing hardware platforms. Where students are studying a number of different subjects, using the same platform for practical work reduces the learning curve of gaining familiarity with the development tools and boards, thus increasing the time available for subject-specific practical work and projects.

We note in particular that Zynq can actually provide a foundation for several related academic subjects, and support practical examples in others, as depicted in Figure 7.1 and discussed later in this section.

7.2.2. Digital Design and FPGA Teaching

As described in earlier chapters, the PL part of the Zynq device is equivalent to FPGA fabric, and can be used in isolation of the PS. This allows the Zynq to be used like a standard FPGA, and thus to support classes in VHDL / Verilog, digital design, and FPGA technology, or for project-based learning in areas such as communications, control systems, signal processing, and robotics.

7.2.3. Computer Science

The dual-core ARM processor in the Zynq can be used in isolation of the FPGA fabric, thus appearing to the user as a standard ARM processor. Hence, Zynq can be used as a practical platform for typical Computer Science teaching, including embedded systems, software design, operating systems, and more challenging applications including homogeneous dual-core Symmetric Multi-Processing (SMP), and Asymmetric Multiprocessing (AMP) — see Chapter 21 for more information on this last topic.

7.2.4. Embedded Systems and SoC Design

Teaching of embedded systems design involves educating students in the concepts, features and operation of embedded systems, and also giving them some practical exposure to embedded systems design.

Students can learn about general embedded systems concepts such as buses, interrupts, processors, peripherals, and memory access, as all of these can be covered using the Zynq platform. It also significant that, due to the composition of Zynq, students gain experience both of the popular ARM processor architecture, and also of FPGA programmable logic.

7.2.5. Algorithm Implementation (e.g. Signal, Image, and Video Processing)

Another candidate learning mode is to provide students with a prepared system, which has one or more 'black boxes' into which they can insert their own custom functionality. This gives students an the opportunity to concentrate on one specific aspect of a system, and gain direct feedback on the impact of that subsystem when integrated into a processing chain.

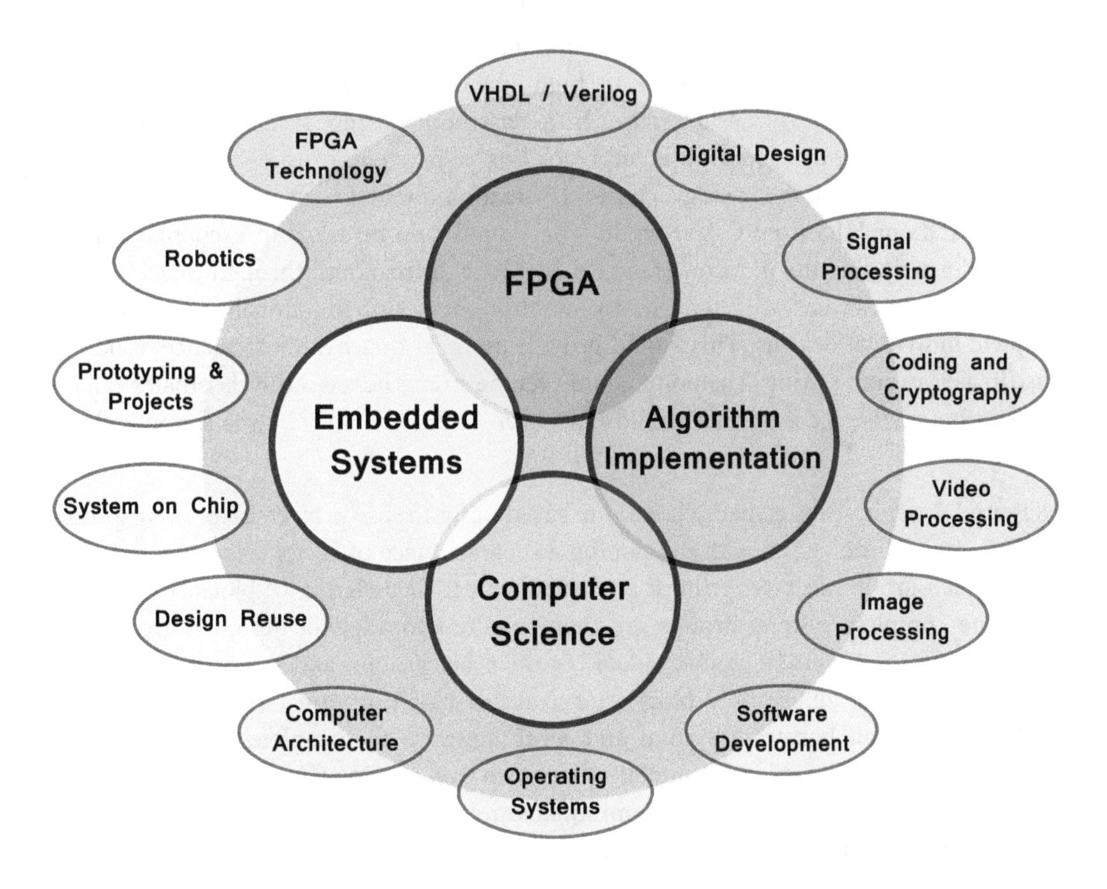

Figure 7.1: Academic subjects to which Zynq is relevant

Examples of this 'black box' approach could include both software and hardware functionality. For instance, a specific filtering stage could be inserted into an SDR, an edge detection algorithm as part of an image processing application, or a software algorithm could be developed to identify particular patterns in a stream of captured sensor data. A good illustrative example would be a real-time video system, which streams data from a camera onto the Zynq device for processing, and then to a video display. Students could be asked to implement an image filtering or feature detection algorithm, and would have the opportunity to gain direct visual feedback on the results. As desired, different design methods and tools could be used to implement algorithm functionality, including HDL, System Generator for DSP, or HLS if targeting the PL part of the Zynq, or software if implementing the functionality using the processor.

7.2.6. Design Reuse

The experience of adding or changing functionality in the manner described in Section 7.2.5 can also be extended, in order to highlight another industry-relevant aspect of embedded systems design: design reuse. In a classroom setting, students (or groups of students) could each develop a different 'black box' subsystem, or interchangeable variations of the same 'black box' using a chosen IP creation method such as System Generator for DSP, HDL, or HLS from C-level code. They could then be asked to exchange designs with classmates using the industry standard IP-XACT format, and through this, either to collectively form an entire system, or to establish the set of functionality necessary to complete individual designs. This would provide an ideal first-hand experience of design reuse in a classroom setting. The point could even be strengthened if the classmates can, in conjunction, realise a design that would have been impossible in the time frame without sharing IP.

Figure 7.2 depicts this style of classroom-based design reuse project. Here, four groups (A, B, C, and D) are tasked with developing a separate piece of IP, for implementation in the PL. Each group must design, test and document their assigned IP block. In order to create the complete system, groups are then required to adopt a design reuse model, wherein they make their IP block available to the other groups, and in return can obtain their IPs, such that all required blocks are available to all groups. Group A must then develop the overall hardware system and write software to complete the design, and likewise, the other groups can complete their own systems in the same way. Note that groups of differing sizes could be accommodated relatively easily, by assigning the larger groups with more complex IPs to develop.

In sharing developed IP, and attempting to reuse the work of others, students can also gain a genuine appreciation of the real engineering importance of testing and documen-

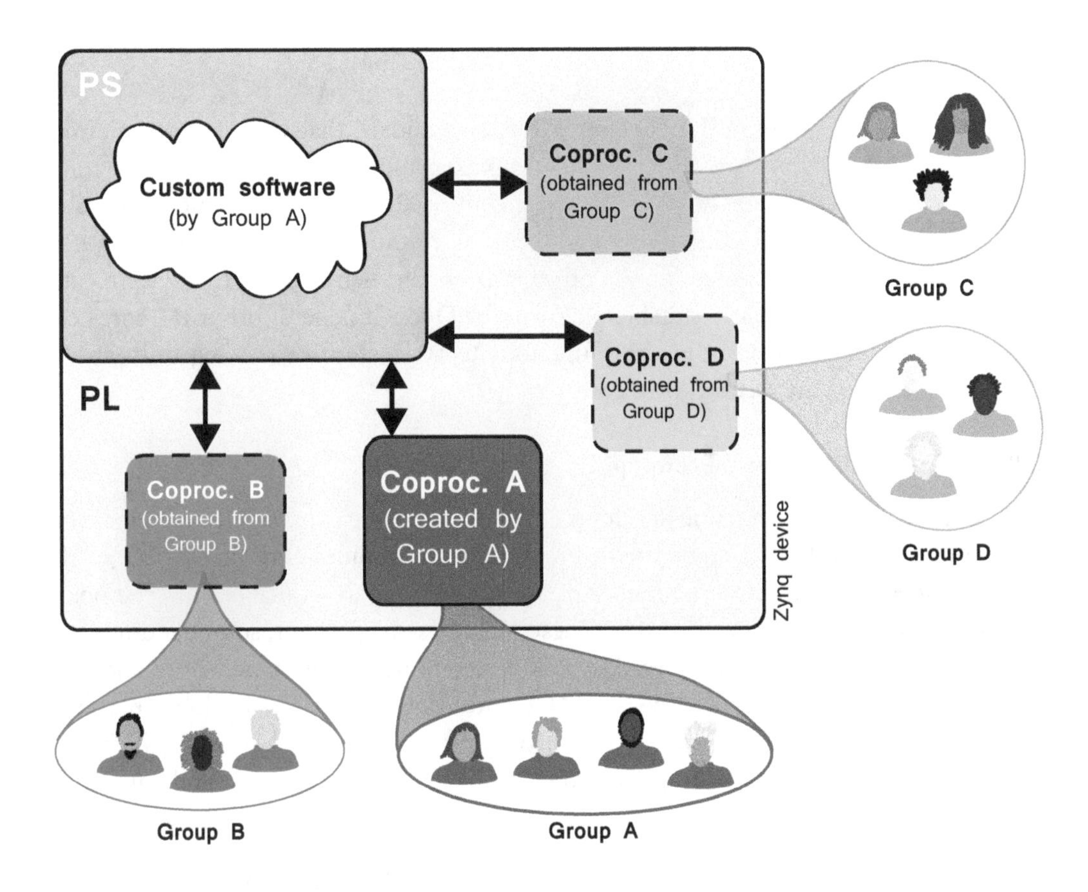

Figure 7.2: Design reuse in the classroom setting (the perspective of Group A)

tation. That is, not to satisfy tutors or gain marks! Rather, to ensure robust operation, and to clearly convey the necessary technical details to other engineers reusing their designs. The experience of being both a producer of IP, and a consumer of others' IP, provides an ideal opportunity to consider these vital aspects of the development process.

7.2.7. New and Emerging Design Methods

Zynq can also be used for teaching and research in new and emerging design methods, languages and techniques, like ***HLS***, ***OpenCV***, and ***OpenCL***.

HLS is the subject of dedicated discussion in Chapters 14 and 15.

OpenCV (Open source Computer Vision) is a free C/C++ library of functions for real-time image processing [16]. OpenCV is available via a Berkeley Software Distribution

(BSD) open source license [17], meaning that it is free for both academic and commercial use.

OpenCL (Open Computing Language) is an open standard that provides a framework for running software across heterogeneous computing units [15]. A set of units could comprise any combination of Central Processing Units (CPUs), DSPs, Graphics Processing Units (GPUs), FPGAs, and other types of processing resource. OpenCL relies on a host processor controlling the execution of other processing elements in the system, and therefore Zynq is a particularly suitable platform for OpenCL: one of the ARM cores can implement the host control, offloading functions to (i) custom processing units in the FPGA fabric, and (ii) the second ARM core.

7.2.8. Sensing, Robotics, and Prototyping

The facilities of Zynq academic development boards include expansion connectors that can be used to extend functionality with Pmod add-on modules and, in the case of the ZedBoard, an FMC connector. A number of Pmod modules are available from the board manufacturer and third parties [5], [12]. These include sensors (light, sound and temperature sensors, proximity detectors, gyroscopes, accelerometers, etc.); actuators and data converters (e.g. stepper motor drivers, DACs and ADCs); communications interfaces (Bluetooth, GPS, WiFi etc.) and visual display outputs (LCD and OLED displays). Pmods are also available for creating simple custom interfaces and test points, and for breadboarding.

Given all of these facilities, there is great scope to develop practical projects based around Zynq development boards. For example, this might include sensing of ambient light and temperature for smart buildings, or the construction of a robotic vehicle, perhaps even including interaction with a mobile phone application via Bluetooth. As an example of this type of application, the 'ZRobot', has been created by Xilinx University Program as a teaching development platform, and is shown in Figure 7.3 [11]. Students can enhance and extend the functionality of the robot by adding further sensors and actuators, or by customising its behaviour via modification of software routines.

7.2.9. An Example Course

An interesting exemplar of Zynq-based teaching was presented by Professor Karl L. Wang of Harvey Mudd College, Claremont, California, USA, in his paper "An Effective Project-Based Embedded System Design Teaching Method", co-authored by colleagues from Digilent [20]. The paper reviews the content and structure of a newly developed course, based around the ZedBoard as the class teaching platform. The 16-week class is

Figure 7.3: The ZRobot, based around a ZedBoard development platform

formed from a series of lectures and structured laboratories, followed by a group-based project phase, in which the students build on their knowledge and skills to address a self-defined problem. The groups appear to have chosen a variety of topics, including a pin-ball game, music synthesiser, and a robot with collision avoidance sensing. The authors of the paper reported a positive response to the teaching methods, and identified some areas for future improvement.

7.3. Projects and Competitions

Student design contests are regularly run by Digilent, the manufacturer of the ZedBoard and other academic development boards. These are open to students all over the world, and are intended to stimulate innovative design ideas and projects based around Digilent products (which includes FPGA, Zynq, and microcontroller boards). In the most recently held design contest, at the time of writing (2013), students addressed problems such as shape recognition image processing, powerline communication, and the implementation of cryptographic algorithms, leveraging the Zynq PL for hardware acceleration.

Xilinx University Program also occasionally organises design contests in regions around the world, with the similar aim of motivating student participation and innovation [24].

7.4. Academic Research

Just as Zynq is relevant to a variety of related topics in the engineering curricula, it has great potential for engineering and computing research. At the time of writing, Zynq is still very 'young' and therefore the scope and diversity of innovation built upon Zynq are still to fully emerge; however, it is useful to survey some of the research activities to date.

It may be expected that research around Zynq will be strongly application-driven, with the areas identified in Chapter 5 providing stimulus; for example, image and video processing, software defined radio, communications networks, etc. Even aside from these key identified areas, Zynq holds wide appeal, and a survey of the papers published to date could provide an early indication of the diversity we might come to expect.

Some of the highlights so far include:

- ***Optical flow accelerators*** — An application in image processing, optical flow is concerned with the movement of features between successive video frames. In [13], researchers report on the implementation of an optical flow algorithm using HLS with OpenCV, and targeting a Zynq device. The benefits of HLS as a design methodology (to be further discussed in Chapter 14) were observed first hand in the context of this project; for example, the ability to scan the design space for the most suitable implementation. The stated outcomes indicate that the developed design achieved performance comparable to a desktop processor, but while consuming only about one seventh of the energy.

- ***Remote health / safety monitoring*** — A paper presenting the design and implementation of a system to monitor vulnerable individuals who may be susceptible to falling and suffering injury [19]. Data is obtained from a medical sensor worn by the patient, which collects movement data from a tri-axial accelerometer. The sensor data is sparse and therefore requires reconstruction, which is achieved using a PL-based accelerator implementing the Orthogonal Matching Pursuit (OMP) algorithm. A further PL accelerator detects falls based on the reconstructed data, while software routines are used to classify falls, and hence determine what action should be taken.

- ***Image processing and embedded vision*** — As discussed in Chapter 5, Zynq is a very suitable platform for image processing. The authors of [9] present a system, prototyped on a ZC702 board, which takes advantage of the ARM processor and PL to support a software framework running on Linux, and image processing hardware accelerators, respectively. The realised functionality includes image stabilisation, contrast normalisation, and identification of moving targets. The paper reports on the achieved resource costs and frame rates, and the extensibility of the system.

- ***'Evolvable' hardware*** —As reviewed in Chapter 5, DPR is a technique that facilitates the reprogramming of defined areas of an FPGA or Zynq PL, while the rest of the device continues to operate unaffected. The implementation of DPR (and related techniques for evolvable hardware) using ARM-based software control is a particularly interesting area, and the authors of [6] and [7] report on such systems for adaptable (or 'evolvable') image filtering. They report on significant performance improvements specifically attributable to the Zynq architecture, and identify areas for further investigation.

- ***Zynq hardware virtualisation*** — In [18], the subject of virtualising the Zynq platform is addressed, with the aim of abstracting detail and facilitating more dynamically adaptable use of the PL in particular. In other words, a scheme where the PL is used not just to host static coprocessors, but to implement functionality that adapts according to processing requirements. The authors assess the potential for Zynq to support this mode of operation using a standard hypervisor application, and demonstrate a framework for software and hardware virtualisation on Zynq.

- ***Gas identification*** — In [1], an SoC design based on Zynq is presented for the identification of gases. Principal component analysis (a matrix-based method) is used for machine learning and thereafter to classify gases. The paper reports that Vivado HLS was successfully used for hardware acceleration of the computationally parallel matrix operations, with directives used to optimise the implementation. The results presented indicate that the Zynq solution was faster than a software-based solution operating on a 64-bit Intel i7 processor.

- ***Biosensing*** — The problem of implementing a flexible biosensing platform for a large number of channels is addressed in [10]. Such platforms are required to support the acquisition of data from a large number of sensors (1000+), with subsequent FFT-based signal processing. The designed system is mapped to the Zynq architecture (on the ZedBoard), with PL-based acceleration of the FFT operations. It is reported that the system supports 8 times more sensors and a sampling frequency

1000 times higher than the previous generation platform. The reconfigurable nature of the Zynq, and the flexibility to extend the design to accommodate more channels, is also noted.

This is not an exhaustive list, and the breadth and depth of research around the Zynq platform can only be expected to increase.

7.5. The Xilinx University Program (XUP)

Academia is supported directly by the XUP group within Xilinx. It is important to raise awareness of the group, and the services it offers, within the academic community [23].

7.5.1. Introducing XUP

XUP provides support to universities around the world in the use of Xilinx tools and technologies, and is managed and delivered by a dedicated global team. XUP was established with the aim of enabling universities to use Xilinx technologies in their curricula and research, and the team undertakes a variety of initiatives to both encourage and facilitate academic activities. These are briefly reviewed over the remainder of this section.

7.5.2. Software Support and Licenses

Academic licenses of Vivado *System Edition* are available at low cost, to facilitate the use of fully functional Xilinx software for classroom teaching and research. Qualifying institutions can purchase licenses from a Xilinx distributor or educational partner in local regions around the world, and in some cases XUP can also make licenses available for free, based on a qualifying application to its donation programme.

Additionally, students can access Xilinx *WebPACK* for personal and home use, meaning that coursework assignments can be completed outside of the lab, and that students can work on their personal laptops if they prefer. WebPACK includes the core functionality of the Vivado tools (FPGA design, software development, and simulation), and is directly downloadable from the internet, free of charge [21].

The latest WebPACK version of the software typically provides support for most past and some current generation Xilinx devices, with the exception of the newest, highest end, and restricted devices in each product family. For example, at the time of writing, the smaller Zynq devices in the family are supported by Vivado WebPACK 2013, but the largest devices are not, and require the full version of Vivado. WebPACK is usually sufficient for academic teaching, or for students and hobbyists to use at home, as most development boards (especially those intended for teaching) tend to use smaller devices in a product family.

Readers may wish to refer to Table 3.1 on page 49 for clarification of the features available in WebPACK, and a comparison with the *Design* and *System* Editions of Vivado.

7.5.3. XUP Development and Teaching Boards

XUP works with partners to produce FPGA and Zynq development boards specifically designed for academia. These tend to feature an array of I/O peripherals including LEDs, switches and buttons, video in/out, audio in/out, ethernet, and expansion ports and pins.

At the time of writing, two XUP Zynq boards are available:

- ***ZYBO*** — A low cost board featuring a XC7Z010 Zynq device and a key set of I/O peripherals. The ZyBo board is physically compact, and well suited to classroom teaching. ZYBO was introduced in Section 3.6.5 on page 69, including a photograph of the board in Figure 3.9.
- ***ZedBoard*** — An enhanced development board based around the XC7Z020 Zynq device, and featuring an expanded set of peripherals. Suitable for classroom use, but also for project work and more demanding applications.

XUP boards are supported by the manufacturer, and also by XUP directly, through the provision of teaching and training material targeted to these boards. Members can access teaching material for Zynq via the XUP website.

The ZedBoard in particular is suitable for research work, although certain topics of investigation may demand larger devices or specific peripheral support, and thus prompt the use of professional grade boards such as the ZC702 or ZC706. In addition to classroom teaching and projects, ZYBO may be useful for certain types of research, for example sensor and network applications.

7.5.4. XUP Workshops and Training Materials

Another of XUP's functions is to develop in-house teaching and training material, including specific support for Zynq, with the aim of educating professors, researchers and teaching assistants in the effective use of Xilinx tools and devices. This material is made available for academics to reuse for their own teaching, whether in its original form, or integrated with other educational materials. Students can also download the material for self directed learning.

These teaching materials form the basis of XUP *Professor Workshops*, which are run by XUP staff and academic partners, and are scheduled regularly at locations around the world. Workshops are free to attend, typically have a duration of 2 days, and are open to professors, teaching staff and researchers. Several of the workshops currently offered focus on Zynq.

7.5.5. Technical Support for Universities

As members of XUP, academics can access technical support, which may be useful in developing courses or supporting projects. Students are asked to first approach their professors and university support staff for technical help, and any unresolved issues can thereafter be channelled to XUP by their professors, as appropriate.

7.5.6. Eligibility

All award-granting universities and some research institutes have institutional eligibility for membership of XUP. Professors, teaching assistants and researchers working at an eligible institution can then apply to become members, and this is achieved by first creating an account on the Xilinx website, and then making an application for membership of XUP.

Once approved, members can access resources and technical support though XUP, download training material, and register to attend Professor Workshops.

7.5.7. Getting in Touch with XUP

Further information about XUP is available via its website [23],

http://www.xilinx.com/university

including the most up-to-date versions of teaching material, current workshop schedules, and contact details.

7.6. Training for Industry

Having discussed support for the academic community earlier in this chapter, it is also important to review sources of training for industry.

7.6.1. Courses and Authorised Training Providers

Xilinx has a network of Authorised Training Providers (ATPs) around the world, who deliver training on Xilinx FPGA technologies and design tools. The core portfolio of courses is developed and designed by Xilinx, and includes courses covering the Zynq architecture, the Vivado Design Suite, together with many other relevant topics. Training is available in three primary modes: 'live' classroom-based courses, and both recorded and live interactive online training.

7.6.2. Other Resources

In addition to formal classes, Xilinx also provides documentation, reference designs, and tutorials, covering FPGA design concepts, use of the Xilinx development tools, and specific application examples. Typically any purchased development board will be accompanied by a set of tutorials, examples and reference designs applicable to the board, providing a basis for the development of user designs. This form of 'out of the box' support is comparable with that supplied along with university development boards, as detailed in Section 7.5.3.

7.6.3. Online Videos

Xilinx also provide a comprehensive set of *QuickTake* training videos, which demonstrate design concepts and techniques, and highlight useful features of the Xilinx development tools. The videos range from a few minutes to an hour in duration, and they provide, as their name suggests, short and focused treatments of specific topics.

To give the reader a taste of the training offered by this mechanism, the *QuickTake* videos are organised by subject, with topics current at the time of writing including: various aspects of the Vivado design flow; applying design constraints; system-level design; programming and debug; Vivado HLS; and several others. As well as being available via the Xilinx website [22], the videos can also be viewed on YouTube [25].

7.7. Chapter Review

In this chapter, we have shown the applicability of Zynq to the engineering curriculum, particularly in the context of networked embedded systems. It has been noted that Zynq and its associated design tools and development boards can provide a basis for several related topics, including FPGA design, computer science, and algorithm implementation, as well as embedded systems design. We have also seen that Zynq is well suited to project-based learning, with design contests providing an additional, external source of motivation.

The potential of Zynq as a platform for research was also considered, and some early examples of research work undertaken based on Zynq were cited. It should be expected that much further research will be emerge in time, as more projects come to completion, are disseminated, and stimulate others.

XUP was introduced and its role as a support mechanism for university teaching and research was explained, together with the types of support provided. Professors, teachers and researchers at eligible institutions can become members of XUP to access specific resources for academia.

Lastly, we reviewed sources of training for industry, including classes delivered by Xilinx ATPs, and online video training. These are in addition to more standard sources of support, such as documentation and tutorials available on the Xilinx website.

7.8. References

NOTE: All URLs last accessed June 2014.

[1] A. A. S. Ali, A. Amira, F. Bensaali and M. Benammar, "Hardware PCA for Gas Identification Systems Using High Level Synthesis on the Zynq SoC", *Proceedings of the 20th IEEE Conferences on Electronics, Circuits and Systems (ICECS)*, December 2013, pp. 707-710.

[2] L. Atzori, A. Iera, G. Morabito, "The Internet of Things: A Survey", *Computer Networks*, Vol. 54, Issue 15, October 2010, pp 2787-2805.

[3] H. Chao, H. Stuttgen, and D. Waddington, "IPv6: The Basis for the Next Generation Internet", Guest Editorial, *IEEE Communications Magazine*, January 2004, pp. 86-87.

[4] Cisco, Inc., "The Zettabyte Era - Trends and Analysis", White Paper, May 2013.
Available: http://www.cisco.com/en/US/solutions/collateral/ns341/ns525/ns537/ns705/ns827/VNI_Hyperconnectivity_WP.pdf

[5] Digilent, Inc., Sensors / Peripherals / Interfaces (Pmods™) webpage,
Available: http://www.digilentinc.com/Products/Catalog.cfm?NavPath=2,401&Cat=9

[6] R. Dobai and L. Sekanina, "Image Filter Evolution on the Xilinx Zynq Platform", *Proceedings of the 2013 NASA/ESA Conference on Adaptable Hardware and Systems,* Torino, Italy, June 2013, pp. 164 - 171.

[7] R. Dobai and L. Sekanina, "Towards Evolvable Systems Based on the Xilinx Zynq Platform", *Proceedings of the 2013 IEEE International Conference on Evolvable Systems*, Singapore, April 2013, pp. 89 - 95.

[8] D. Evans, "The Internet of Things: How the Next Evolution of the Internet is Changing Everything", White Paper, Cisco, April 2011.
Available: http://www.cisco.com/web/about/ac79/docs/innov/IoT_IBSG_0411FINAL.pdf

[9] E. Gudis et al, "An Embedded Vision Services Framework for Heterogeneous Accelerators", *Proceedings of the IEEE Conference on Computer Vision and Pattern Recognition Workshops*, Portland, Oregon, USA, June 2013, pp. 598 - 603.

[10] J. Leitão, J. Germano, N. Roma, R. Chaves and P. Tomás, "Scalable and High Throughput Biosensing Platform", *Proceedings of the 23rd International Conference on Field Programmable Logic and Applications (FPL)*, September 2013, pp. 1 - 6.

[11] P. Lysaght, "All Programmable Technologies in Academia", keynote presentation, *International Symposium on Applied Reconfigurable Computing*, March 2013, Los Angeles, USA.
Available: http://www.isi.edu/events/arc2013/Xilinx-ARC2013-Invited-Lysaght.pdf

[12] Maxim Integrated, *Pmod-Compatible Plug-in Peripheral Modules* webpage.
Available: http://www.maximintegrated.com/en/design/design-technology/fpga-design-resources/pmod-compatible-plug-in-peripheral-modules.html

[13] J. Monson, M. Wirthlin, and B. L. Hutchings, "Implementing High-Performance, Low-Power FPGA-Based Optical Flow Accelerators in C", *Proceedings of the 24th IEEE International Conference on Application-Specific Systems, Architectures and Processors (ASAP)*, Washington DC, USA, June 2013, pp. 363-369.

[14] D. Nagel, "212 Billion Devices to Make Up 'The Internet of Things' by 2020", *THE Journal online*, October 2013.
Available: http://thejournal.com/articles/2013/10/07/212-billion-devices-to-make-up-the-internet-of-things-by-2020.aspx

[15] OpenCL website,
Available: http://www.khronos.org/opencl/

[16] OpenCV website,
Available: http://opencv.org/

[17] Open Source Initiative, "Open Sources Licenses By Category" webpage,
Available: http://opensource.org/licenses/category

[18] K. D. Pham, A. K. Jain, J. Cui, S. A. Fahmy, and D. L. Maskell, "Microkernel Hypervisor for a Hybrid ARM-FPGA Platform", *Proceedings of the 24th IEEE International Conference on Application-Specific Systems, Architectures and Processors (ASAP)*, Washington DC, USA, June 2013, pp. 219-226.

[19] H. Rabah, A. Amira, and A. Ahmad, "Design and Implementation of a Fall Detection System Using Compressive Sensing and Shimmer Technology", *Proceedings of the 24th International Conference on Microelectronics (ICM)*, pp. 1 - 4, December 2012.

[20] K. L. Wang, C. S. Cole, T. Wang, and J. Harris, "An Effective Project-Based Embedded System Design Teaching Method", *Proceedings of the 120th American Society for Engineering Education (ASEE) Annual Conference and Exposition*, Atlanta, USA, June 2013.

[21] Xilinx, Inc., *Vivado Design Suite Evaluation and WebPACK* webpage,
Available: http://www.xilinx.com/products/design_tools/vivado/vivado-webpack.htm

[22] Xilinx, Inc., *Vivado Video Tutorials* webpage,
Available: http://www.origin.xilinx.com/training/vivado/

[23] Xilinx, Inc., *Xilinx University Program* website,
Available: http://www.xilinx.com/university

[24] Xilinx, Inc., Xilinx University Program Design Contests webpage,
Available: http://www.xilinx.com/support/university/design-contests.html

[25] YouTube, *Xilinx Channel*,
Available: http://www.youtube.com/user/XilinxInc?feature=watch

8

First Designs on Zynq

This is the first of the practical chapters of The Zynq Book, which can be identified by the green chapter numbers and piping on the first page. In these chapters, which are placed at appropriate points within the book, the corresponding tutorials and practical exercises that are provided on the accompanying website will be introduced, along with a brief overview. It is not the intention to provide the practical exercises in the book itself; due to the ever evolving state of software tools and design flows, keeping the detailed practical instructions online permits better concurrency with the design tools.

As this is the first of the practical chapters, the accompanying exercises will be relatively short and simple, with the main aim being to familiarise the reader with the required software tools and the corresponding Zynq design flow.

Although not a design tutorial, the first practical that is available on the website is a guide which provides a step-by-step walkthrough of the installation of the required tools and ZedBoard device drivers. It is recommended that even if you already have the software tools installed, that you read through this guide to ensure that all of the tools are setup as required for the future practical exercises, and that the required working directory structure is in the correct location on your host machine.

The first practical tutorial focuses on introducing the Zynq design flow, guiding you through the creation of a new design project, where a hardware design will be created, before exporting the design for the development of a software application.

Before starting, it is useful to provide an overview of The Zynq Book website.

8.1. Software Installation Guide

A comprehensive guide for the installation of the Vivado Design Suite is provided in the "*Vivado Design Suite User Guide: Release Notes Installation and Licensing*", which is available from Xilinx [1]. The two chapters of relevance are:

- **Chapter 3**, which details how to download and install a copy of the Vivado Design Suite; and
- **Chapter 5**, which details how to obtain and manage a product license.

8.2. Aims and Outcomes

The general aim of this initial set of practical exercises is to introduce the design flow and associated tools. In doing so, a simple design will be constructed, targeting the ZedBoard.

After completion of this tutorial you will be able to:

- Create a new Zynq hardware project.
- Be able to configure a Zynq PS targeting the ZedBoard.
- Create and connect an interconnect between the Zynq PS and PL.
- Implement an IP module in the Zynq PL.
- Generate HDL files for the Zynq hardware design, and create a bitstream hardware description for the Zynq PL.
- Create a simple software application which will execute on the Zynq PL and communicate with IP implemented in the PL.

The main aim of this tutorial is to provide an introduction to the Zynq design flow, and the most important outcome is that you are familiar with the software tools required to develop Zynq-based systems.

8.3. Overview of Exercise 1A

The first practical exercise involves the creation of a new hardware project in Vivado IDE, targeting the ZedBoard, and will guide you through the process step-by-step. The required steps are:

1. Launch Vivado IDE for the first time.
2. Invoke the New Project Wizard.
3. Specify the required working directory for the design.
4. Target the ZedBoard as the default part for project.

This is a very simple exercise, with the sole objective being that your first Vivado IDE project is correctly setup to allow you to complete the further exercises successfully.

Exercise 1A is available on the website: www.zynqbook.com

8.4. Overview of Exercise 1B

Building upon Exercise 1A, the next stage is to introduce the Vivado IDE environment, focussing on the default layout, before creating a simple Zynq embedded system which will first configure the Zynq PS for operation on the ZedBoard before implementing a GPIO controller in the Zynq PL. The Vivado IP Integrator tool will be used to create the system using the provided graphical environment, and the provided Designer Assistance tools will be utilised for the automatic configuration and connection of an AXI interconnect which connects the Zynq PS with the IP module in the PL. The GPIO controller will be connected to the LEDs that are available on the ZedBoard.

The steps involved in this Exercise are:

1. Introduce the Vivado IP working environment and features.
2. Create a new IP Integrator block design.
3. Add and configure a Zynq Processing System module which is targeted at the ZedBoard.
4. Add and configure a GPIO controller to connect to the LEDs on the ZedBoard.
5. Use the IP Integrator Designer Assistance tools to create and configure an AXI interconnect to connect the Zynq PS and GPIO controller.
6. Generate HDL files for the hardware design and create a bitstream hardware description file.
7. Export the finalised hardware design to the SDK.

An overview of the hardware design that will be created in this exercise is provided in Figure 8.1

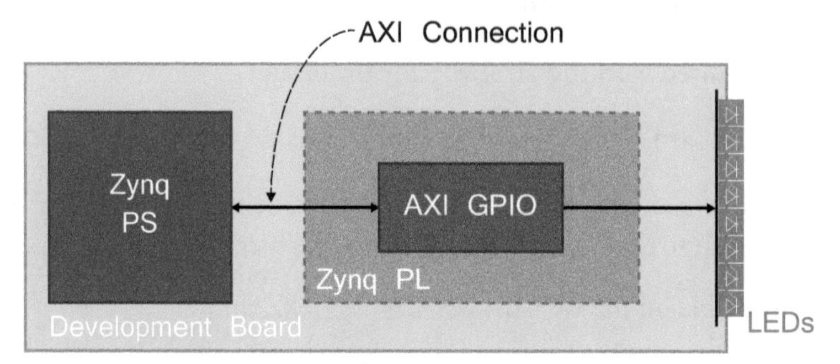

Figure 8.1: Exercise1A Zynq Hardware Design

Exercise 1B is available on the website: www.zynqbook.com

8.5. Overview of Exercise 1C

The final exercise in this first tutorial introduces the software creation side of the Zynq design process, having already created the hardware design in Exercise 1B. A software application will be created which will control the LEDs on the ZedBoard. The created application will run on the Zynq PS and communicate with the GPIO controller in the PL. The software drivers that are created by IP integrator for communication between the software and the PL implemented hardware modules will be explored before building and executing the software application on the ZedBoard.

The steps involved in this Exercise are:

1. Create an empty application project.
2. Add source code and build the application project.
3. Explore the generated software drivers and the corresponding functions which allow the software to communicate with the implemented hardware.
4. Program the Zynq PL with the bitstream that was generated in Exercise 1B.
5. Execute the software application on the hardware and confirm that it controls the LEDs on the ZedBoard as expected.

Exercise 1C is available on the website: www.zynqbook.com

8.6. Possible Extensions

Having completed Exercise 1C, there are some possible variations you can introduce to personalise the developed system. For example, you could:

- Change the LED blink rate.
- Create a custom pattern for the LED blinking.
- Add a further GPIO controller to the hardware design which connects to the DIP switches on the ZedBoard. Use the GPIO driver functions to take input from the DIP switches to control the output of the LEDs.
  Hint: As this extension requires hardware changes, the system would need to be re-exported from Vivado IDE before making additional software changes in the SDK.

8.7. What Next?

This set of practical exercises concludes Part A of the book, "*Getting to Know Zynq*".

Next, we move onto Part B, which looks in more detail at creating the hardware component of a Zynq SoC. Further practical exercises will follow, illustrating these concepts.

8.8. References

NOTE: All URLs last accessed June 2014.

[1] Xilinx, Inc, "Vivado Design Suite User Guide: Release Notes, Installation and Licensing", UG973, June 2014.
Available:
http://www.xilinx.com/support/documentation/sw_manuals/xilinx2014_2/ug973-vivado-release-notes-install-license.pdf

PART B

Zynq SoC & Hardware Design

9

Embedded Systems and FPGAs

The term 'embedded system' has become widely used when describing a large number of different applications, with a large number of platforms now available. These range from small microcontrollers and DSPs, to large FPGAs and GPPs, and defining what exactly comprises an embedded system is growing more and more challenging.

The aim of this chapter is to introduce the concept of embedded systems and to provide some examples of where they would be deployed in terms of real-life applications. The scope will then be narrowed to focus on embedded systems on FPGAs, and the generic embedded system architecture will be explored. Individual components of the architecture will be detailed, starting with the processor, and a high-level overview of some of the fundamental operations, such as interrupts and execution cycles, will be discussed. Some basic bus operations and properties, such as bus arbitration, memory transfers and bandwidth, will also be covered.

9.1. What is an Embedded System?

An embedded system is a specialised computing system that is optimised to carry out a single, or very few, dedicated functions. Embedded systems form part of larger devices, or pieces of equipment, with the purpose of controlling a dedicated function within those machines. The pre-determined functionality of an embedded system contrasts with that of a GPP, such as those found in a Personal Computer (PC), whereby a single processor will perform a large number of very varied functions.

A GPP is not designed with any specific task in mind. A desktop computer, for example, can be used to perform a myriad of tasks including document creation and editing, a home entertainment system, an image and video editing system, or an internet terminal. An embedded system is application-specific, and as such can be finely optimised to deliver the chosen characteristics of a given application. Those characteristics could be to provide very high performance in a given area, or low power consumption. It is only by making the system application-specific that allows an embedded system to achieve such performance. A GPP may be able to perform the functions of an embedded system, but is unlikely to achieve the same system performance, and/or power consumption requirements. The generic programmability of a GPP also comes at a cost, with GPPs generally being far more expensive than those used in embedded systems.

9.1.1. Applications

Embedded systems are deployed in a large number of devices across a breadth of fields. An overview of some of the more prominent uses for embedded systems is provided in Figure 9.1.

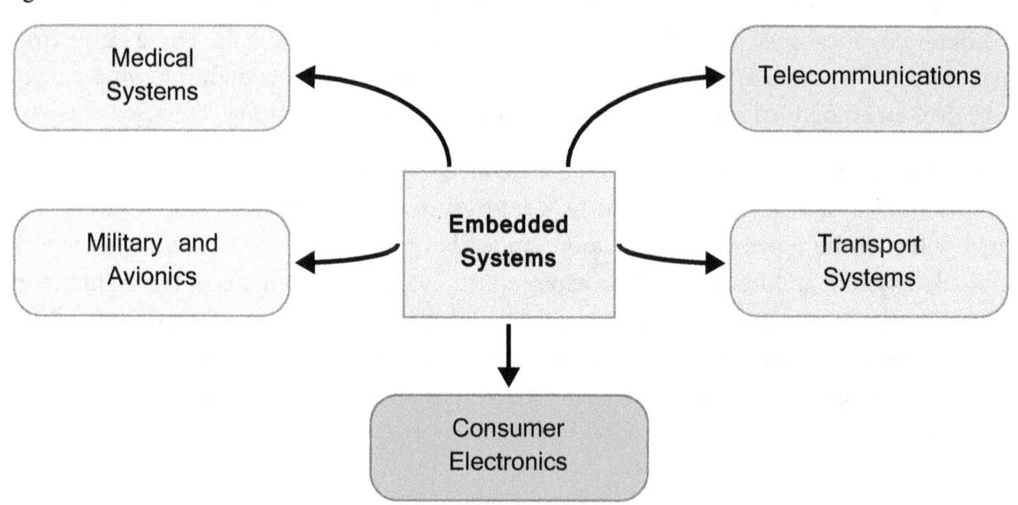

Figure 9.1: Embedded system application areas

A number of more specific applications from each of the highlighted fields are provided as follows:

- **Telecommunications** — Mobile phones, routers, consumer radio and television.

- **Medical Electronics** — Body scanning devices such as MRI machines, electronic stethoscopes and pacemakers.

- **Consumer Electronics** — Digital cameras, video games consoles and washing machines.

- **Military and Avionics** — RADAR and SONAR, guided missile systems, satellite stations and flight navigation systems.

- **Transport Systems** — Anti-Lock Braking Systems (ABS), air bag deployment, on-board entertainment, GPS and satellite navigation systems.

9.1.2. Generic Embedded System Architecture

In general, the architecture of an embedded system follows the architecture outlined in Figure 9.2.

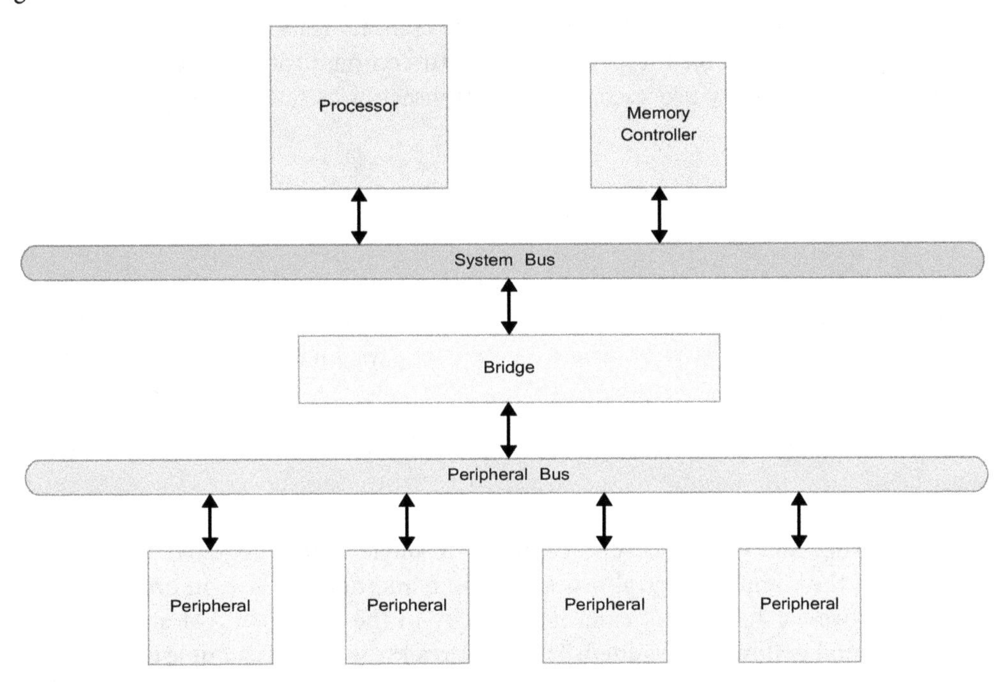

Figure 9.2: Generic embedded system architecture

The individual components in Figure 9.2 are defined as follows:

- **Processor** — This is the 'brains' of the system. It is programmed to perform the tasks specific to the application of the embedded system.

- **Memory Controller** — Memory controllers manage the reading and writing of data to and from main memory in an embedded system. Memory controllers reside in on-chip soft cores, providing an interface between the system memory and all other parts of the system.

- **Peripherals** — These are the components around the central processing unit. Peripherals can be implemented as individual integrated circuits, be contained on-chip with the processor, or may reside in an area of programmable logic such as an FPGA.

- **System Bus** — In an embedded system with multiple buses, the system bus connects the processor, memory controller and other high-speed devices together. As such, it is the bus with the greatest bandwidth in the system.

- **Peripheral Bus** — The second bus creates two separate sections of the system, allowing two arbiters to control and manage communications across the two buses. This allows devices on the peripheral bus to communicate with each other, even when a high-priority processor-memory transaction is taking place on the system bus.

9.2. Processors

A processor is the main controlling unit within an embedded system. It controls and orchestrates the system, supports software and co-ordinates exchanges with peripheral components. In embedded systems where an operating system is used for the governing of the system, it will run on the processor.

There are a variety of processors which can be used in an embedded system, as outlined below:

- **Microprocessor** — A microprocessor is a single chip integrated circuit which contains the complete central processing unit, and nothing else. In order to make a microprocessor functional, external memory in the form of RAM and ROM, and other peripherals must be added. The CPU in a PC is a good example of a microprocessor.

- **Microcontroller** — A microcontroller contains an entire computing system on a single chip. Unlike a microprocessor, a microcontroller comprises of a CPU with a fixed amount of RAM/ROM and peripherals all within a single Integrated Circuit (IC).

- **Digital Signal Processor** — A digital signal processor is a processor which has been designed specifically for the task of digital signal processing, with an instruction set that is targeted accordingly. DSPs are designed to perform fast arithmetic operations, and are capable of performing a multiply-accumulate operation in a single clock cycle. This makes a DSP very efficient, both in terms of performance and power consumption, when used for specific audio/video tasks, but very poor when used for other tasks due to the restricted instruction set.

- **Embedded Processor** — An embedded processor is one which is physically embedded within the programmable fabric of an FPGA device. Embedded processors come in two varieties — *hard* and *soft* processors. Hard processors are those which are built from dedicated silicon outwith the general-purpose logic of the FPGA device, whereas a soft processor is built using the general-purpose logic of the FPGA. A soft processor must be synthesised to fit into the FPGA fabric. In the case of all embedded processors, both hard and soft, the local memory, bus interconnects, memory controllers and internal peripherals must be realised in the FPGA general-purpose logic.

9.2.1. Co-processors

A coprocessor is a processing core that supplements the functionality of the primary processor, and is optimised for a single, specific task. By off-loading computations from the main processor to one or more co-processing units, the overall system performance can be accelerated.

Whereas the main processor may be used for a variety of different tasks, co-processors are generally used to perform dedicated tasks. Examples of tasks which may be performed on dedicated co-processors are:

- High-speed arithmetic
- Image and video processing
- Digital signal processing
- Data encryption

With reference to FPGA based embedded systems, the programmable logic provides a perfect platform in which to create co-processing cores due to the ability to perform parallel execution. This means that complex tasks that would require a large number of sequential CPU clock cycles to compute can be executed far quicker in a PL based co-processor. These are known as *soft coprocessor* cores. Other forms of acceleration can be implemented using dedicated hard processing cores, that do not reside in the FPGA fabric. Collectively, this task off-loading processing is known as *hardware acceleration*.

9.2.2. Processor Cache

A cache is a small memory located between the CPU and main memory. It has a lower access time than main memory and the cache is not accessible via the system bus. Cache is used to store data that is frequently accessed by the processor from main memory. Operations which make use of cached data are therefore much faster than those where the data is in main memory only.

As processors typically read data at a much faster rate than the system's main memory, the processor speed is constrained to that of the memory. By including cache memory in a system — which contains the most frequently accessed data that can be read at a higher rate than main memory — the processor is no longer constrained to the speed of the main memory. This leads to an increased efficiency in terms of data access. The processor speed is still limited, however, in the event of a cache miss — whereby the processor fails in its attempt to read or write data in the cache. This results in the data being read/written to main memory, and increased access latency.

A number of different levels of cache can exist within a system. These are outlined in Figure 9.3, and expanded upon below.

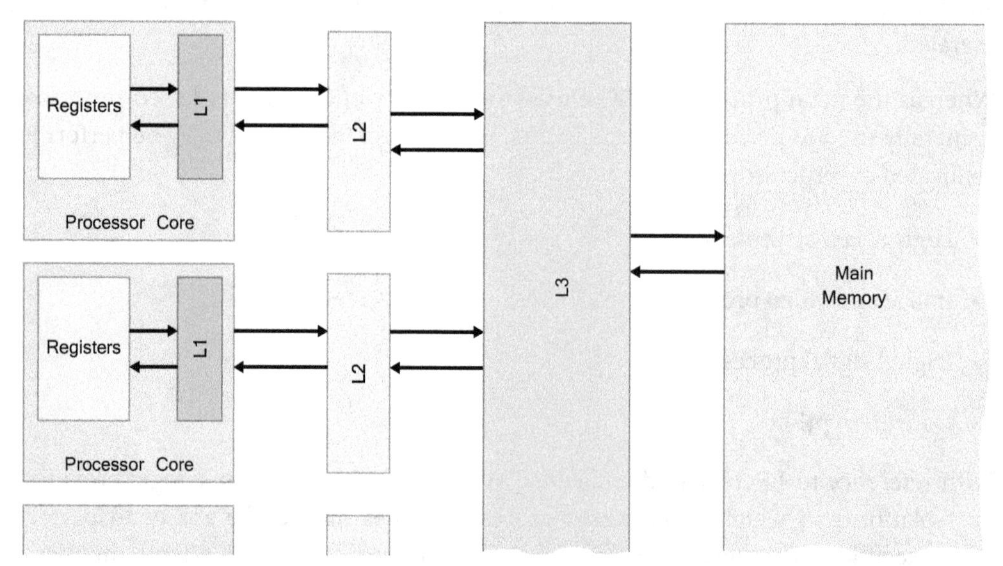

Figure 9.3: Cache levels and their locations relative to the processor cores and main memory

It is useful to first introduce the two types of memory which are used — Dynamic RAM (DRAM) and Static RAM (SRAM) — before going on to discuss the different levels of cache.

Dynamic RAM (DRAM)

DRAM is the most common form of memory that is used in computing systems. A DRAM chip consists of a number of *memory cells* which can each hold 1-bit of data, stored in a capacitor. Every memory cell also features a transistor, which acts as a switch to allow the control circuitry to read or write the state of the capacitor. As both the capacitor and transistor are extremely small, millions of individual memory cells can fit in a single DRAM chip.

Due to the fact that capacitors leak electric charge, the state of the bit of information held by each memory cell will eventually fade unless the charge of the capacitor is periodically refreshed by the memory controller. The memory controller does this by reading the state of each memory cell and then writing the state back again. This is where dynamic RAM gets its name.

Static RAM (SRAM)

SRAM uses a different technology to store information than DRAM. Whereas each bit of memory in DRAM is stored in a capacitor, SRAM uses latches to store the data. Each memory cell requires 4 or 6 latches to store a single bit of data, and therefore requires much more space on a chip than DRAM, making it more expensive.The upside to SRAM is that it does not need to be refreshed, thus making it much faster than DRAM. Due to the expense of SRAM, it is usually only used in high-speed, low-capacity memory chips.

Level 1 (L1) Cache

L1 cache is the smallest form of cache memory, the size of which is typically 8 to 128 KB. It is implemented in the form of SRAM which is built into the fabric of the processor core and, as such, operates at the same clock rate. L1 cache typically consists of two sections: *data* and *instruction* caches.

L1 cache is used to store a local copy of frequently accessed data and instructions, enabling instant access by the processor.

Level 2 (L2) Cache

L2 cache is usually external to individual processing cores, but is located extremely close by. It is larger than L1 cache, typically in the region of 256 to 1024 KB, but has slower access speeds. L2 cache is in the form of DRAM and is unified in a single section (unlike L1 which is split into two sections). Larger quantities of data are constantly read in by L2 cache from main memory before being fed to L1.

Level 3 (L3) Cache

Level 3 (L3) cache is shared among all processor cores. It is also implemented in DRAM, and is the largest form of cache memory, typically in sizes of 2 MB and upwards.

9.2.3. Execution Cycles

In order for a program that is stored in memory to be executed by the processor it must go through an *instruction execution cycle.* This is the process by which a system retrieves an instruction from memory, determines the required actions for that instruction, and executes those actions. Each instruction execution can be divided into three unique parts, as shown in Figure 9.4. For obvious reasons, this process is sometimes referred to as the *fetch-and-execute* cycle or the *fetch-decode-execute cycle.*

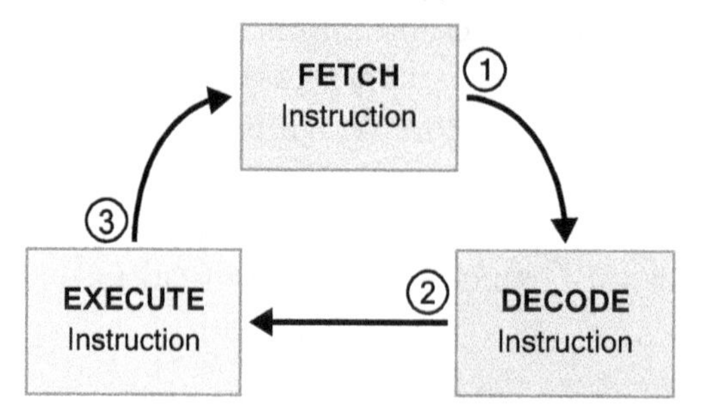

Figure 9.4: Steps of the instruction execution cycle

Before taking a closer look at each of the stages of the instruction execution cycle in detail, it is useful to understand some of the terms that will be used:

- **Machine Code** — When writing a software application, it is often in a high-level language (such as C/C++) which is easily understood by the developer, and human-readable. Code in this form is meaningless to a processing system, and must be converted, or compiled, into a form that the processor can understand. The final, low-level, output that is readable by the processor is known as machine code - a stream of binary data relating to the program that the processor is able to interpret and process.

- **Opcodes** — An opcode is an operation code which uniquely defines a function to be performed — a machine code representation of a processor instruction. A processor can execute many different operations, so each instruction is assigned a unique numeric code.

- **CPU Instruction Set** — The instruction set for a given processor is the basic set of commands which a processor can understand. It contains a specification of each of the opcodes and native commands that can be performed by that processor.

Fetch Instruction

The fetch instruction is the first stage of the instruction execution cycle. The flow of the operation is outlined in Figure 9.5.

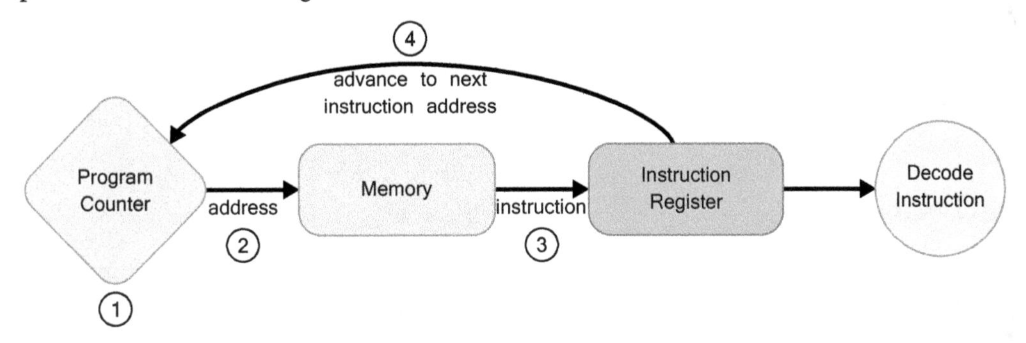

Figure 9.5: Fetch instruction flow

With reference to Figure 9.5, the fetch instruction flow is as follows:

1. The program counter register contains a value that corresponds to an address in memory containing the next instruction to be processed.

2. This value is transferred to the memory address register where the control unit checks the value and fetches the corresponding instruction from memory.

3. That instruction is then stored in the memory buffer register before being transferred to the instruction register.

4. The program counter register is advanced by the control unit so that the value matches the memory location of the next instruction to be executed.

Decode Instruction

Once the instruction has been fetched from memory, the next step is to get the instruction into a form that the processor can understand. This is the decode process. At this point, the instruction that is stored in the instruction register is checked by the control unit. This detects the opcode and addressing mode which have been used, and consequently what actions must be performed to correctly execute the instruction.

There are three main addressing modes:

- **Immediate Addressing** — This requires no lookup of data as all of the required data is located within the operands of the instruction. This makes immediate addressing the fastest, but least flexible mode.

- **Direct Addressing** — Operands of instruction contain the memory address of the required data. Required data must be fetched from this address.

- **Indirect Addressing** — Operands of instruction contain a memory address. Within this memory address is a pointer to the address where the required data is stored. This makes indirect addressing the most flexible, but slowest due to the two data lookups.

Execute Instruction

Depending on the opcode that was determined by the decode process, a number of different actions can be performed at this stage. In general, there are four main groups of actions:

- Transfer of data between the processor and memory.

- Transfer of data between processor and I/O devices.

- Processing data, such as using the Arithmetic Logic Unit (ALU).

- Control operations that change the sequence of subsequent operations. These can be conditional based on the values in a certain flag register.

Once the instruction has been executed, the next instruction to be processed is fetched.

The flow of an execute operation cycle is shown in Figure 9.6.

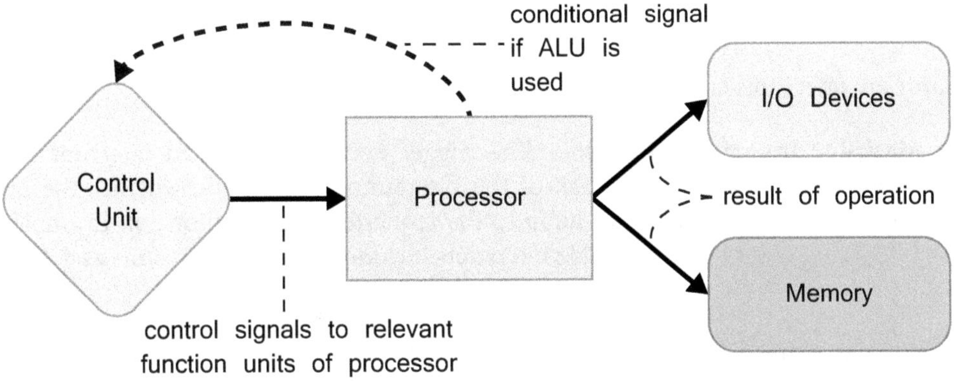

Figure 9.6: Execute instruction flow

With reference to Figure 9.6, the flow is as follows:

1. The control unit passes decoded information from the decode process (as a sequence of control signals) to the required functional units of the processor. This could be, for example, to read values from registers, or writing to a register.

2. The processor performs the required operations, and the output is either stored in memory or sent to an output device.

3. If the ALU is used, a conditional signal is relayed back to the control unit. As an example, the signal could update the address of the program counter so that the next instruction is fetched.

9.2.4. Interrupts

An interrupt is a signal which is generated to indicate to the processor that its attention is required. Interrupts can be generated by hardware processing units and peripherals, and also within the software itself. In hardware, an interrupt signal is an asynchronous signal that is generated by a processing unit to indicate the need for attention from the processor. In software, an interrupt is also an asynchronous event which indicated to the processor that a change in execution is needed. The process of polling the generated interrupts, however, is synchronous.

When the processor receives an interrupt it will halt the task that is currently being processed, and jump to the one which has requested attention. This is in contrast with the process of polling, which is the synchronous sampling of a device's status by the software program. Instead of having the processor constantly poll the I/O ports of a device to see if its attention is needed, the device will interrupt the processor.

Hardware interrupts can be further categorised into the following types:

- **Maskable Interrupts (IRQs)** — The trigger event of a masked interrupt is not always important. It is the task of the programmer to decide whether the event should cause the program to jump to the requested execution or not. Examples of devices which may use maskable interrupts include timers, comparators and ADCs.
- **Non-Maskable Interrupts (NMIs)** — These are interrupts which should never be ignored, and are therefore deemed much more important than maskable interrupts. Events which require NMIs include power-on, external reset (from a physical button) and serious device faults.
- **Inter-Processor Interrupts (IPIs)** — In multiple processor systems, one processor may need to interrupt the operation of another processor. In this situation, an IPI will be generated.

9.3. Buses

A bus provides the interconnection by which processors interface with other processors and peripherals. Processors, memory controllers and peripherals connect to the bus via a standard bus interface. The particular bus interface will depend on the particular bus architecture that is in use, but at a basic level a bus consists of address, control (read/write requests and acknowledgements) and data signals [2]. This is detailed in Figure 9.7, where the three main bus signals are highlighted:

- The **address line** transfers the memory addresses, or target port identification numbers, to I/O devices.
- The **control line** governs the control and timing signals of a system, synchronising operations and transmitting control signals such as interrupt requests and acknowledgements.
- The **data line** is responsible for the transfer of data.

Data transfer occurs in a *bus cycle.*

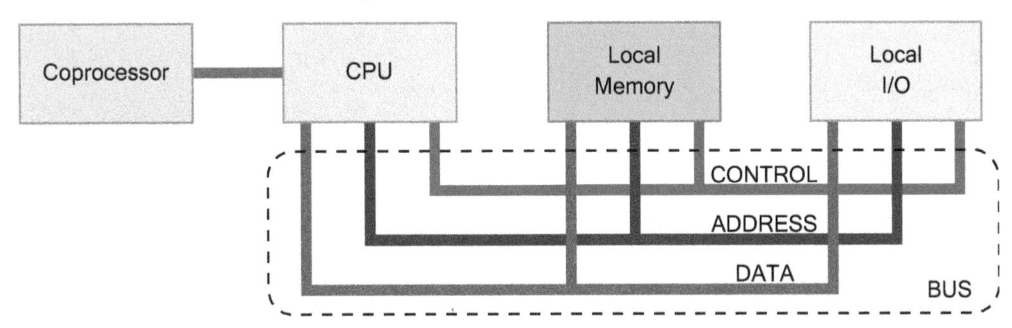

Figure 9.7: Bus comprised of control, address and data lines

Any coprocessors that are present in an embedded system are directly connected, or closely coupled, to the main processor via the processor bus. In a multi-processor system, individual processor subsystems are connected by the *system bus.*

Buses also contain a bus arbiter which controls access to the bus. Not all connected modules can request access to the bus, and they are categorised accordingly as *bus masters* and *bus slaves* [2].

9.3.1. System and Peripheral Buses

In larger embedded system designs, it may be appropriate to use multiple buses to provide sufficient bandwidth for communication between all processing and peripheral cores. In such systems, the two main types of bus are the *system bus* and the *peripheral bus.*

System Bus

In an embedded system design, the system bus is the main method of providing communication between the various peripheral and processing cores. In smaller embedded system designs, the system bus may be the only bus that is present within the design. In larger, multiple-bus designs, the system bus will be the bus with the largest bandwidth that connects the memory controller and the processor, as well as any high-speed devices. All remaining peripherals, which do not require such high-speed access to the processor and memory controller, will be connected via the peripheral bus.

Peripheral Bus

In order to split the embedded system design into separate domains, a second bus — known as the peripheral bus — can be added. This can be for a number of reasons, such as making the distinction between low-speed and high-speed devices or to provide dedicated bandwidth for the communication between a group of peripheral cores. This allows a set of peripherals to communicate between themselves in parallel with any communication on the system bus, such as a processor-memory access.

Bus Bridge

In order to allow the communication between devices on separate buses, such as the processor (on the system bus) requesting data from a peripheral core (on the peripheral bus), a *bridge* between the two buses is required. A bridge is connected to both buses and communicates requests between them. On one bus the bridge is connected as a bus master, while having a slave connection on the other.

9.3.2. Bus Masters and Slaves

The modules that connect to the buses in an embedded system can be split into two distinct categories:

- **Bus Masters** have the ability to request access to the bus and, as such, are responsible for initiating data transfers by driving the address and control signals. Bus cycles can be initiated, and other bus modules informed of the bus cycle type.
- **Bus Slaves** do not possess the ability to initiate bus cycles, and instead only monitor bus activity. Signals on the address and control lines are decoded, and when addressed, can either place, or accept, data on/from the bus.

9.3.3. Bus Arbitration

As a bus is shared amongst all devices in an embedded system, there is a requirement to determine which bus master device is permitted to use the bus at any one time. The method for determining this access is called *bus arbitration*. If more than one master device requests access to the bus at the same time, it is the job of the bus arbiter to decide which device should be granted access first. In simple terms, the requesting master device with the highest priority (lowest assigned number) will be granted access to the bus first, provided a bus cycle is not already in progress. Lower priority masters (higher assigned numbers) will be placed in a waiting queue, and processed upon completion of the higher priority requests.

9.3.4. Memory Access

The way in which memory controllers are accessed in an embedded system can have a huge effect on overall performance. Even if a very efficient type of memory and memory controller is used, the system performance could suffer from poor memory access control. It is important that the system is structured, and accessed, in a way which will maximise the memory bandwidth whilst keeping the required resources to a minimum.

Programmable Input/Output (I/O)

One way of controlling the movement of data between the memory controller and other peripherals is to route all data transfers via the processor. This method of memory transfer is known as programmable I/O, and allows the system to process memory transfers with a minimum amount of resources. This approach requires that the peripheral and the processor are situated on the same bus, and the processor acts as a central point of communication between all peripherals and the memory. If the number of memory transfer requests between peripherals and memory is high, the processor could spend a lot of time performing memory transfers and less time performing other computations [2].

Programmable I/O is an effective way of managing memory transactions, whilst using a minimum amount of resources, provided that the system implements most of the functionality in the programmable logic. If, however, the processor is required to perform a large number of other computations, other methods may be preferable [2].

Direct Memory Access (DMA)

One way of reducing the burden on the processor is to use Direct Memory Access (DMA) to perform memory transfers. Using this approach, the processor issues a memory transfer request to the DMA controller, which will then perform the memory transaction. This allows the processor to perform other tasks while the DMA controller performs the transfer. In this situation, the DMA controller acts as both a bus master and a bus slave. As a master, the DMA controller communicates with the memory controller whilst also arbitrating for the bus. While acting as a slave, the DMA controller sets up memory transfers by responding to requests from bus masters (the processor, in most cases). In order to initiate the transaction, the DMA controller must be provided with the following information [2]:

- **Source address** — the address where the data will be read from.
- **Destination address** — the location where the data should be written to.

- **Transfer length** — the number of bytes that should be transferred.

An overview of a DMA memory transfer is provided in Figure 9.8.

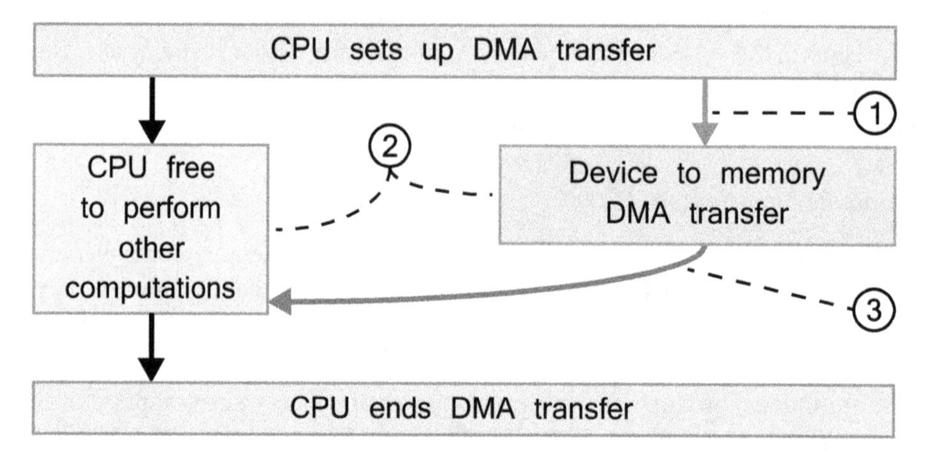

Figure 9.8: DMA memory transfer operation

With reference to Figure 9.8, the DMA memory transfer process is:

1. The processor sets up the device wishing to use the DMA to transfer data to memory by issuing a DMA command and disables all DMA interrupts.

2. The DMA controller transfers data from the peripheral device to memory leaving the CPU free to perform other computations.

3. Following completion of the data transfer, an interrupt is sent to the CPU to inform it that it can close the DMA transfer.

9.3.5. Bus Bandwidth

Bus bandwidth is the total amount of data that can be transferred on a bus in a given unit of time. The value for bus bandwidth depends on two parameters:

- **Bus data width** — this is the number of physical lines over which the bus transfers data simultaneously. A bus with 32 individual data lines can transmit 32 bits of data simultaneously.

- **Bus frequency** — this is the speed at which the bus operates. This refers to the number of data bits which can be transmitted/received per second. This is defined in Hertz (Hz).

The relationship between these parameters and bus bandwidth is given in Equation (1).

$$Bus\ Bandwidth\ (\text{Mbits/s}) = Bus\ Width\ (\text{bits}) \times Bus\ Frequency\ (\text{MHz}) \qquad (1)$$

As an example, a bus with a data width of 32 bits operating at 10 MHz has a bus bandwidth and, therefore, a maximum throughput of:

$$Bus\ Bandwidth = 32\ \text{bits} \times 10\ \text{MHz} = \mathbf{320\ Mbits/s\ (40\ MBytes/s)}$$

Obviously, the more peripherals that are connected to a bus, the higher the required throughput on the bus will be. Therefore, it is important that the bus bandwidth is sufficiently high in order to prevent the system from saturating the bus.

9.4. Chapter Review

In this chapter, the concept of an embedded system has been introduced and a generic embedded system architecture has been explored. The role of processors within embedded systems has been discussed, and some of the processing features, such as processor cache and execution cycles have been covered. The function of coprocessors was also introduced as well as the use of software/hardware interrupts.

Communication between all components within an embedded system is reliant on the inclusion of a bus system, and their functionality has been discussed in this chapter. The requirement for multiple bus systems was introduced, and the distinction between bus master and slave devices was reviewed. Bus arbitration and memory access techniques were summarised, and the importance of bus bandwidth was discussed.

9.5. References

[1] D. Liu, "Introduction" in *Embedded DSP Processor Design: Application Specific Instruction Set Processors*, Morgan Kaufmann, 2008, pp 1 - 46.

[2] R. Sass and A. G. Schmidt, "Managing Bandwidth" in *Embedded Systems Design with Platform FPGAs: Principles and Practices*, 1st. Ed, Morgan Kaufmann, 2010, pp 295 - 346.

[3] R. Sass and A. G. Schmidt, "System Design" in *Embedded Systems Design with Platform FPGAs: Principles and Practices*, 1st. Ed, Morgan Kaufmann, 2010, pp 115 - 196.

10

Zynq System-on-Chip Design Overview

In this chapter we will examine all aspects of the Zynq SoC design. We will take a closer look at certain aspects of the ARM processing cores and their operating modes, as well as laying down the foundations for some of the later chapters in this book. To be specific, the various interconnects available, interfacing signals, interrupts and the memory facilities on a Zynq device will all be introduced in this chapter, before being covered in greater detail as the book progresses.

One of the main objectives of this chapter is to detail the connections and shared resources that exist between the PS and the PL. These include the interconnect system, memories and a variety of different interfaces. Further, this chapter also describes the various operating modes of the Zynq SoC and the configurations of the PS and PL which can be used for different tasks and purposes.

As all Zynq-7000 AP devices feature the same dual-core ARM Cortex-A9 based PS, the information covered in this chapter can be applied across the entire Zynq range, unless otherwise stated. There are some small differences in certain models, mainly pertaining to the maximum operating frequency of the PS, and these will be explicitly mentioned in the relevant sections.

10.1. Interfacing and Signals

In this section the user-visible interfaces and signals of the Zynq-7000 AP SoC device will be detailed. Special attention is given to the interfaces between the PS and PL. The major groups of signals and interfaces are highlighted in Figure 10.1.

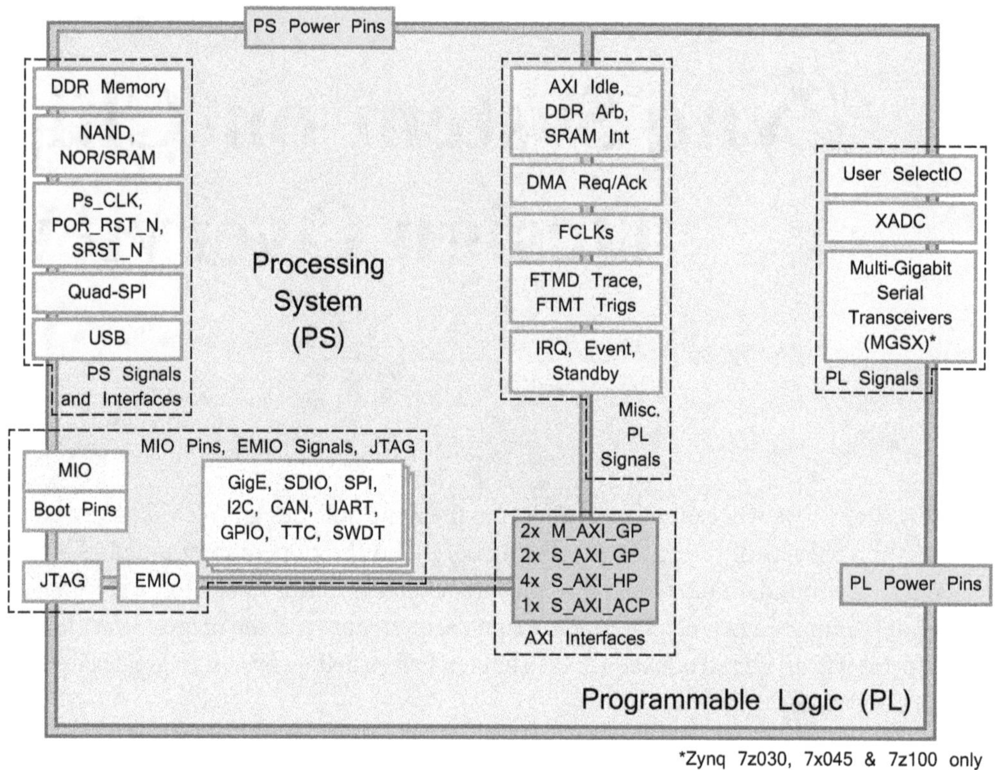

Figure 10.1: Zynq-7000 AP SoC interfaces, signals and pins

10.1.1. PS-PL AXI Interfaces

The main form of connection between the PS and PL elements of Zynq is via AXI interfaces, which provide high bandwidth, low latency links between both parts of the device. Each AXI interface on the PS side consists of multiple AXI channels; over a thousand signals are used to implement the nine PL interfaces. Theses are detailed in Table 10.1.

Table 10.1: PS-PL AXI interfaces

Interface Name	Interface Description	Master	Slave	Signals
M_AXI_GP0	General purpose ports with a path to the OCM and access to a port on the DDR memory controller.	PS	PL	See Section 10.2.6 AXI_GP Interfaces.
M_AXI_GP1		PS	PL	
S_AXI_GP0		PL	PS	
S_AXI_GP1		PL	PS	
S_AXI_ACP	Accelerator Coherency Port (ACP), cache-coherent transaction and path to L2 cache.	PL	PS	See Section 10.2.5 AXI_ACP Interface.
S_AXI_HP0	High performance ports with read/write FIFOs, a path to the OCM and two dedicated memory ports on the DDR controller. AXI_HP interfaces are also know as AFI.	PL	PS	See Section 10.2.4 AXI_HP Interfaces.
S_AXI_HP1		PL	PS	
S_AXI_HP2		PL	PS	
S_AXI_HP3		PL	PS	

10.1.2. PL Co-Processing Interfaces

This section provides an overview of the interfaces that are available for communication between the PL and the PS.

Accelerator Coherency Port (ACP) Interface

The ACP is a 64-bit slave interface on the SCU which provides an asynchronous cache-coherent access point from the PL to the PS. The ACP is accessible by a number of PL masters to provide access to the memory subsystem in the same manner as the APU processors. This has the effect of increasing overall performance, improving power consumption and simplifying software. The ACP interface behaves as a standard AXI slave interface and supports most standard read and write transactions without any additional coherency requirements placed on the PL components. The ACP therefore provides cache-coherent access from the PL to the CPU caches while any memories local to the PL are non-coherent with the CPUs [2].

Any read transactions through the ACP to a coherent section of memory go through the SCU to check if any of the required data is currently held within CPU L1 cache. If data is held within the L1 cache, the requested data is returned directly to the requesting component. In the case that the data is not found in the L1 cache, the L2 cache is also checked before the data request is passed to the main memory. [2]

Write transactions to coherent memory regions have coherence enforced by the SCU, before the write is forwarded to main memory. Optionally, write transactions can also allocate into L2 cache, which removes the performance and power impact of writing to off-chip memory [2].

ACP Usage

The ACP supplies a low latency path between the accelerators implemented in the PL and the PS. The steps involved in a communication between the PS and a PL accelerator are summarised as follows [2]:

1. Input data for the accelerator is prepared with the CPU's local cache space.
2. A message is sent to the accelerator from the CPU via one of the AXI General Purpose (AXI_GP) master interfaces to the PL.
3. The PL accelerator fetches data via the ACP. It is processed and a result is returned via the ACP.
4. A flag is set by the accelerator by writing to a known location which indicates that the data processing is complete. The status of the flag can be polled by the PS or an interrupt can be generated.

In comparison to a tightly-coupled co-processor, ACP has relatively high access latencies. As a result, ACP is not recommended for fine-grained instruction-level acceleration. Due to the relatively small transaction overhead when compared to the transaction time, ACP does not have a clear advantage over traditional memory mapped PL acceleration for coarse-grain acceleration (such as video frame-level processing) either. ACP is therefore best used for medium-grain acceleration, such as a block-level cryptography accelerator. [2].

ACP read and write behaviour based on the current cache status is detailed in Table 10.2. From this, it is clear that access latencies are low when cache hits occur [2].

Table 10.2: ACP read and write behaviour based on current cache status [2]

Action	Description
ACP Read — I (Invalid)	Data fetched from external memory by the SCU via one of two AXI master interfaces and forwarded to the ACP directly. CPU L1 cache state is not affected.
ACP Read — M (Modified)	Data fetched from the L1 cache with M status by the SCU. CPU L1 cache state is not affected.
ACP Read — S (Shared)	Data fetched from the L1 cache with S status by the SCU. CPU L1 cache state is not affected.
ACP Read — E (Exclusive)	Data fetched from the L1 cache with E status by the SCU. CPU L1 cache state is not affected.
ACP Write — I (Invalid)	Data is written to external memory through one of two AXI master interfaces. CPU L1 cache state is not affected.
ACP Write — M (Modified)	Data in the CPU L1 cache is flushed out to external memory before ACP data is written into the external memory interface. L1 cache status is changed from M to I. L1 cache flush step is skipped if the SCU overwrites the entire cache line.
ACP Write — S (Shared)	Data is written to external memory through one of two AXI master interfaces. L1 cache status is changed from S to I.
ACP Write — E (Exclusive)	Data is written to external memory through one of two AXI master interfaces. L1 cache status is changed from S to I.

ACP Limitations

A number of limitations of the ACP exist [2]:

- Locked access is not allowed for coherent memory.
- Exclusive access is not allowed for coherent memory.

- Accesses from other AXI masters can be starved by continuous access to the OCM over the ACP. ACP bandwidth should be moderated to less than the peak OCM bandwidth to allow access from other masters. This can be achieved by regulating burst sizes to less than eight 64-bit words.

- Write transactions with AWLEN=3, AWSIZE=3 and WSTRB ≠ `11111111` can cause the cache line in the CPUs to be corrupted.

- Modules that prioritise write requests over read requests, such as PCI Express (PCIe), should not be connected to the ACP as they can create deadlock. They should be connected to the AXI GP or HP ports to avoid this.

10.1.3. Interrupt Interface

Interrupts between the PS and PL are controlled by the Generic Interrupt Controller (GIC), which supports 64 interrupt lines. Six interrupts are driven from within the APU, including the L1 parity fail, L2 interrupt and Performance Monitor Unit (PMU) interrupt.

The interrupt output from the GIC drives either IRQ or Fast Interrupt ReQuest (FIQ) signals as inputs to the CPUs. Selection of the processor target for the interrupt is via an SCU register within the APU. Interrupts specific to the APU are outlined in Table 10.3 [2].

Table 10.3: APU-specific interrupts [2]

Interrupt	Description
32	Errors from CPU0 including L1 data cache, L1 instruction cache, Translation Look-aside Buffer (TLB), Global branch History Buffer (GHB), and Branch Target Address Cache (BTAC) parity errors.
33	Errors from CPU1 including L1 data cache, L1 instruction cache, TLB, GHB, and BTAC parity errors.
34	Any errors from the L2 cache controller, including parity errors.
92	Parity errors from the SCU generate a third interrupt.
37	PMU of CPU0.
38	PMU of CPU1.

10.2. Interconnects

The interconnect that is located within the PS facilitates the communication of read, write and response transactions between master and slave clients, and is comprised of multiple switches to connect system resources using AXI point-to-point channels. The interconnect, belonging to the ARM AMBA family of buses, implements a large array of interconnect communication capabilities and overlays for Quality-of-Service (QoS), debug and test monitoring. Multiple outstanding transactions are managed by the interconnect, which is designed for low-latency paths for ARM CPUs. In terms of the PL master controls, the interconnect is capable of high-throughput and cache coherent data paths [4].

10.2.1. Interconnect Features

The AXI interconnect system is the primary mechanism for data communications on Zynq devices. A summary of the interconnect features is provided below [4]:

AXI high performance datapath switches:

- Snoop control unit
- L2 cache controller
- Interconnect switches based on ARM NIC-301:
  - Central interconnect
  - Master interconnect
  - Slave interconnect
  - Memory interconnect
  - OCM interconnect
  - AMBA Advanced High-Performance Bus (AHB) and Advanced Peripheral Bus (APB) bridges

PS-PL Interfaces:

- AXI_ACP, one cache coherent master port for the PL
- AXI_HP, four high performance/bandwidth master ports for the PL
- AXI_GP, four general purpose ports. Two master ports and two slave ports.

10.2.2. Interconnects, Masters and Slaves

The overall interconnect structure features a number of different individual interconnect switches, as well as two classes of connections: master and slave. These are outlined in this section.

Interconnect Switches

The various interconnect switches available on the interconnect are summarised as follows [4]:

- ***Central Interconnect*** — The central interconnect is the core of the ARM NIC-301-based interconnect switches.

- ***Master Interconnect*** — The master interconnect controls the switching of the low-to-medium speed traffic from AXI_GP ports, Device Configuration (DevC) and Device Access Port (DAP) to the central interconnect.

- ***Slave Interconnect*** — The slave interconnect controls the switching of low-to-medium speed traffic from the central interconnect to I/O peripherals, AXI_GP and other blocks.

- ***Memory Interconnect*** — The memory interconnect controls the switching of high-speed traffic from the AXI_HP ports to DDR DRAM and OCM (via the OCM interconnect).

- ***OCM Interconnect*** — The OCM interconnect controls the switching of high-speed traffic between the OCM and the central and memory interconnects.

- ***SCU***— The SCU functions like a switch from the perspective of traffic to its AXI master ports and from its AXI slave ports, due to the address filtering feature.

- ***L2 Cache Controller*** — The L2 cache controller functions like a switch from the perspective of traffic to its AXI master ports and from its AXI slave ports, due to the address filtering feature.

Examples of the two types of interconnect connections are provided as follows.

Interconnect Masters

Interconnect masters include [4]:

- CPUs
- ACP
- High performance PL interfaces, AXI_HP{3:0}
- General purpose PL interfaces, AXI_GP{1:0}
- DMA controller
- AHB masters — I/O Peripherals (IOPs) with local DMA units
- Device Configuration (DevC) and Debug Access Port (DAP)

Interconnect Slaves

Interconnect slaves include [4]:

- General purpose PL interfaces, M_AXI_GP{1:0}
- OCM
- DDR DRAM
- Global Programmers View (GPV) — programmable registers of the interconnect
- AHB slaves — IOPs with local DMA units
- APB slaves — programmable registers in various modules

10.2.3. Connectivity

The interconnect does not offer a full cross-bar structure, whereby all masters can communicate with all slaves. Which master can access which slave is detailed in Table 10.4.

Table 10.4: Interconnect master-slave access [4] ('X' denotes access)

Master / Slave	OCM	DDR Port 0	DDR Port 1	DDR Port 2	DDR Port 3	M_AXI_ GP	AHB Slaves	APB Slaves	GPV
CPUs	X	X	----	----	----	X	X	X	X
AXI_ACP	X	X	----	----	----	X	X	X	X
AXI_HP{0,1}	X	----	----	----	X	----	----	----	----
AXI_HP{2,3}	X	----	----	X	----	----	----	----	----
S_AXI_GP{0,1}	X	----	X	----	----	X	X	X	----
DMA Controller	X	----	X	----	----	X	X	X	----
AHB Masters	X	----	X	----	----	X	X	X	----
DevC, DAP	X	----	X	----	----	X	X	X	----

10.2.4. AXI_HP Interfaces

There are four AXI_HP interfaces which provide high bandwidth datapaths from the PL bus masters to the OCM and DDR memories. Two FIFO buffers are included on each interface for read and write traffic. The interconnect that connects the PL to memory routes high-speed AXI_HP ports to two DDR memory ports, or the OCM. In some Xilinx documentation, the AXI_HP interfaces are also referred to as the AXI FIFO Interface (AFI), in reference to their buffering capabilities.

Features of the AXI_HP interfaces include [4]:

- 32- or 64-bit data master interfaces, independently programmed on a per port basis.
- Automatic expansion of transfer size from 32-bits to 64-bits for unaligned 32-bit transfers.
- Programmable release threshold of write commands.
- Asynchronous clock frequency domain crossing for all PL-PS interfaces.
- Read and write FIFOs.
- Command and Data FIFO fill-level counts are PL-viewable.

PL ports offer QoS signalling.

A diagram of the AXI_HP interfaces is provided in Figure 10.2.

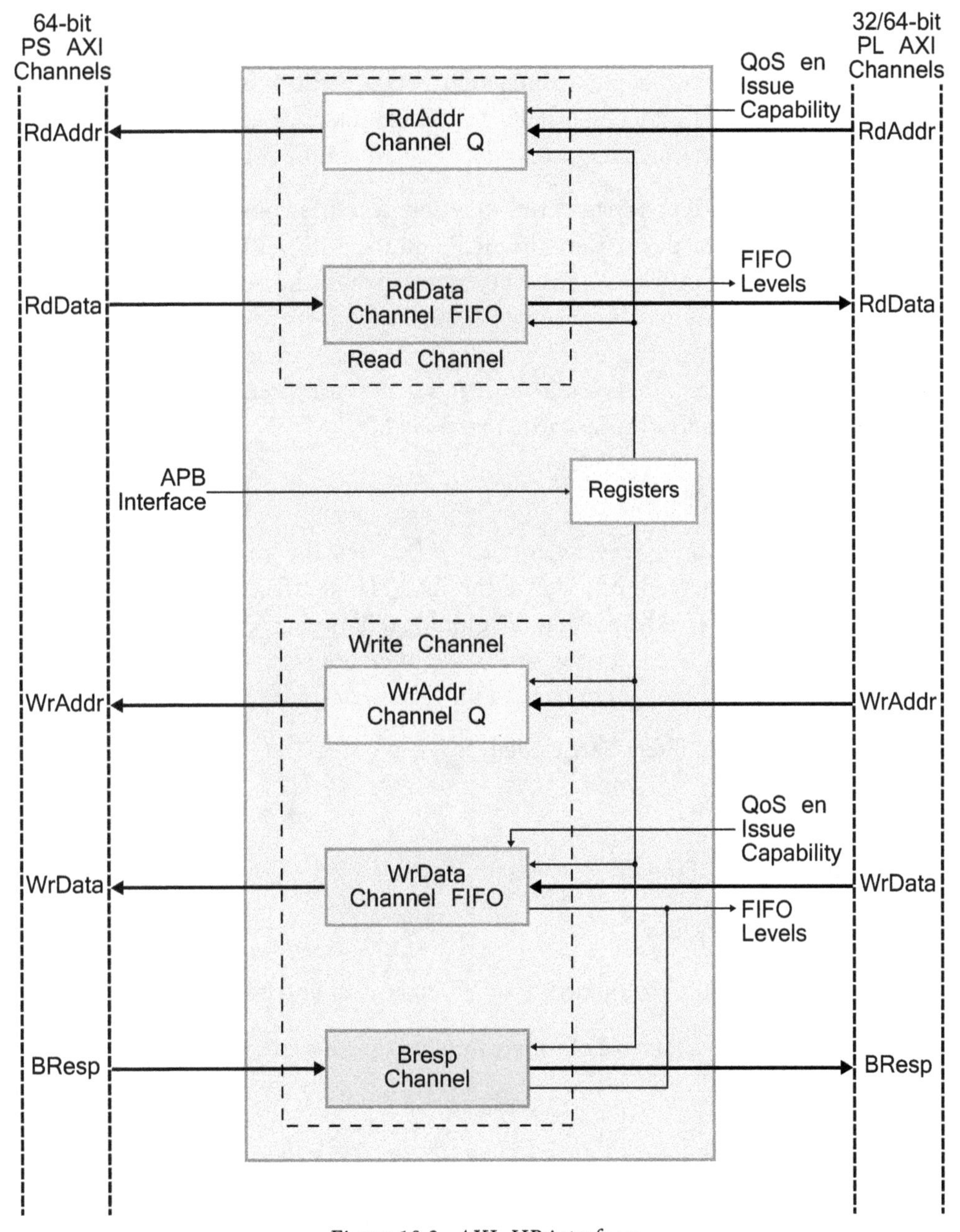

Figure 10.2: AXI_HP interfaces

Further information on the AXI_HP interface is available in Chapter 5 of the Zynq-7000 All Programmable SoC Technical Reference Manual [4].

10.2.5. AXI_ACP Interface

The ACP provides a low-latency connection between the PS and PL, with optional coherency with L1 and L2 caches [4]. It is a 64-bit interface which allows the PL to implement an AXI master which has access to the OCM and L2 cache.

From the perspective of the system, the ACP interface has comparable connectivity as the CPUs in the APU. For this reason, the ACP and the APU CPUs directly compete for resources outside of the APU [4]. This means that when the ACP interface is in use, sections of the cache space will be occupied by the co-processing task. Due to this, CPU processes which rely on CPU caches for high performance, or even real-time performance, may not be able to meet required deadlines. If this is the case, then it may be preferable to use the AXI_HP interface to store task data in the OCM.

10.2.6. AXI_GP Interfaces

The AXI_GP interfaces are directly connected between the ports of the master interconnect and the slave interconnect. Unlike the AXI_HP interfaces which feature a 1 KB data FIFO for buffering [4], AXI_GP has no additional buffering. Performance is therefore constrained by the ports of the master and slave interconnects. These interfaces are for general-purpose only and should not be used for high performance tasks.

Features of the AXI_GP interface include [4]:

- 32-bit data bus width.
- 12-bit master port ID width.
- 6-bit slave port ID width.
- Master and slave port acceptance capability of 8 reads and 8 writes.

The AXI_GP interface is capable of supporting multiple peripherals on each port.

10.3. Memory

Zynq-7000 AP devices feature a number of different types of memory and memory interfacing facilities. In this section these memory facilities will be introduced.

10.3.1. Memory Interfaces

The memory interface unit on all Zynq-7000 AP devices includes a dynamic memory controller and static memory interface modules. The dynamic memory controller is compatible with the following types of memories: DDR3, DDR3L, DDR2 and LPDDR2. The static memory controllers support a NAND flash interface, a Quad-SPI flash interface, a parallel data bus and parallel NOR flash interface [9].

Dynamic Memory Interface

The multi-protocol Double Data Rate (DDR) memory controller consists of three major modules: a core memory controller and scheduler (DDRC), an AXI memory port interface (DDRI) and a digital PHY and controller (DDRP) [3].

The DDR memory controller can be configured to operate in either 16-bit or 32-bit mode, offering access to a 1 GB address space with a single rank DRAM memory configuration of 8-, 16- or 32-bit. EEC memory is supported, but only in 32-bit bus access mode. A maximum speed of 1333 Mb/s is supported when DDR3 is used [9].

Shared access to common memory for the PS and PL is supported through the multi-ported DDRI, which features Four AXI slave ports to accommodate this [9]:

- PL access via two dedicated 64-bit ports (AXI_HP).
- One 64-bit port dedicated to the ARM CPU(s) via the L2 cache controller. This port can be configured for low latency.
- All other AXI masters share the remaining port via the central interconnect.

Each AXI interface features a dedicated transaction FIFO.

The DDRP PHY processes read/write requests from the controller and, within the timing constraints of the timing DDR memory, translates them into specific signals. The PHY uses signals from the controller to create internal signals that connect to the DDR pins via the digital PHYs. PCB signals connect the DDR pins to the DDR device(s).

A block diagram of the DDR memory controller is provided in Figure 10.3.

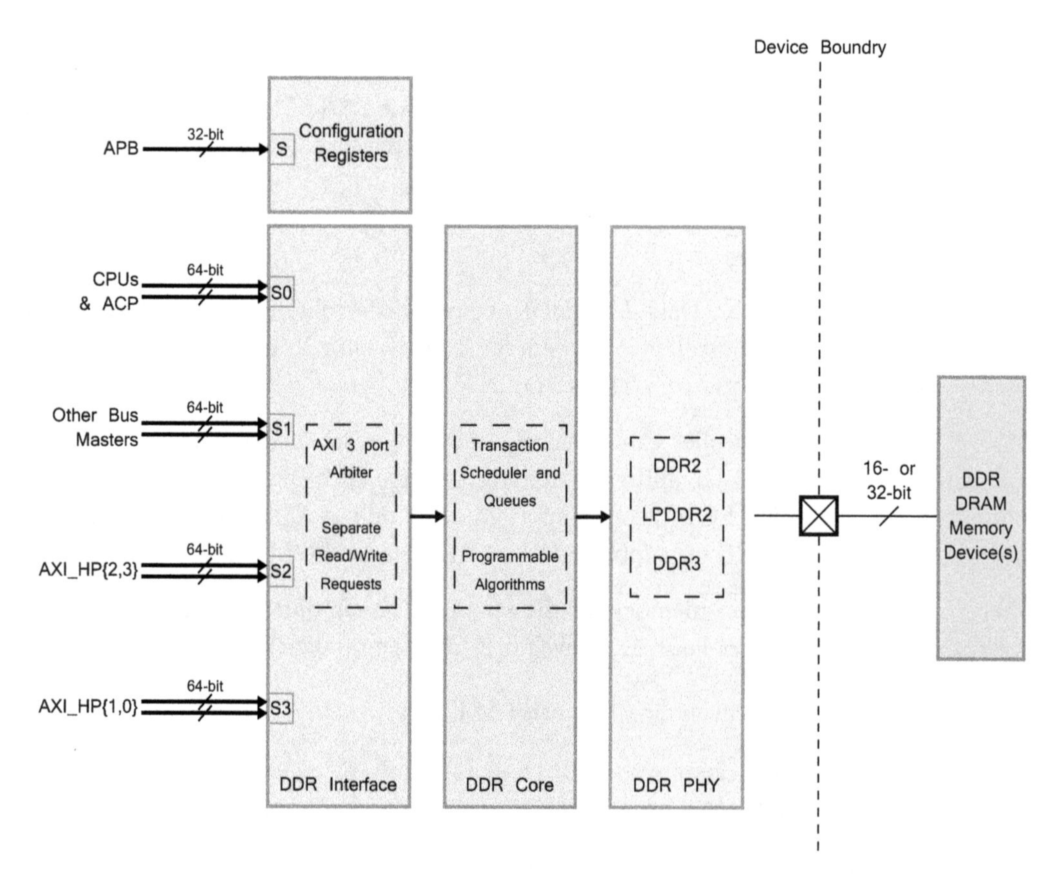

Figure 10.3: DDR memory controller block diagram

The features of the DDR memory controller are extensive and for the purposes of this book, not all will be covered. For further information on the DDR memory controller, please refer to Chapter 10 of the Zynq-7000 All Programmable SoC Technical Reference Manual [3].

The devices supported by the DDR memory controller are subject to the conditions outlined in Table 10.5.

Table 10.5: Memory connectivity limitations [9]

Parameter	Value	Notes
Maximum total memory density	1 GB	1 GB of address map is allocated to DRAM
Total data width (bits)	16, 32	ECC can only use a 32-bit configuration: 16 data bits, 10 check bits
Component data width (bits)	8, 16, 32	4-bit devices are not supported
Maximum ranks	1	----
Maximum row address (bits)	15	----
Maximum bank address (bits)	3	----

A list of example memory configurations for Zynq devices is provided in Table 10.6.

Table 10.6: Possible Zynq-7000 SoC memory configurations

Memory Type	Component Configuration	Number of Components	Component Density	Total Width	Total Density
DDR3/ DDR3L	x16	2	4 Gb	32	1 GB
DDR2	x8	4	2 Gb	32	1 GB
LPDDR2	x32	1	2 Gb	32	256 MB
LPDDR2	x16	2	2 Gb	32	512 MB
LPDDR2	x16	1	2 Gb	16	256 MB

Static Memory Interface

The Static Memory Controller (SMC) can be used as either a NAND flash controller, or a parallel port memory controller. The SMC supports the following types of memories [8]:

- NAND flash
- NOR flash
- Asynchronous SRAM

All addresses, commands, data and memory device protocols are handled by the SMC, allowing the user to access the controller by reading or writing into the operational registers. The optional registers of the SMC are configured through an APB interface. The SMC is based on the ARM PL353 static memory controller.

The features of both the NAND flash and Parallel (SRAM/NOR) interfaces are outlined in Table 10.7.

Table 10.7: SMC interface features [8]

NAND Flash Interface	Parallel (SRAM/NOR) Interface
Up to a 1 GB device	8-bit bus width
Programmable I/O cycle timing	Programmable I/O cycle timing on a per chip select basis
16-word read and 16-word write data FIFOs	16-word read and 16-word write data FIFOs
8-word command FIFO	8-word command FIFO
8/16-bit I/O width with single chip select	One chip select with up to 26 address signals
Open NAND Flash Interface (ONFI) Specification 1.0	Two chip selects with up to 25 addresses (32 + 32 MB)
Asynchronous memory operating mode	Asynchronous memory operating mode
1-bit ECC hardware with software assist	----

A block diagram of the SMC is provided in Figure 10.4.

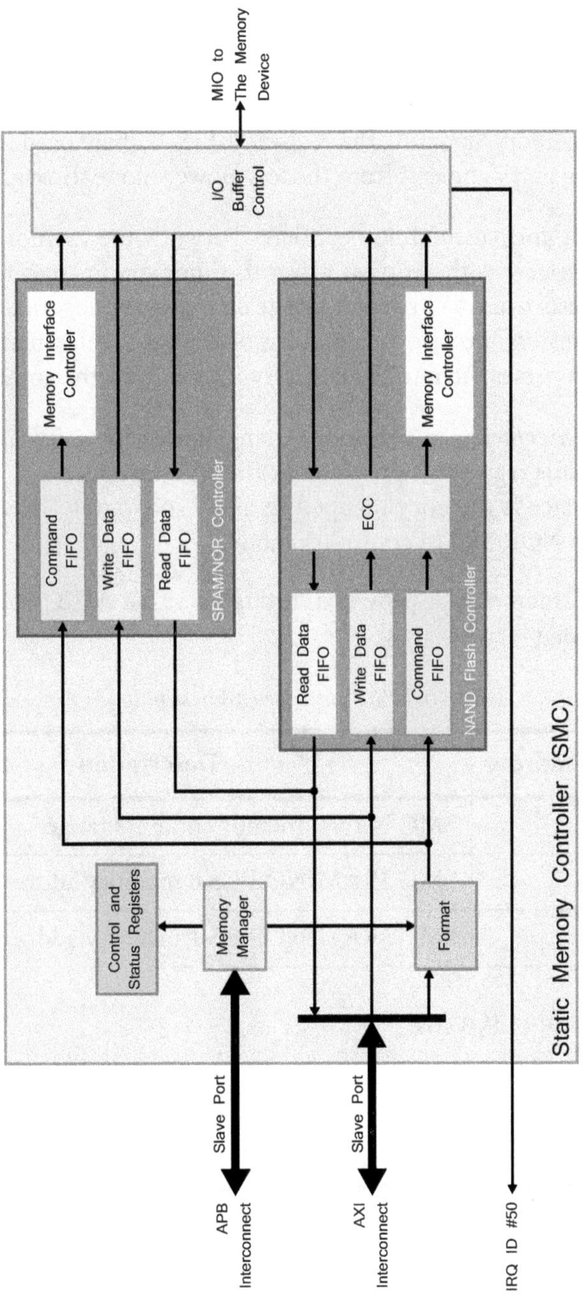

Figure 10.4: Static Memory Controller (SMC) block diagram

With reference to the block diagram in Figure 10.4, the function of some of the modules is described below [8]:

- ***Memory Manager*** — The memory manager controls and tracks the current state of the CPU_1x domain state machine. It is responsible for controlling direct commands issued to the memory, updating the register values that are used in the memory clock and controlling entry and exit from the low-power mode through the APB interface.

- ***Format*** — The format module negotiates between the memory manager and the AXI slave interface, with requests from the memory manager having the higher priority. Requests from AXI read and write channels are negotiated on a round-robin basis. AXI transfers are also mapped onto the applicable channels by the format module, which passes them to the memory interface via the command FIFO.

- ***Interconnect Interfaces*** — A memory mapped area for software to read and write control and status registers is provided by the APB interface.
  The AXI interface is memory mapped to allow software to read and write to/from the memory in NOR/SRAM controller mode.

Access to the SMC memories is provided through a 32-bit AHB bus. The SMC memory address map is provided in Table 10.8.

Table 10.8: SMC memory address map [8]

Register Base Address	Description
`0xE100_0000`	SMC NAND memory address range
`0xE200_0000`	SMC SRAM/NOR CS 0 memory address range
`0xE300_0000`	SMC SRAM/NOR CS 1 memory address range

10.3.2. On-Chip Memory (OCM)

The on-chip memory contains 256 KB of RAM and 128 KB of ROM — this is where the BootROM resides. The OCM supports two 64-bit AXI slave interface ports — one port is dedicated for CPU/ACP access via the APU SCU and the other is shared by all other bus masters within the PS and PL. The BootROM is not visible to the user as it is reserved for exclusive use by the boot process [6].

By implementing RAM as a double-wide memory (128 bits) the OCM is able to support high AXI read and write throughput for RAM access. In order to make full use of the high

RAM access throughput, user applications must use 128-bit aligned addresses and even AXI burst sizes [6].

The TrustZone security feature is supported for memory blocks of 4 KB; the 256 KB of RAM can be divided into 64 blocks of 4 KB, with each assigned independent security attributes.

A block diagram of the OCM is provided in Figure 10.5.

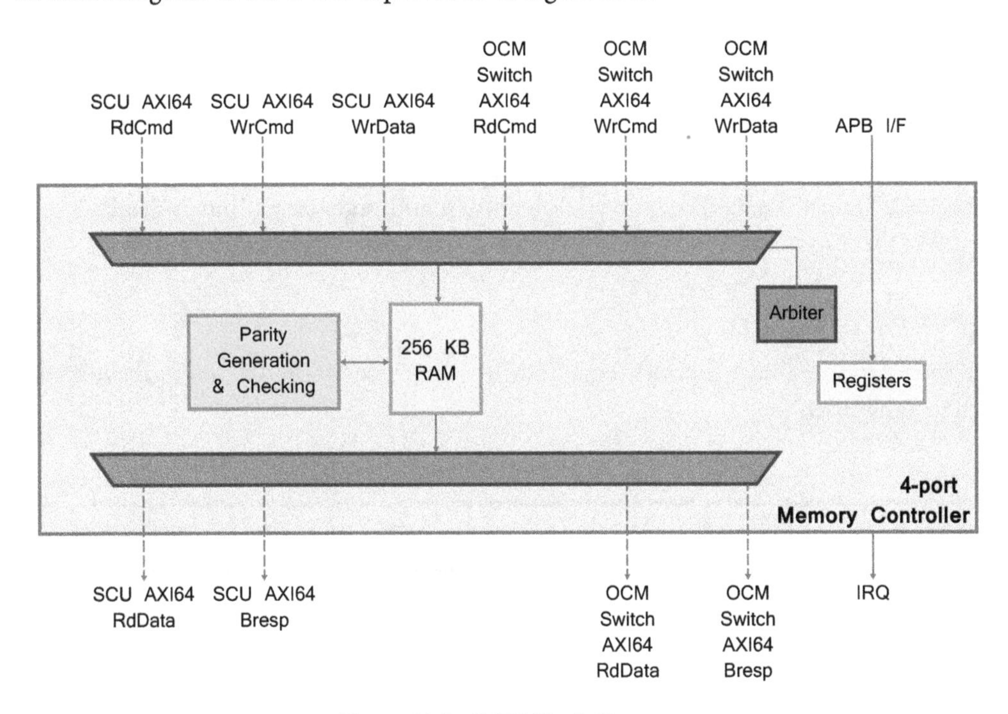

Figure 10.5: OCM block diagram

With reference to the block diagram in Figure 10.5, there are 10 AXI channels associated with the OCM [6]:

- 5 AXI channels for the CPU/ACP (SCU) port
- 5 AXI channels for the other PS/PL masters (OCM switch port)

Arbitration of the read and write channels of the SCU and OCM switch ports is controlled within the OCM module. Only RAM accesses receive parity generation and checking. A register access APB port and interrupt signal (IRQ) are the other main interfaces to/from the OCM module [6].

Key features of the OCM module include [6]:

- 256 KB on-chip RAM
- 128 KB on-chip BootROM (not visible to user)
- Two AXI 3.0, 64-bit slave interfaces
- Support for full AXI 64-bit bandwidth for simultaneous read/write commands on the OCM interconnect port (optimal address alignment restrictions apply)
- TrustZone security support for on-chip RAM (4 KB page granularity)
- Byte-wise parity generation, checking and interrupt support for RAM

Further details on the OCM memory address map and register settings is available in Chapter 29 of the Zynq-7000 All Programmable SoC Technical Reference Manual [6]

10.3.3. Memory Map

Zynq-7000 AP SoCs have support for an address space of 4 GB. The memory map is provided in Table 10.9

Table 10.9: Zynq-7000 SoC memory map [9]

Start Address	Size	Description
0x0000_0000	1024 MB	DDR DRAM and OCM
0x4000_0000	1024 MB	PL AXI slave port 0
0x8000_0000	1024 MB	PL AXI slave port 1
0xE000_0000	256 MB	IOP devices
0xF000_0000	128 MB	Reserved
0xF800_0000	32 MB	Programmable registers access via AMBA APB
0xFA00_0000	32 MB	Reserved
0xFC00_0000	64 MB — 256 KB	Quad-SPI linear address base (except top 256 KB which is OCM), 64 MB reserved, only 32 MB is currently supported
0xFFFC_0000	256 KB	OCM when mapped to high address space

10.4. Interrupts

The interrupt structure of Zynq devices is closely linked with the dual-core ARM A9 based PS, which also incorporates a GIC PL390 interrupt controller. This section describes the functions of the interrupt controller and the system-level interrupt environment, of which an overview is provided in Figure 10.6

The rest of this section will cover a number of key topics, including:

- Functionality of the GIC
- Private, shared and software interrupts
- Interrupt prioritisation and handling

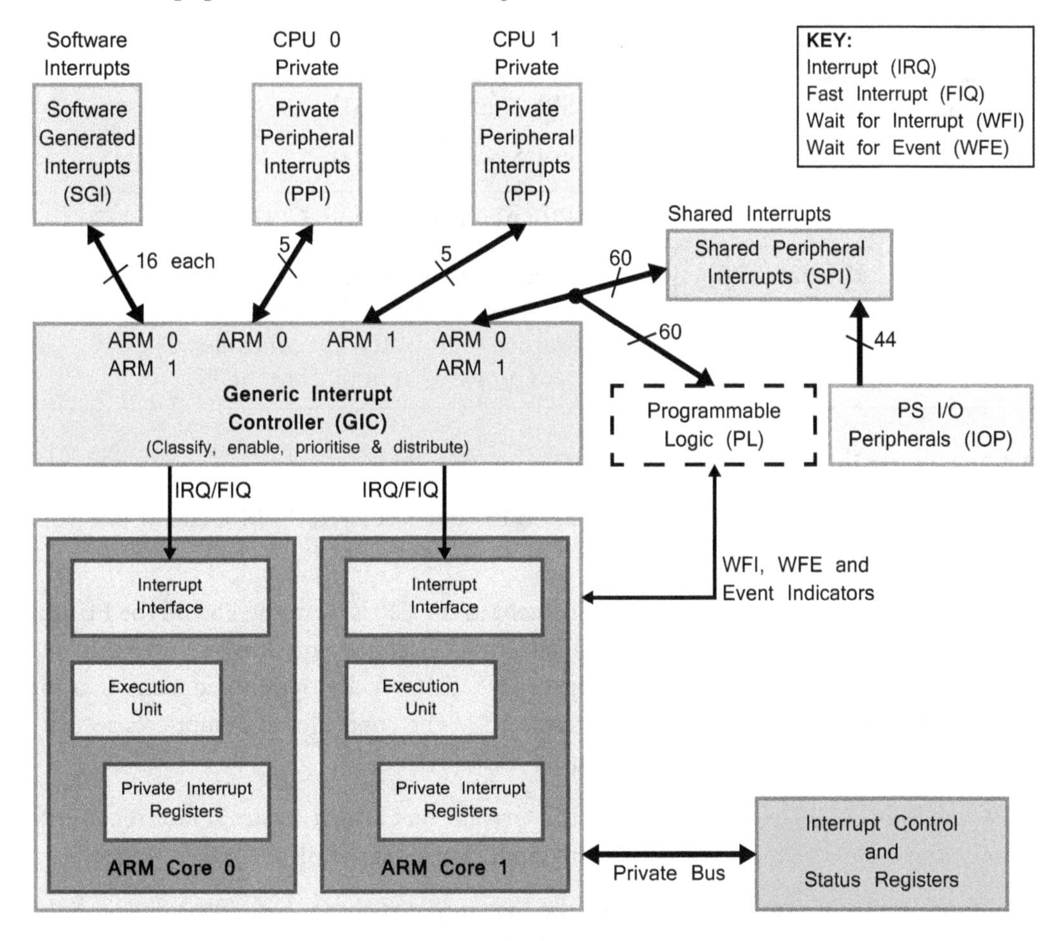

Figure 10.6: System-level interrupt environment

10.4.1. Interrupt Signals

The PS IOP interrupt signals are routed to the PL and are asserted asynchronously to the FCLK clocks. Going the other way, the PL can assert up to 20 interrupts asynchronously to the PS, with up to 16 of the interrupt signals mapped to the interrupt controller as a peripheral interrupt. Each interrupt signal is mapped to one or both of the CPUs and is set to a priority level. The four remaining PL interrupt signals are inverted and directly routed to the nIRQ and nFIQ signals of the Private Peripheral Interrupt (PPI) unit of the interrupt controller. Each CPU has its own nIRQ and nFIQ signal. The relevant interrupt signals between the PS and PL are summarised in Table 10.10 [5].

Table 10.10: PL interrupt signals [5]

Type	PL Signal Name	I/O	Destination
PL to PS Interrupts	IRQF2P[7:0]	I	SPI: Numbers [68:61]
	IRQF2P[15:8]	I	SPI: Numbers [91:84]
PS to PL Interrupts	IRQP2F[19:16]	I	PPI: nFIQ, nIRQ (both CPUs)
	IRQP2F[27:0]	O	Programmable logic. These signals are received from the I/O peripherals and are forwarded to the interrupt controller. They are also provided as outputs to the PL.

10.4.2. Generic Interrupt Controller (GIC)

The general interrupt controller is based on the non-vectored ARM General Interrupt Controller Architecture v1.0.

The controller manages interrupts that are sent to the CPUs from the PS and the PL. It is a centralised resource, and is capable of enabling, disabling, masking and prioritising interrupt sources, sending them to the appropriate CPU(s) in a programmed manner as the next interrupt is accepted by the CPU interface [5]. The controller also supports security extension for the implementation of a security-aware system [5].

GIC registers are accessed via the CPU private bus which ensures fast read/write response times by avoiding bottlenecks and temporary blockages in the interconnect [5].

All interrupt sources are centralised by the interrupt distributor before the one with the highest priority is dispatched to the individual CPUs. The GIC also ensures that an interrupt that targets more than one CPU can only be taken by a single CPU at a time. A unique interrupt ID number identifies each interrupt source, and have their own configurable priority and list of targeted CPUs [5].

Further information on the GIC can be obtained from the ARM Generic Interrupt Controller Architecture Specification [1].

10.4.3. Interrupt Sources

An overview of the individual interrupt sources is provided in this section, covering:

- CPU Private Peripheral Interrupts (PPIs)
- PL and PS Shared Peripheral Interrupts (SPIs)
- Software Generated Interrupts (SGIs)

A block diagram of the interrupt sources is outlined in Figure 10.7.

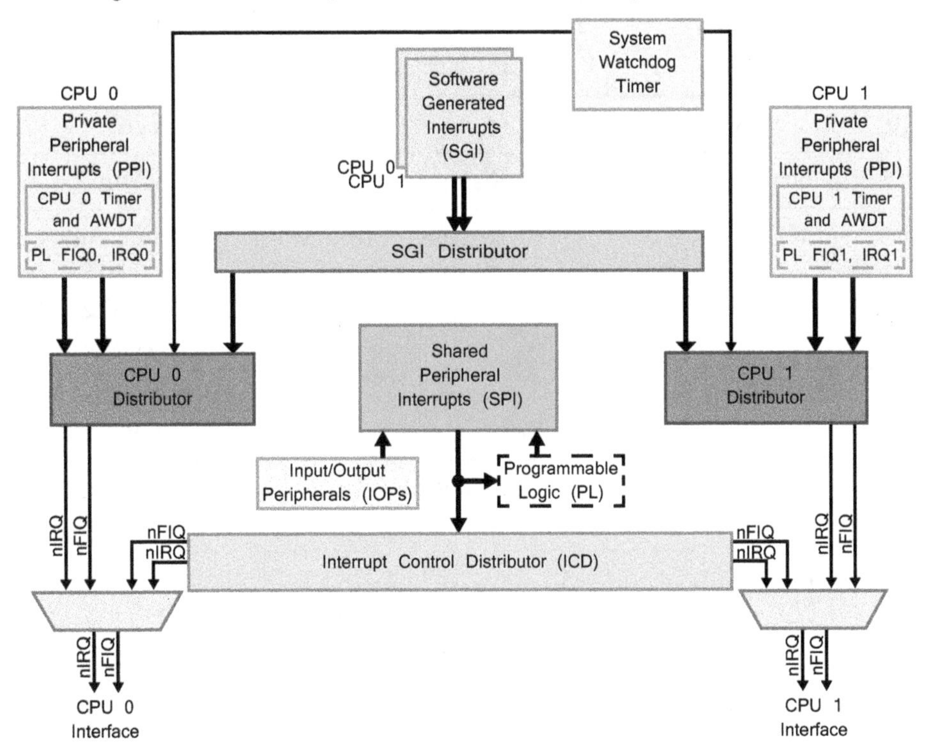

Figure 10.7: Interrupt controller block diagram

CPU Private Peripheral Interrupts (PPI)

Each CPU core connects to a private set of five peripheral interrupts, which are summarised in Table 10.11 [5].

The GIC must be programmed to accommodate the fact that all PPI sensitivity types are fixed by the requesting sources and cannot be changed. These registers are not programmed by the Boot ROM, and therefore the GIC must be programmed to accommodate these sensitivity types by the SDK device drivers [5].

It is worth noting that both the interrupt (IRQ) and fast interrupt (FIQ) signals from the PL are inverted when transferring to the PS, before being sent to the interrupt controller. The signals are therefore active High in the PS at the PS-PL interface before being inverted to active Low level in the PL.

Table 10.11: Private Peripheral Interrupts (PPI) [5]

IRQ ID #	**Name**	**PPI #**	**Type**	**Description**
26:16	Reserved	-----	----	Reserved
27	Global Timer	0	Rising Edge	Global timer
28	nFIQ	1	Active Low level in PL (Active High at PS-PL interface)	Fast interrupt signal from the PL CPU0: IRQF2P[18] CPU1: IRQF2P[19]
29	CPU Private Timer	2	Rising Edge	Interrupt from private CPU timer
30	AWDT{0, 1}	3	Rising edge	Private watchdog timer for each CPU
31	nIRQ	4	Active Low level in PL (Active High at PS-PL interface)	Interrupt signal from the PL CPU0: IRQF2P[16] CPU1: IRQF2P[17]

Shared Peripheral Interrupts (SPI)

A group of roughly 60 interrupts from various modules can be routed to the PL or one or both of the CPUs. The prioritisation and reception of those of the interrupts that are destined for the CPUs is managed by the interrupt controller [5].

As with PPIs, the interrupt sensitivity types of SPIs are fixed by the requesting sources and cannot be changed. The SDK drivers must program the GIC to accommodate the sensitivity types.

The various SPI interrupts are summarised in Table 10.12.

Table 10.12: PS and PL Shared Peripheral Interrupts (SPI) [5]

Source	Interrupt Name	IRQ ID#	Status Bits	Required Type	PS-PL Signal Name	I/O
APU	CPU 1, 0 (L2, TLB, BTAC)	33:32	spi_status_0[1:0]	Rising Edge	----	----
	L2 Cache	34	spi_status_0[2]	High Level	----	----
	OCM	35	spi_status_0[3]	High Level	----	----
Reserved	----	36	spi_status_0[4]	----	----	----
PMU	PMU [1,0]	38, 37	spi_status_0[6:5]	High Level	----	----
XADC	XADC		spi_status_0[7]	High Level	----	----
DVI	DVI	40	spi_status_0[8]	High Level	----	----
SWDT	SWDT	41	spi_status_0[9]	Rising Edge	----	----
Timer	TTC 0	43:42	spi_status_0[11:10]	High Level	----	----
Reserved	----	44	spi_status_0[12]	----	----	----
DMAC	DMAC Abort	45	spi_status_0[13]	High Level	IRQP2F[28]	O
	DMAC [3:0]	49:46	spi_status_0[17:14]	High Level	IRQP2F[23:20]	O
Memory	SMC	50	spi_status_0[18]	High Level	IRQP2F[19]	O
	Quad SPI	51	spi_status_0[19]	High Level	IRQP2F[18]	O
Debug	CTI	----	----	High Level	IRQP2F[17]	O
IOP	GPIO	52	spi_status_0[20]	High Level	IRQP2F[16]	O

Table 10.12: PS and PL Shared Peripheral Interrupts (SPI) [5]

Source	Interrupt Name	IRQ ID#	Status Bits	Required Type	PS-PL Signal Name	I/O
IOP	USB 0	53	spi_status_0[21]	High Level	IRQP2F[15]	O
	Ethernet 0	54	spi_status_0[22]	Rising Edge	IRQP2F[14]	O
	Ethernet 0 Wakeup	55	spi_status_0[23]	Rising Edge	IRQP2F[13]	O
	SDIO 0	56	spi_status_0[24]	High Level	IRQP2F[12]	O
	I2C 0	57	spi_status_0[25]	High Level	IRQP2F[11]	O
	SPI 0	58	spi_status_0[26]	High Level	IRQP2F[10]	O
	UART 0	59	spi_status_0[27]	High Level	IRQP2F[9]	O
	CAN 0	60	spi_status_0[28]	High Level	IRQP2F[8]	O
PL	FPGA [2:0]	63:61	spi_status_0[31:29]	High Level	IRQF2P[2:0]	I
	FPGA [7:3]	68:64	spi_status_1[4:0]	High Level	IRQF2P[7:3]	I
Timer	TTC 1	71:69	spi_status_1[7:5]	High Level	----	----
DMAC	DMAC [7:4]	75:72	spi_status_1[11:8]	High Level	IRQP2F[27:24]	O
IOP	USB 1	76	spi_status_1[12]	Rising Edge	IRQP2F[7]	O
	Ethernet 1	77	spi_status_1[13]	Rising Edge	IRQP2F[6]	O
	Ethernet 1 Wakeup	78	spi_status_1[14]	High Level	IRQP2F[5]	O
	SDIO 1	79	spi_status_1[15]	High Level	IRQP2F[4]	O
	I2C 1	80	spi_status_1[16]	High Level	IRQP2F[3]	O
	SPI 1	81	spi_status_1[17]	High Level	IRQP2F[2]	O
	UART 1	82	spi_status_1[18]	High Level	IRQP2F[1]	O
	CAN 1	83	spi_status_1[19]	High Level	IRQP2F[0]	O
PL	FPGA [15:8]	91:84	spi_status_1[27:20]	High Level	IRQF2P[15:8]	I
SCU	Parity	92	spi_status_1[28]	Rising Edge	----	----
Reserved	----	95:93	spi_status_1[31:29]	----	----	----

Software Generated Interrupts (SGI)

Each CPU core is capable of interrupting both CPUs, the other CPU or itself using a SGI. There are 16 SGIs, which are summarised in Table 10.13 [5]. An SGI is generated by writing the SGI interrupt number to the Software Generated Interrupts Register (ICDSGIR) and specifying the target CPU, or CPUs. The write is carried out on the CPU private bus of the source CPU. A separate set of SGI registers is available to each CPU, allowing them to generate one or more of the 16 software generated interrupts. The interrupts are cleared by either reading the Interrupt Acknowledge Register (ICCIAR) or by writing a value of '1' to the corresponding bits of the Interrupt Clear-Pending Register (ICDICPR) [5].

All SGIs are edge triggered, and the sensitivity types for them are fixed and cannot be changed.

Table 10.13: Software Generated Interrupts (SGI) [5]

IRQ ID #	Name	SGI #	Type	Description
0	Software 0	0	Rising Edge	A set of 16 interrupt sources that are private to each CPU can be routed to up to 16 common interrupt destinations. Each destination can be one or more CPUs.
1	Software 1	1	Rising Edge	
⋮	⋮	⋮	⋮	
15	Software 15	----	Rising Edge	

10.4.4. Interrupt Prioritisation and Handling

This section will briefly touch upon the way in which interrupts are prioritised and handled by Zynq devices.

Interrupt Prioritisation

All interrupt requests, whether they are PPI, SGI or SPI, are assigned a unique ID number which is used by the interrupt controller to arbitrate. A list of pending interrupts for each CPU is held by the interrupt distributor, which will select the highest priority interrupt before asserting it to the CPU interface. If two interrupts of equal priority arrive at the same time, the one with the lowest interrupt ID is issued first [5].

Prioritisation logic exists for each CPU, enabling the highest priority interrupt to be selected for each CPU. The central list of interrupts, processors and activation information is held by the interrupt distributor which is then responsible for triggering software interrupts to the CPUs [5]. In order to provide a separate copy for each processor, the SGI and PPI distributor registers are banked. Logic ensures that an interrupt targeting more than one CPU can only be taken by one CPU at a time [5].

Upon transmission of the highest pending interrupt to the CPU interfaces, the interrupt distributor will receive back an acknowledgement from the CPU allowing it to change the status of the corresponding interrupt. The interrupt can only be ended by the CPU that acknowledges the interrupt [5].

Interrupt Handling

The interrupt distributor operates a state machine for each supported interrupt on each CPU interrupt. The possible interrupt states are [1]:

- inactive
- pending
- active
- active and pending

When the interrupt controller receives an interrupt request, it will mark the state of that request as *pending*. The regeneration of a pending interrupt does not affect the state of that interrupt [1]. Once the interrupt has been acknowledged, the status changes from *pending* to *active and pending* if the pending state of the interrupt persists after the interrupt has become active, or the interrupt is generated again. Otherwise, the status is changed from *pending* to *active* [1]. When the interrupt controller receives confirmation from the processor that the interrupt handling has been completed, the status of the interrupt is changed from *active* to *inactive*, provided that the interrupt has not been generated again. Otherwise, it is changed from *pending and active* to *pending*.

Further information on the handling of interrupts by the GIC is covered in Chapter 3 *Interrupt Handling and Prioritization* of the ARM Generic Interrupt Controller Architecture Specification [1]

10.4.5. Further Reading

It is impossible to cover all aspects of interrupts within the confines of this chapter; the ARM specification of the GIC is over 150 pages long! With that being the case, a few of the topics that can be found in the ARM Generic Interrupt Controller Architecture Specification [1] are highlighted below:

- GIC security extensions
- Interrupt handling state machines
- Programmers model interface to the GIC
- Details of GIC registers
- Distributor and CPU interfaces

10.5. Chapter Review

In this chapter, some of the features of the Zynq SoC have been examined in detail. The methods of interfacing between the L1 cache and the PS have been introduced and discussed, along with relevant signals. Specific attention was given to the ACP access point which provides cache-coherent data transfer and requests between the PL and the PS. The interrupt interface between the PL and the PS was also introduced.

The AXI interconnect system, as well as the various datapath and interconnect switches, were discussed and the available interconnect masters and slaves identified. Further information on the AXI interconnect can be found in Chapter 19.

The final sections of the chapter detail the various memory interfaces and controllers that feature in the Zynq SoC, as well as the interrupt systems.

10.6. References

NOTE: All URLs last accessed June 2014.

[1] ARM, "ARM Generic Interrupt Controller: Architecture Specification", v1.0, September 2008. Available: http://infocenter.arm.com/help/topic/com.arm.doc.ihi0048a/IHI0048A_gic_architecture_spec_v1_0.pdf

[2] Xilinx, Inc, "Application Processing Unit" in *Zynq-7000 All Programmable SoC Technical Reference Manual*, UG585, v1.5, February 2014, pp. 60-111. Available: http://www.xilinx.com/support/documentation/user_guides/ug585-Zynq-7000-TRM.pdf

[3] Xilinx, Inc, "DDR Memory Controller" in *Zynq-7000 All Programmable SoC Technical Reference Manual*, UG585, v1.5, February 2014, pp. 278-314. Available: http://www.xilinx.com/support/documentation/user_guides/ug585-Zynq-7000-TRM.pdf

[4] Xilinx, Inc, "Interconnect" in *Zynq-7000 All Programmable SoC Technical Reference Manual*, UG585, v1.5, February 2014, pp. 117-146. Available: http://www.xilinx.com/support/documentation/user_guides/ug585-Zynq-7000-TRM.pdf

[5] Xilinx, Inc, "Interrupts" in *Zynq-7000 All Programmable SoC Technical Reference Manual*, UG585, v1.5, February 2014, pp. 213-224.
Available: http://www.xilinx.com/support/documentation/user_guides/ug585-Zynq-7000-TRM.pdf

[6] Xilinx, Inc, "On-Chip Memory (OCM)" in *Zynq-7000 All Programmable SoC Technical Reference Manual*, UG585, v1.5, February 2014, pp. 707-717.
Available: http://www.xilinx.com/support/documentation/user_guides/ug585-Zynq-7000-TRM.pdf

[7] Xilinx, Inc, "Signals, Interfaces and Pins" in *Zynq-7000 All Programmable SoC Technical Reference Manual*, UG585, v1.5, February 2014, pp. 42-59.
Available: http://www.xilinx.com/support/documentation/user_guides/ug585-Zynq-7000-TRM.pdf

[8] Xilinx, Inc, "Static Memory Controller" in *Zynq-7000 All Programmable SoC Technical Reference Manual*, UG585, v1.5, February 2014, pp. 315-324.
Available: http://www.xilinx.com/support/documentation/user_guides/ug585-Zynq-7000-TRM.pdf

[9] Xilinx, Inc, "Zynq-7000 All Programmable SoC Overview", DS190, v1.6, December 2013.
Available: http://www.xilinx.com/support/documentation/data_sheets/ds190-Zynq-7000-Overview.pdf

11

Zynq System-on-Chip Development

In this chapter we will examine all aspects of the Zynq design flow that concentrate on software development. We will take a closer look at certain aspects of software development and partitioning for Zynq.

The various sections of this chapter explore the important concepts of hardware/software partitioning, Zynq software development and profiling. This ties together the software development cycle on Zynq devices from the early conception stages, through to those of testing and verification.

11.1. Hardware/Software Partitioning

Hardware/software partitioning, also known as hardware/software co-design, is an important stage in the design of embedded systems and, if well executed, can result in a significant improvement in system performance. The process of hardware/software partitioning, as the name implies, involves deciding which system components should be implemented in hardware and which should be implemented in software. The impetus behind the partitioning process is that hardware components, such as those which reside in FPGA programmable logic fabric, are typically much faster due to the parallel processing nature of FPGA devices. However, it also tends to be more expensive. Software components, on the other hand, implemented on a GPP or a microprocessor, are cheaper to both create and maintain, but are also slower due to the inherent sequential processing. In order to achieve a good trade-off between performance and cost, high-performance components can be implemented in hardware, while less intensive processes can be implemented in software.

Traditionally the partitioning process was carried out manually by the systems designer, who would decide which of the design modules would be implemented in hardware, and which could be realised as software. More recently, a number of algorithms and techniques have been developed which enable the automation of the partitioning decision process for a variety of different design environments [1]. The advent of techniques such as HLS also had a significant effect on the partitioning process, allowing software algorithms to be directly converted into RTL for hardware implementation.

An overview of hardware/software partitioning is provided in Figure 11.1.

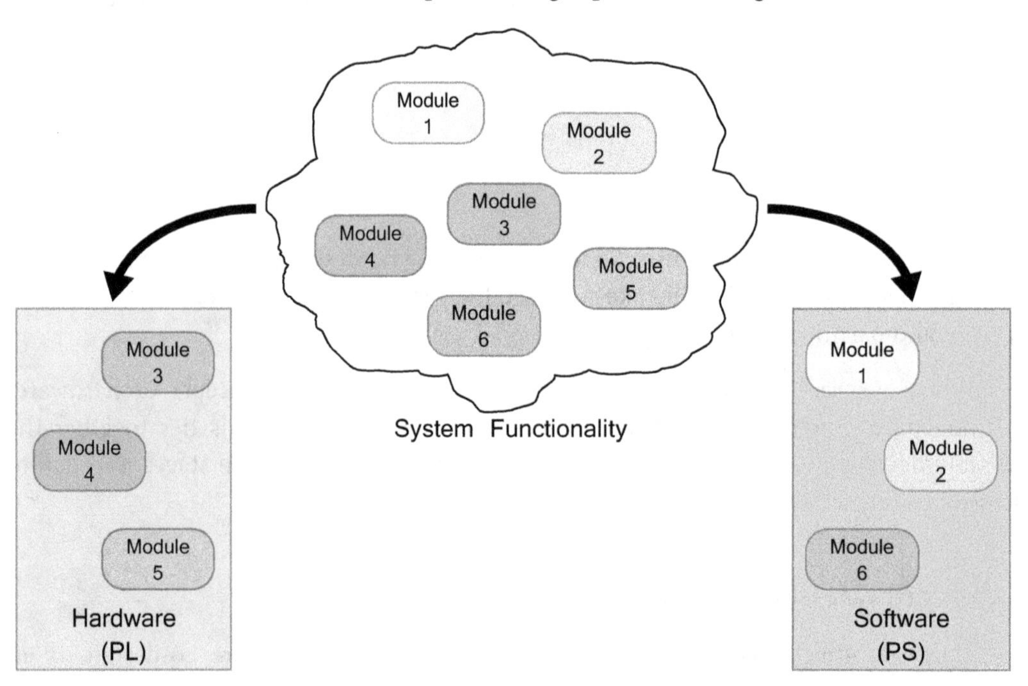

Figure 11.1: Hardware/software partitioning overview

FPGA programmable logic fabric is a good target for the implementation of problems which can be efficiently divided into multiple, parallel tasks. Due to the inherent parallel execution of the programmable logic, multiple operations can be processed simultaneously to calculate the final result in a shorter time than if processed sequentially. Example applications for FPGA implementation include digital filtering operations, beamforming and image processing. Traditionally these tasks are repetitive and quite static in nature. On the other hand, problems exist which are more dynamic and unpredictable, and these are better suited to implementation on a processor based system.

Another factor to consider when deciding whether a process should be implemented in hardware or software, is the number format which will be used. Traditionally, processors have had much better support for floating point calculations due to specially developed vector math engines and dedicated floating point units. FPGAs can support floating point calculations, but a large amount of logic is required to implement them. This is similarly true of high-precision fixed point calculations. Now, with the increased size of FPGAs, high-precision calculations are becoming more common as logic footprint required for their implementation, with respect to the size of the device, reduces. Therefore, if an application requires high-precision floating point calculations, it is best suited to either a processor-based implementation, or that of a large scale FPGA.

It is important to properly exploit the parallel execution of the programmable logic for the implementation of strict timing-driven functions. Figure 11.2 demonstrates the advantage of parallel hardware execution. Whereas the software execution would require 12 clock cycles (sequential execution) to produce the output, **G**, the parallel implementation would only require 2 clock cycles (parallel execution) to produce the same result. If, however, the inputs and outputs of the operation are to be transmitted between the PS and

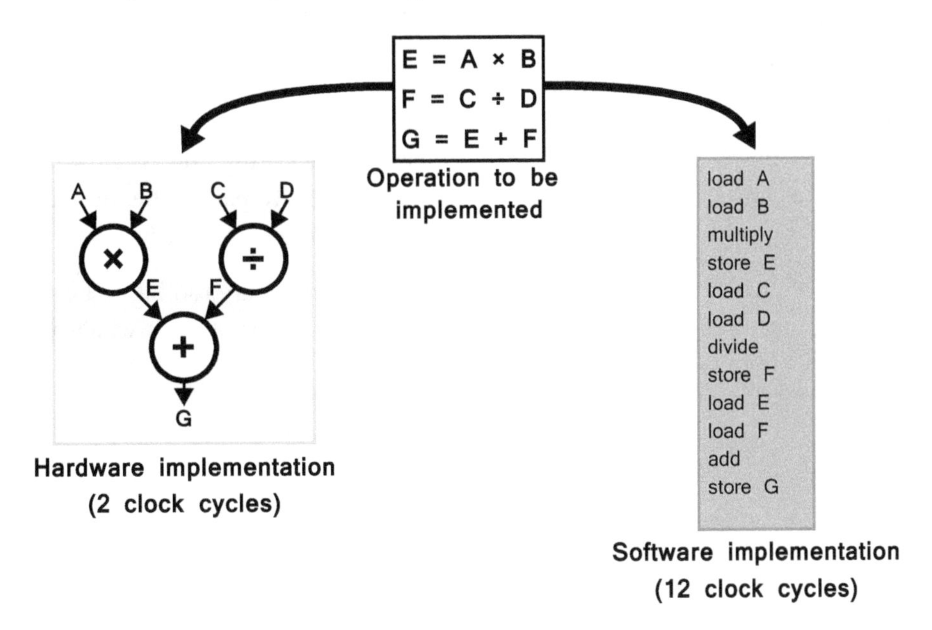

Figure 11.2: Parallel vs sequential implementation

the PL, there will be a communication overhead. With the closely coupled PS and PL of the Zynq-7000 SoCs and the high-speed interconnects, this is less of a problem.

11.2. Profiling

Profiling is a form of program analysis that is used to aid the optimisation of a software application. It is used to measure a number of properties of application code, including:

- Memory usage
- Execution time of function calls
- Frequency of function calls
- Instruction usage

Profiling can be performed statically (without executing the software program) or dynamically (performed while the software application is running on a physical or virtual processor). Static profiling generally performed by analysing the source code, or sometimes the object code, whereas dynamic profiling is an intrusive process whereby the execution of a program on a processor is interrupted to gather information.

The use of profiling allows you to identify bottlenecks in the code execution that may be a result of inefficient code, or poor communication between function interactions with a module in the PL or another function within software. It could also be the case that the algorithm or routine may be inherently more suitable for implementation in hardware. Once identified, the bottlenecks can be optimised by rewriting the original software function or by moving it to the PL for acceleration. Alternatively, part of the function could remain in software, while the problematic section could be moved to hardware.

Profiling is also useful when analysing large programs which are too big to be analysed by reading the source code. The use of profiling can help you spot bugs which you may otherwise not have noticed.

The execution flow of various functions is shown in Figure 11.3, with the number of clock cycles required to execute a given function highlighted.

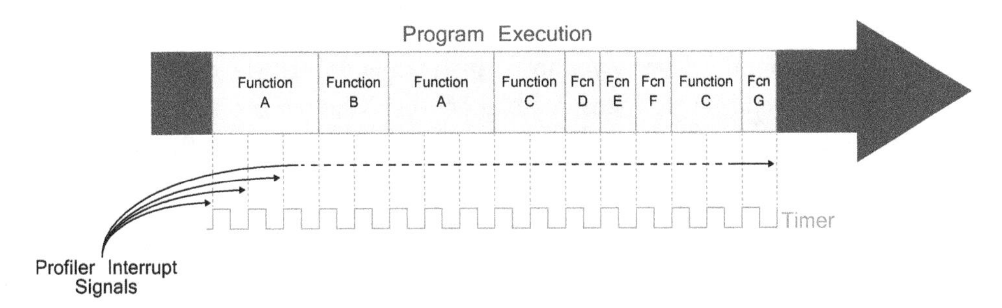

Figure 11.3: Profiling

By profiling a program, and thus determining the number of clock cycles required to execute each individual function, you can determine whether a function needs optimised. Having written a software function, the engineer should have a good idea of roughly how long a function should take to execute on a given PS. This estimate can be compared to the results of the profiling and large discrepancies investigated. An example of the profiling output for the execution flow in Figure 11.3 is provided in Figure 11.4.

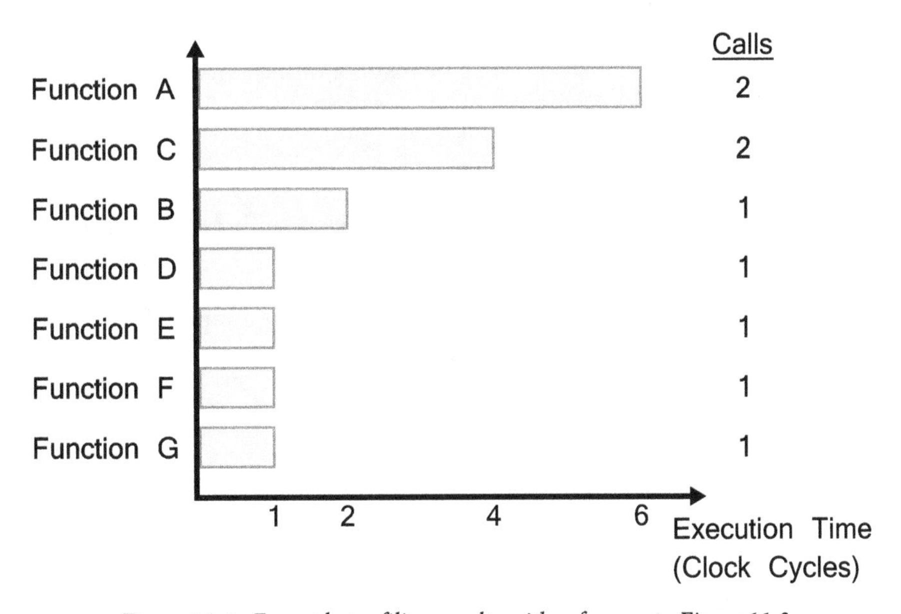

Figure 11.4: Example profiling results with reference to Figure 11.3

11.3. Software Development Tools

Software application development flows for the Zynq-7000 AP SoC devices allow the user to create software applications using a unified set of Xilinx tools, as well as utilising a wide range of tools from third-party vendors which target the ARM Cortex-A9 processors [2]. Although this section will focus mainly on the Xilinx tools and tool flows, the discussed concepts are broadly applicable to third-part tools. Those familiar with software development will recognise familiar components such as the GNU compiler toolchain and Eclipse-based IDE.

11.3.1. Software Tools

Xilinx provides design tools for the development and debugging of software applications for Zynq-7000 AP SoC devices. Provided software includes [2]:

- **A software IDE** - This is an integrated design environment for the development of software applications for execution on the PS.
- **GNU-based compiler toolchain** - Used to convert the source code of your application into an executable program. It is a collection of programming tools based on those produced by the GNU Project, including a GCC compiler, GNU Debugger (GDB) debugger, utilities and libraries.
- **JTAG debugger** - Allows for hardware debugging of the software application whilst running on the target device via a JTAG connection.
- Various other associated utilities

The provided software tools allow the user to develop both bare-metal applications, which run directly on the Zynq-7000 device without an OS, and Linux applications.

Third-party software tools are also available which provide support for the Cortex-A9 processors. They include [2]:

- Software IDEs
- Compiler toolchains
- Debug and trace tools
- Embedded OS and software libraries
- Simulators

- Models and virtual prototyping tools

The level of direct support and integration varies for Zynq-7000 AP SoC devices by third-party solutions. Tool support for targeting Linux Kernel development is not provided by Xilinx, but is provided by third-party vendors.

11.3.2. Hardware Configuration Tools

Xilinx provides two hardware configurations tools which provide support for the Zynq-7000 AP SoC devices. These are:

- **Vivado IDE Design Suite IP Integrator** - IP integrator provides a graphical interface which allows the user to create a block diagram of a system which incorporates both the PL and the PS.
- **ISE Design Suite Embedded Development Kit (EDK) Xilinx Platform Studio (XPS)** - XPS captures information about the PS and any peripherals, including configuration settings, register memory address map and hardware bitstream for the initialisation of the PL.

Vivado IP Integrator

Vivado IDE Design Suite IP Integrator provides a block-based design environment for the construction of Zynq-7000 AP SoC systems. The block diagram allows you to configure both PS and PL components. Configuration information is stored in an XML file and other INIT files which are then used by software design tools to infer compiler parameters, define JTAG settings, create and configure BSP libraries and automate other hardware related operations [2].

More information on Vivado IP Integrator can be found in Chapter 18 — IP Reuse and Integration.

Xilinx Platform Studio

ISE Design Suite EDK Xilinx Platform Studio stores configuration information in an XML file and other INIT files which are then used by software design tools to infer compiler parameters, define JTAG settings, create and configure BSP libraries, program the PL and automate other hardware related operations [2].

11.3.3. Software Development Kit (SDK)

The Xilinx SDK provides an environment whereby fully functioning software applications can be created, compiled and debugged all within one tool. The SDK includes GNU-based compiler toolchain (GCC compiler, GDB debugger, utilities and libraries), JTAG debugger, flash programmer, drivers for Xilinx IPs and bare-metal BSPs and middleware libraries for application-specific functions [2]. All of the features that have been mentioned are accessible from within the Eclipse-based IDE, which incorporates the C/C++ Development Kit (CDK) [2].

Features of the SDK include [2]:

- Project management
- Error navigation
- C/C++ code editing and compilation environment
- Application build configuration and automatic `makefile` generation
- Integrated environment for debugging and profiling embedded targets
- Additional functionality through third-party plug-ins, such as source code and version control

The SDK is available as part of the Xilinx Vivado IDE, the ISE Design Suite and EDK, or as a standalone application. An application template for creating a First-Stage Boot Loader (FSBL), as well as a graphical interface for building a boot image, is also included in the SDK. Documentation and reference material is available directly from the SDK help system [2].

11.3.4. Microprocessor Debugger

The XMD is a command line driven JTAG debugger that can be used to download, debug and verify programs. It includes a Tcl interface that supports scripting for repetition of tasks. XMD serves as the GDB server for GDB and SDK when debugging bare-metal applications [2].

When debugging Linux applications, the SDK interacts with a GDB server running on the target platform. Debuggers can connect to XMD running on the host computer or on a remote machine running on the same network [2].

11.3.5. Sourcery CodeBench Lite Edition for Xilinx Cortex-A9 Compiler Toolchain

In addition, the SDK also includes the Sourcery CodeBench Lite Edition for the Xilinx Cortex-A9 compiler toolchain. This is used for Linux application and bare-metal Embedded Application Binary Interface (EABI) development [2].

The Xilinx Sourcery CodeBench Lite Edition in the SDK comprises of identical GNU tools, libraries and documentation as the standard editions, but with the following additions [2]:

- Default toolchain settings for the Xilinx Cortex-A9 processors
- Bare-metal (EABI) start up support and default linker scripts for the Xilinx Cortex-A9 processors
- Vector Floating Point (VFP) and NEON optimised libraries

11.3.6. Logic Analysers

Analysis tools allow you to analyse internal signals and ports of any custom logic which is running in the PL. Two analysis tools are provided by Xilinx: ChipScope™ Pro Analyzer and Vivado Logic Analyzer. Both tools perform similar functions, but operate within different environments; ChipScope Pro Analyzer is a standalone application whereas Vivado Lab Tool operates within the Vivado IDE.

ChipScope Pro comprises of two tools: ChipScope Pro Core Inserter and ChipScope Pro Analyzer. Analyzer inserts logic analyzer, system analyzer and virtual I/O cores into the PL custom logic designs, while Analyzer allows you to analyse and view the internal signals. All signals are captured at the operating speed of the PL and can be displayed and analysed in the ChipScope debugging tools. Capturing of signal data, and software debugger break-points, can be triggered by events in the custom logic [2].

Vivado Logic Analyzer operates in a similar manner to ChipScope, but accessed from within the Vivado IDE rather than as a standalone application [2].

11.3.7. System Generator for DSP

System Generator for DSP is a block-based development environment that allows users to create DSP hardware co-processor designs within the MATLAB/Simulink environment. It supports rapid development and simulation of DSP hardware, whilst enabling automated generation of the co-processor which reduces the overall system development time. The co-debug feature of the SDK allows users to run and debug programs running on the processor while retaining control over the hardware under development in System Generator [2].

11.4. Chapter Review

In this chapter the concept of hardware/software partitioning has been introduced, whereby the system components are divided between software (running on the PS) and hardware (implemented in the PL). The process of profiling was also discussed, which can aid with identifying software-based bottlenecks within the system. Such software functions could benefit from hardware acceleration, and thus implementation in the PL. The Xilinx Vivado HLS tool can aid with the transition from software to hardware by automating the conversion from C/C++/SystemC algorithms to RTL code.

The Xilinx tools provided for Zynq software development have also been introduced, both in terms of Linux and baremetal development. The hardware configuration tools, such as Vivado IDE IP Integrator and Xilinx Platform Studio were also introduced.

11.5. References

NOTE: All URLs last accessed June 2014.

[1] M. López-Vallejo and J. C. López, "On the Hardware-Software Partitioning Problem: System Modeling and Partitioning Techniques" *ACM Transactions on Design Automation of Electronic Systems (TODAES)*, vol. 8, no. 3, pp. 269-297, July 2003.

[2] Xilinx, Inc, "Zynq-7000 All Programmable SoC Software Developers Guide", UG821, v8.0, April 2014. Available: http://www.xilinx.com/support/documentation/user_guides/ug821-zynq-7000-swdev.pdf

12

Next Steps in Zynq SoC Design

This practical chapter and its corresponding tutorial seeks to build on the basics learned in the previous chapter, expanding on existing skills in developing a Zynq system.

In particular, this chapter will focus on adding further IP to the basic Zynq system, using IP Integrator to handle interrupts generated by a timer and the GPIO on the Zedboard. You will be guided through the process of connecting this additional IP to the system and details of the connections made explained at each stage.

Following generation of a hardware design, it will be exported to the SDK to develop a software application that makes clear use of interrupts and effectively conveys the function.

12.1. Prerequisites

Before beginning this exercise it is recommended you have read up to and including Chapter 10, entitled *Zynq System-On-Chip Development*, and have successfully completed the previous practical exercise *First Designs on Zynq* as this exercise will expand upon the previously created Zynq system.

12.2. Aims and Outcomes

The aim of this practical exercise is to introduce adding further IP to a basic Zynq system and making the appropriate connections. To achieve this, a simple design constructed and targeted to the Zedboard will have additional IP added and connected, before creating a software application that clearly exhibits the operation of the system.

After completion of this tutorial you will be able to:

- Add additional IP to an existing, basic Zynq hardware project.
- Be able to configure the IP to utilise interrupts.
- Amend your design to include multiple sources of interrupts.
- Use an Interrupt Controller to connect hardware interrupts between the IP and the Zynq PS.
- Create more advanced software applications which will execute on the Zynq PL, utilising interrupts and several types of IP.

The main aim of this tutorial is to extend basic skills in developing systems for Zynq, further familiarising the reader with the available software tools and flow, as well as the creation of more complex software applications.

12.3. Overview of Exercise 2A

The first practical exercise is identical to that of the previous tutorial in the creation of a simple Zynq system for the ZedBoard. This is repeated to develop familiarisation with the procedure.

Exercise 2A is available on the website: www.zynqbook.com

12.4. Overview of Exercise 2B

This practical exercise involves the expansion of the previously created hardware project targeting the ZedBoard in Vivado IDE. The required steps are:

1. Add additional IP (GPIO push buttons) to generate hardware interrupts.
2. Connect hardware interrupts to the interrupt controller in the Zynq PS.

3. Generate HDL files for the hardware design and create a bitstream hardware description file.

4. An overview of the hardware design that will be created in this exercise is provided in Figure 12.1

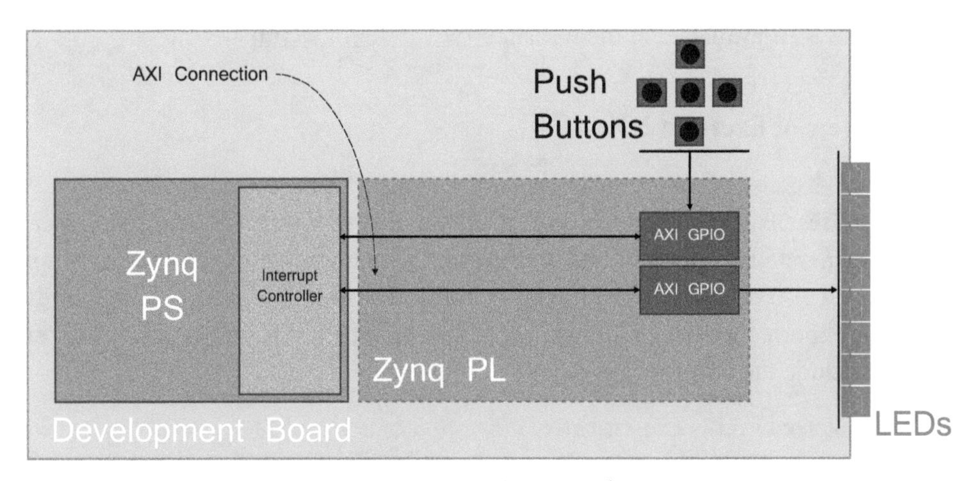

Figure 12.1: Exercise 2B Zynq Hardware Design

With our hardware design from Exercise 1 augmented to include additional IP, the next exercise will focus on the creation of a software application that reveals the operation of a single interrupt in a Zynq system.

Exercise 2B is available on the website: www.zynqbook.com

12.5. Overview of Exercise 2C

Following creation of a hardware design in Exercise 2B, the next step is to export the design to the Xilinx SDK. A software application will then developed, that uses presses of the push buttons to generate an interrupt. This interrupt is handled by the AXI interrupt controller and passed to the Zynq PS. This will increase the value of a variable, creating a counter, the value of which will be displayed in binary on the LEDs by passing and writing the variable value to the AXI GPIO instance,

The steps involved in this Exercise are:

5. Export the finalised hardware design to the SDK.

6. Create a simple software application that uses interrupts from the push buttons to increment a counter, the value of which is displayed on the LEDs in binary.

7. Program the FPGA and confirm that the software application operates as expected, showing the function of interrupts in a Zynq system.

   Exercise 2C is available on the website: www.zynqbook.com

12.6. Overview of Exercise 2D

In this final exercise of the tutorial, we will expand the hardware design created and tested in Exercises 2B and 2C to include a further source of interrupts. An AXI Timer will be added to generate an interrupt and increment the counter on expiration. Appropriate connections will also be made to the interrupt controller. Finally, the augmented hardware design will be exported to the SDK and the software application from the previous exercise modified to include the new functionality.

The steps involved in this Exercise are:

1. Open the previously created project in Vivado IDE.

2. Add an AXI Timer source using IP Integrator.

3. Make appropriate connections between the timer and the hardware and interrupt controller in the PS.

4. Generate HDL files for the hardware design and create a bitstream hardware description file.

5. An overview of the hardware design that will be created in this exercise is provided in Figure 12.2

6. Export the finalised hardware design to the SDK.

7. Modify the software application of the previous exercise to include new interrupt functionality.

8. Program the FPGA and execute the software application on the hardware, confirming that it operates as expected.

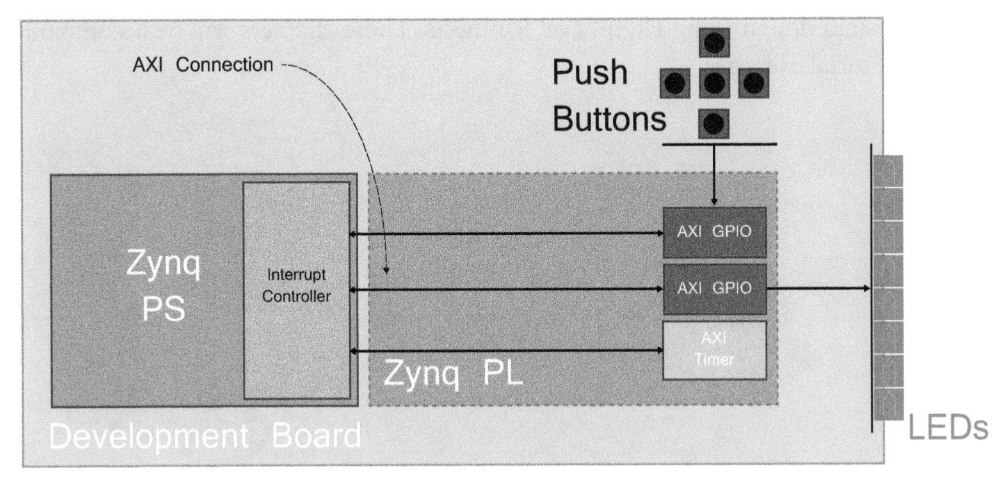

Figure 12.2: Exercise 2D Zynq Hardware Design

Exercise 2D is available on the website: www.zynqbook.com

12.7. Possible Extensions

Having completed Exercise 2D, there are some possible variations you can introduce to personalise the developed system. For example, you could:

- Alter the timer expiration value or push button incremental values.
- Add a further GPIO controller to the hardware design which connects to the DIP switches on the ZedBoard. Use the GPIO driver functions to take input from the DIP switches to control the following options:
  - Pause or resume the timer operation
  - Whether the counter increments or decrements
  - Selection between different timer expiration values

12.8. What Next?

The next few chapters of this book will focus on the creation of custom IP blocks and abstracted system design using High Level Synthesis. These chapters will be accompanied by relevant tutorial exercises.

13

IP Block Design

IP plays a very important role in today's FPGA and embedded system industry, allowing system designers to pick-and-choose from a wide array of pre-developed design blocks. This has a number of advantages in terms of development time, as well as providing guaranteed functionality without the need for extensive testing.

In this chapter we will take a look at what intellectual property means in terms of the Zynq-7000 platform, the industry trends and what sources of IP are available. We will also look at the Xilinx design flows that are available for the creation and maintenance of your own IP catalogue.

13.1. Overview

A sensible starting point when dealing with IP is to define what we mean when we use the terms "*IP block*" and "*IP core*". An IP core, or IP block, is a hardware specification that can be used to configure the logic resources of an FPGA or, for other silicon devices, physically manufacture an integrated circuit [3]. In terms of IP cores, there are two types: hard IP cores and soft IP cores.

Soft IP cores allow the end user to customise the IP to a certain degree. The extent of the customisation is dependent on the exact format that the IP is delivered in. The highest level of customisation is available when the soft core is provided as synthesisable RTL, i.e. the source HDL code is provided. For those familiar with software programming, this is equivalent to a computer program being provided in high-level source code, such as C/C++. The user is free to make modifications to the HDL source code before synthesising (or compiling, to continue the software comparison) it for implementation on a target device. It is worth noting that most IP vendors will offer no support or warranty for IP designs which have been modified. In order to provide protection against modification, RTL can be provided in encrypted source form. Although the source files can not be modified, the IP can still be customised through the use of parameters.

Another format which soft cores can be provided in, is as a gate-level netlist. This means that the IP vendor has partially synthesised the individual IP component and will provide you with a logic gate implementation of the IP's functionality. This, although still customisable, makes any changes to the functionality of the IP harder to implement. As such, IP delivered as a netlist provides the IP vendor with a certain degree of protection from revealing the underlying algorithms and processes. This is somewhat analogous to a software developer compiling the C source code and providing an assembly code listing; the code is still modifiable, but it must be done at a lower level.

Both of these IP delivery methods are considered to be *soft* as they allow the user to control the synthesis, place and route design flow when targeting the end device. Therefore, ultimate control over the placement of the IP functionality, and the amount of hardware resources used on the target device lies with the end user.

Hard IP cores, on the other hand, are provided in formats that permit no realistic method of modification by the end user, and in some cases the functionality of the IP is realised in the silicon of a device at manufacture.

One method of delivery for hard IP which is targeted at FPGA/ASIC implementation, is to provide a design which has already undergone full synthesis and place and route design flows. This is also sometimes referred to as *firm IP.* As the IP has already undergone the place and route stage of the design flow, each individual IP must be targeted at a specific end devices, or family of devices, and cannot be easily ported to another device. There is also the disadvantage that a specific area of the device must be used for the IP implementation. Although this has the disadvantage that the functionality of the IP cannot be customised in any way, it also brings with it the advantage that there is a high level of predictability in terms of timing performance and required hardware area.

Another method of delivery for the implementation of hard IP in silicon is to provide a transistor-layout. This is a format which is only compatible with the process design rules of a specific foundry, and therefore hard IP delivered for use in one foundry cannot be used in another without undergoing a difficult porting process. Some larger foundry operators even go as far as to provide hard IP cores which specifically target their own foundry process design rules in order to ensure that customers must use their services. This form of hard IP delivery is not applicable to FPGAs.

Xilinx provides an extensive catalogue of soft IP cores for the Zynq-7000 AP family which are optimised to provide efficiency in terms of both performance and hardware area. Functionality ranges from building blocks, such as FIFOs and arithmetic operators, all the way up to fully functional processor blocks, such as the MicroBlaze processor core.

Third-party IP is also available, both commercially and from the open-source community. Third-party IP comes in two forms: vendor-specific and generic. Vendor-specific IP should require little-to-no modification to integrate with your chosen Zynq device. Generic IP, however, may require modifications in order to make them conform with certain HDL naming conventions.

The final option for obtaining IP for your design is to create it yourself. The traditional method of IP generation is for it to be developed in HDLs, such as VHDL or Verilog. More recently, other methods of IP creation have been introduced to the Xilinx tool suite, such as System Generator model-based design or Vivado HLS. Other third-party tools also exist.

13.2. Industry Trends and Philosophy

In recent years, with the demand for more complex SoCs growing steadily, system designers are being forced to incorporate more and more IP into their products. With the more complex systems, this can be anything up to hundreds of individual IP cores. With this many discrete pieces of IP in a single product, the ability to manage each block individually becomes much harder. The solution to this problem is to combine IP cores of a similar key functionality to make a single IP subsystem.

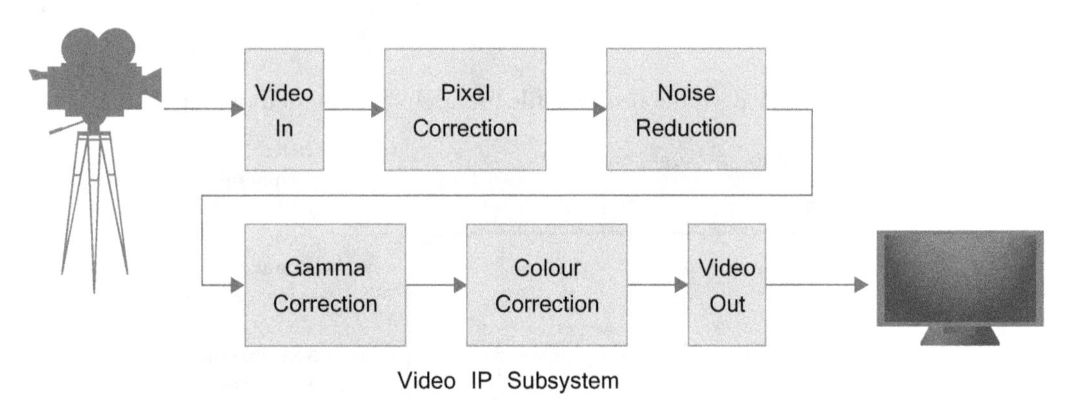

Figure 13.1: Video IP subsystem

A good example of an IP subsystem is a video processing pipeline, as shown in Figure 13.1. The individual video IP blocks — video in, pixel correction, noise reduction blocks etc — are combined together to create a single video IP subsystem which can be integrated into an embedded system design as a single IP block, rather than 6 individual IP cores.

13.3. IP Core Design Methods

Now that we have introduced the concept of IP cores and the types of IP that are available, we will look at the various ways in which we can make our own IP. Xilinx provide a number of tools which enable the creation of custom IP blocks for use in your own embedded system designs.

13.3.1. HDL

Hardware description languages, such as VHDL and Verilog, are specialised programming languages which are used to describe the operation and structure of digital electronic circuits. The ability to create IP cores in HDL allows you the maximum control over the functionality of your peripheral. If you already have an existing HDL design, the ability to turn it into an IP core allows you to take advantage of the benefits of an IP module without having to redesign or use a third-party equivalent.

When designing IP cores in HDL which are to communicate via an AXI interface, care must be taken to conform to the Xilinx IP Packager peripheral signal naming conventions. The top-level HDL file for an IP core defines the design interface and lists the default connections and ports for the bus interface. It also lists any generics and specifies their default values.

An example of the HDL peripheral source file hierarchy is provided in Figure 13.2.

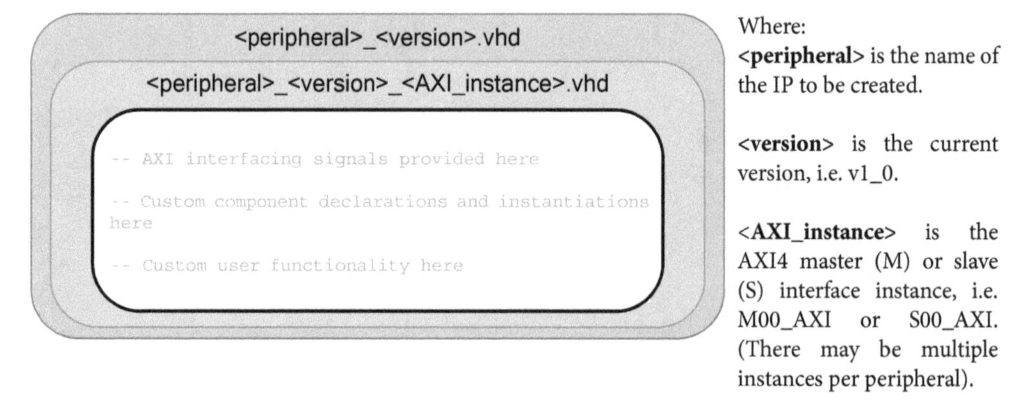

Figure 13.2: HDL peripheral source file hierarchy

The major drawback to creating IP in HDLs, is that complex designs require an experienced engineer to produce an optimal solution. The process can be very time consuming in terms of both development and testing, leading to greater time-to-market. However, when looking for IP which requires very tight timing or hardware constraints, or if the functionality is complex, HDL is usually the best option.

13.3.2. System Generator

System Generator, typically used for DSP design, is a design tool that utilises the Mathworks Simulink design platform for FPGA system design. It provides a high-level, model-based environment for the creation of hardware designs.

With the introduction of the Vivado Design Suite to the Xilinx product catalogue, a new compilation target was introduced to System Generator. IP Packager compilation allows you to package your System Generator design into an IP module which can be included in the Vivado IP Catalog. The System Generator design can then be used like any other module from the IP Catalog and instantiated into a Vivado user design [4]. Further, System Generator is fully integrated into the Vivado design flow, allowing an IP Integrator IP block to be created directly from Vivado.

System Generator first generates an HDL Netlist from the user designed model. Any included Vivado IP modules are automatically copied across to subfolder named *"IP"*, before all RTL design files and Vivado IP design files are packaged into a ZIP file, which is placed in a subfolder named *"ip_packager"*.

Testbench and HTML documentation files can also be automatically generated by the IP catalog compilation tool. The testbench file allows the IP to be fully simulated from a Vivado IDE project. The documentation file includes information about the IP and how it should be interfaced in Vivado.

System Generator offers an intuitive environment for the design of IP where blocks can be connected together to create designs quickly and easily. A large number of blocks are available from simple mathematical operators to complex DSP operations. Functionality is not always implemented in the most readable form in the HDL code, and this can make some designs hard to follow. Depending on the options chosen within the System Generator design, the IP implementation may not be as efficient as if hand coded in HDL.

13.3.3. HDL Coder

HDL Coder is a tool from MathWorks which enables the generation of synthesisable HDL code (both VHDL and Verilog) from MATLAB functions and Simulink models. It provides a workflow which will analyse a MATLAB/Simulink model and automatically convert the system from floating point to fixed point, providing a high level of abstraction. This allows the user to concentrate on the development of algorithms and models without lower-level intricacies of HDL design.

The HDL Coder workflow also provides tools for HDL code verification, allowing the generated HDL code to be tested alongside the original MATLAB/Simulink model. This allows any errors that may be present in the generated HDL code to be easily to identified and fixed.

The HDL code optimisation provided in the workflow provides the option of specifying the target FPGA device in order to offer a large amount control over the implementation allowing you to highlight critical paths, control the HDL architecture and generate hardware resource utilisation estimation.

Once generated by HDL Coder, the HDL code can be used to create an IP core in the same way as detailed in Section 13.3.1.

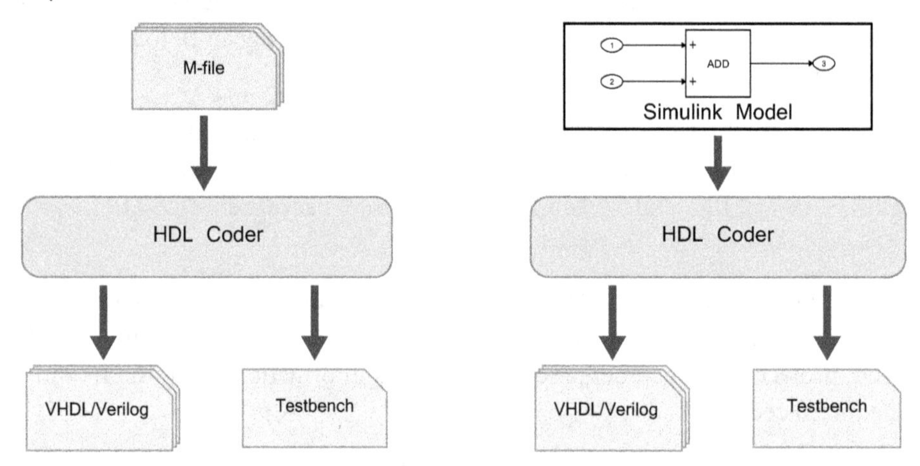

Figure 13.3: HDL Coder flow

While HDL Coder allows you to create IP from existing MATLAB/Simulink models quickly and with minimum design changes, there are some downsides. At present, not all MATLAB functions and Simulink blocks support HDL generation. This means that some of the functionality may have to be reworked with supported functions or blocks. Another point worth noting is that, even though a large amount of customisation in terms of hardware implementation and optimisation is available, the HDL code that is generated may not be as efficient as if it were created from hand coded HDL. The generated code can sometimes be complicated to follow in terms of readability, and in some cases hard to customise.

13.3.4. Vivado High-Level Synthesis

Vivado HLS is a tool provided by Xilinx, as part of the Vivado Design Suite, which is capable of converting C-based designs (C, C++ or SystemC) into RTL design files (VHDL/Verilog or SystemC) for implementation of Xilinx All Programmable devices. An overview of the Vivado HLS design flow is provided in Figure 13.4.

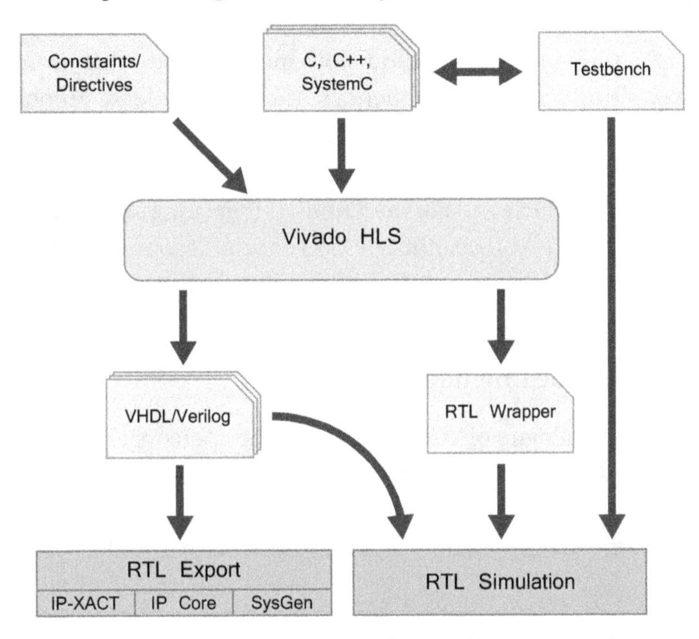

Figure 13.4: Vivado HLS flow

From Figure 13.4 we can see that there are three distinct forms of RTL output available from the Vivado HLS flow. These are outlined below [6]:

1. **IP-XACT** — IP-XACT is a public specification for documenting design IP which was developed by the SPIRIT Consortium. It is a widely-adopted XML schema for the description of IP which is tool-independent and machine-readable [1].
   IP-XACT is the option that should be chosen to allow your IP design to be imported into the Vivado IP Catalog [6].
   IP-XACT is covered in greater detail in Chapter 18 — IP Reuse and Integration.

2. **IP Core** — When this option is selected, your IP will be exported in a format which can be exported into XPS.

3. **SysGen** — This option allows you to export the resulting RTL files as a block which can be used in System Generator designs.

Vivado HLS will be covered in greater detail in Chapter 14 — Spotlight on High-Level Synthesis.

13.3.5. Choosing the Right IP Creation Method

When choosing an IP creation method you must decide what is most important: achieving the maximum operating frequency; utilising the least amount of hardware resources; minimising development time; or a middle ground. You must also consider what methods you or your team have most experience with, and the learning curve required with using a method with which you are not familiar. If existing simulation models exist in a certain format, be it MATLAB/Simulink or C/C++, chances are you may want to utilise them and use a method which can create IP from them such as HDL Coder or Vivado HLS.

13.4. Simulation and Documentation

When building up a catalogue of custom IP, for either personal or commercial use, it is important to ensure that each module functions correctly; the last thing you want is for one of your customers to contact you to inform you that the IP you sold them doesn't work! The best way to ensure that everything is performing as it should is through extensive simulation.

It is also important that each IP module is properly documented so that when a customer wants to include your IP in a design, they know exactly how it works. This, of course, also applies to you using your own IP a few years down the line; thorough documentation acts as a good aid to remind you how your own IP works!

In this section we will cover the simulation and documentation of IP with the tools available in the Vivado Design Suite, as well as relevant third-party tools.

13.4.1. Simulation

Simulation of your IP is extremely important at a number of different points in the design stage. If creating your IP from a non-RTL source, you must first ensure that your design source is performing correctly. Of course, the simulation process will differ depending on the source of your IP. Simulation of each of the IP sources that have been covered in this chapter will be detailed in this section, along with the tools that will help you to validate the final RTL IP files.

RTL Simulation

There are a number of options available to you when simulating RTL files, the first of which is to use the built-in simulator in Vivado. The Vivado IDE includes a HDL simulator which allows you to simulate your design at various points in the Zynq-7000 design flow.

The Vivado simulator supports the following languages [5]:

- Verilog IEEE-STD-1364-2001
- VHDL IEEE-STD-1076-1993
- Standard Delay Format (SDF) version 2.1
- VITAL-2000

Simulation of SystemC files is not available within the Vivado IDE, but instead can be performed by third-party RTL simulation products or in the Vivado HLS tool which will be covered separately in the *Vivado HLS* section.

The following features are supported by the Vivado simulator [5]:

- Source code debugging
- Value Change Dump (VCD) file dumping
- SDF annotation
- Native support for Hard IP blocks
- SAIF dumping for power analysis and optimisation
- Multi-threaded compilation
- Mixed language simulation (VHDL and Verilog)
- Real-time waveform update
- Built-in Xilinx simulation libraries

As well as the built-in Vivado simulator, Xilinx supports the follwing third-party simulators:

- Mentor Graphics QuestaSim/ModelSim (integrated in Vivado IDE)
- Cadence® Incisive® Enterprize Simulator (IES)
- Synopsys VCS® and VCS MX
- Aldec Active-HDL and Rivera-PRO (compatible but not officially supported by Xilinx Technical support.)

An example of the Vivado RTL simulation flow is provided in Figure 13.5.

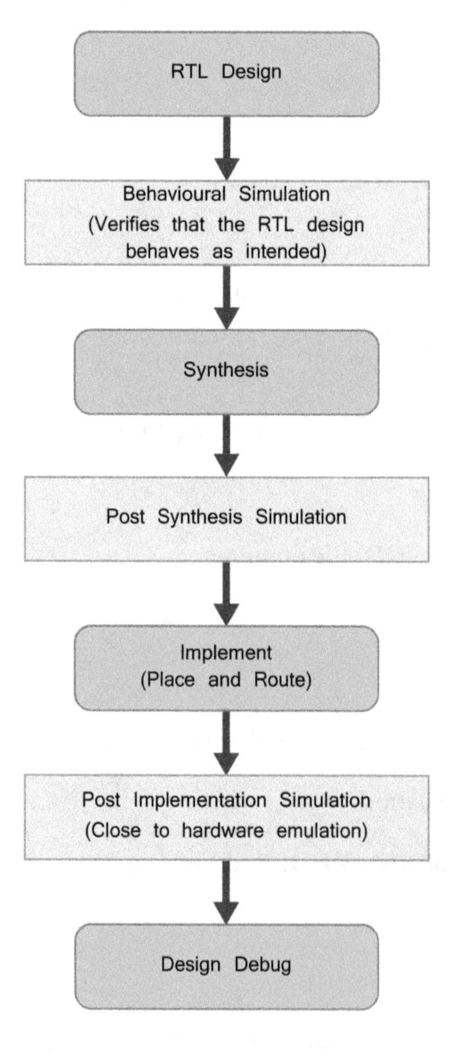

Figure 13.5: Vivado RTL simulation flow

System Generator

As System Generator is based on the Simulink model-based platform, simulation of System Generator designs is a relatively simple procedure. Once all the blocks are connected and the timing parameters have been set, a simulation can be executed by simply clicking the "run" button. Simulation data can be presented in various ways, such as the in-built *Scope* blocks which can display both time- and frequency-domain plots. Data can also be exported to the underlying MATLAB framework for further analysis and display using the powerful plotting tools.

Once RTL code for your IP has been generated, the System Generator tools provide the ability to perform hardware co-simulation for the validation of the design. This allows you to incorporate a design running on an FPGA directly into a System Generator simulation, allowing you to provide the same stimulus to both the System Generator model, and the implemented RTL code, simultaneously. The output from both the simulation and the hardware co-simulation can then be gathered and compared, allowing you to check the validity of the implemented RTL design.

HDL Coder

As HDL Coder works within MATLAB and Simulink, the simulation of all IP designs is identical to the process of simulating a M-code file or a Simulink model. This allows you to present your simulation results in the same way as any other MATLAB/Simulink simulation, using the extensive plotting tools.

With RTL code generated by HDL Coder, the functionality can be compared to that of the original IP model using the HDL Verifier tool. Multiple techniques for verification are supported, including co-simulation of the design by interfacing with HDL simulators (such as ModelSim and QuestaSim from Mentor Graphics and Incisive from Cadence) as well as providing hardware co-simulation via FPGA-in-the-loop. The FPGA-in-the-loop simulation is carried out by automating a process that creates and runs an FPGA implementation on a development board. The process of FPGA-in-the-loop verification is similar to that described in the System Generator simulation section.

Using the HDL Verifier HDL co-simulation and hardware co-simulation allows you to provide both the MATLAB/Simulink model and the generated RTL code with the same input stimulus. Outputs from both designs can then be compared in order to validate the design and identify any errors.

Vivado HLS

HLS automatically creates the scripts that are required to perform co-simulation of the generated design with the original C testbench file [6]. This allows both the original C-based design and the generated RTL files to be tested with the same input stimuli, and the corresponding outputs compared in order to find any discrepancies. This removes the need to create an RTL testbench file.

The following RTL simulators are supported by HLS [6]:

- ModelSim
- VCS
- Open SystemC Initiative (OSCI)
- NCSim
- XSim
- ISim
- Riviera

The VCS, NCSim and Riviera HDL simulators are only supported on the Linux operating system [6]. If SystemC code is generated, the built-in SystemC kernel can be used for verification.

Fixed point data types are supported by C++ and SystemC simulations, and when used simulation results will match that of the implemented RTL files. This allows the effects of bit-accuracy, quantisation and overflow to be analysed with fast C-level simulation [6].

As part of the RTL co-simulation process, HLS generates a SystemC wrapper that creates adapters around the RTL modules. The C code wrapper is then instantiated into the existing C testbench file. An example of a co-simulation wrapper is shown in Figure 13.6.

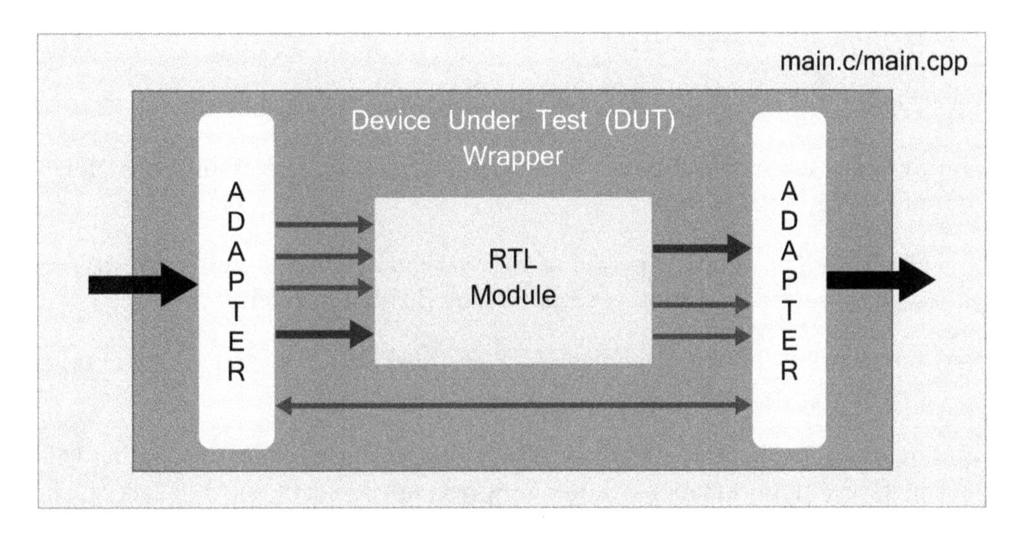

Figure 13.6: RTL co-simulation wrapper

13.4.2. Documentation

Documentation is an important component of any IP, whether it be commercial, open-source or internal. Good IP documentation should allow a user who has never used a specific IP module to understand, connect and implement it within their own design without any further help. Documentation also acts as a good reminder for a user, even if they have created the IP themselves.

In this section, the documentation features of the IP creation tools and methods that have been detailed in this chapter will be reviewed.

System Generator

System Generator has an option to create interface documentation when generating IP. This is an HTML document which specifies the interface to the generated IP design, and comprises of the following sections:

- **Introduction** — Provides a brief introduction to the document.
- **Port Interface** — This section details the port interface of the generated IP. A table lists all of the input and output ports of the design with the following detail:
  - ***Name*** — Top-level port names.

- ***Direction*** — Port direction, e.g. in, out or inout.
- ***HDL Type*** — Port HDL type, e.g. std_logic, std_logic_vector.
- ***Type*** — The connected signal type, e.g. data, clock, clock enable.
- **System Generator Type** — The signal type within System Generator, e.g. ufix16_15, sfix12_11 etc.
- ***Period*** — The sampling period of a given signal. Information of multi-rate realisation is provided in another section of the HTML document.
- ***Comment*** — Further comments are provided here if included in the original System Generator design.

- **Multi-Rate Realization** — If a multi-rate design is generated, this section will be populated with information regarding the network of clock enable signals which are created to control the various clock signals throughout the design. A generic timing diagram is also included to help explain the realisation of the different clock domains.

- **Design Files** — This section provides a list of all of the HDL files that were created by System Generator during the IP generation process. The list is displayed in the order of the top-level module down to the lowermost-level, in order to aid with design compilation. A brief description of each HDL file is also provided, along with a simple block diagram of the design.

- **Design Statistics** — A table is presented in this section which contains the settings that were specified in the System Generator block for design generation. The settings listed include:
  - ***Compilation target*** — Target for the compilation, e.g. IP Catalog, HDL Netlist or Hardware Co-simulation.
  - ***Part*** — The target Xilinx FPGA/Zynq part.
  - ***Synthesis Tool*** — Target synthesis tool, e.g. Vivado or ISE.
  - ***Multirate Implementation*** — Method of multirate implementation, e.g. clock enables.

- **Tools** — A list of the tools and their versions used to generate the design is provided here.

HDL Coder

When generating an IP core from either a MATLAB algorithm or Simulink model, HDL Coder provides the option to generate an *HDL Code Generation Report*. This is an HTML report that includes the following sections [2]:

- **Summary** — This provides a brief overview of the basic settings used to generate the IP and information on the generated IP. Information provided here includes [2]:
    - IP core name and version.
    - Target IP directory.
    - Target language, e.g. VHDL or Verilog.
    - Source model name and version.
- **Target Interface Configuration** — This section contains information Processor/ FPGA synchronisation mode (either free running or coprocessing) and the various interfaces that were specified when creating the IP. This details [2]:
    - ***Port name*** —The names of the input and output ports of the IP.
    - ***Port type*** — Port direction type, e.g. input or output.
    - ***Port data type*** — The data type used within MATLAB/Simulink, e.g. ufix16_15, sfix12_11 etc.
    - ***Target platform interfaces*** — The interface type, e.g. AXI4-Lite, AXI4-Stream or external port.
    - ***Bit Range / Address / FPGA pin*** — Depending on the selection for Target platform interfaces, this section contains the bit range, address of processor-accessible registers or the specified FPGA pin for external ports.
- **Register Address Mapping** — Depending on the interface that was specified for the IP module, this section contains details of any embedded processor-accessible registers in the design. This section includes any register names, their corresponding address offsets from the base address and a brief description of their operation.
- **IP Core User Guide** — This section provides a brief overview of the operation and connectivity of the generated IP. It comprises three subsections [2]:

- ***Theory of Operation*** — Provides a brief overview of the operation of the device, including the master-slave control of the IP and the register-accessible reset and enable signals.

- ***Processor/FPGA Synchronisation*** — A description of the FPGA synchronisation mode (free running or co-processing) is provided here, as well as the data read/write procedure between the IP core and the processor.

- ***Environment Integration*** — Information on how to integrate the newly generated IP into the Xilinx development environment is provided here.

- **IP Core File List** — Hyperlinks to the various files that are generated by HDL Coder are provided here along with a hyperlink to the IP core directory.

13.5. Chapter Review

In this chapter we have introduced the concept of intellectual property and the use of IP blocks, as well as the recent industry trend of IP subsystems. The various ways in which the Vivado Design Suite enables you to create and maintain your own IP catalogue have been introduced, including HDL, Vivado HLS and System Generator. The third-party tool, MathWorks HDL coder has also been discussed.

Importance has been placed on the need for proper documentation, simulation and testing of all IP created, and the various simulation processes for all of the previously introduced IP creation tools have been highlighted. The provided documentation creation options incorporated in corresponding tools have also been detailed.

13.6. References

NOTE: All URLs last accessed June 2014.

[1] "IEEE Standard for IP-XACT, Standard Structure for Packaging, Integrating, and Reusing IP within Tool Flows", IEEE Standard 1685-2009, February 2010.

[2] MathWorks, "HDL Coder User's Guide", R2014a, March 2014. Available via: http://www.mathworks.co.uk/help/

[3] R. Sass and A. G. Schmidt, "Introduction" in *Embedded Systems Design with Platform FPGAs: Principles and Practices*, 1st. Ed, Morgan Kaufmann, 2010, pp 1 - 42.

[4] Xilinx, Inc, "Vivado Design Suite User Guide: Model Based DSP Design using System Generator", UG897, v2014.1, April 2014.

Available: http://www.xilinx.com/support/documentation/sw_manuals/xilinx2014_2/ug897-vivado-sysgen-user.pdf

[5] Xilinx, Inc, "Vivado Design Suite User Guide: Logic Simulation", v2014.2, UG900, June 2014. Available: http://www.xilinx.com/support/documentation/sw_manuals/xilinx2014_2/ug900-vivado-logic-simulation.pdf

[6] Xilinx, Inc, "Vivado Design Suite User Guide: High-Level Synthesis", UG902, v2014.1, May 2014. Available: http://www.xilinx.com/support/documentation/sw_manuals/xilinx2014_2/ug902-vivado-high-level-synthesis.pdf

14

Spotlight on High-Level Synthesis

One of the significant trends in digital systems design is the move to accelerate development cycles, crucially without compromising on verification. This is inevitably driven by commercial factors, and in particular the need to take new products to market as quickly as possible. Of course, any strategies that reduce the design effort are also attractive due to the implied development cost savings.

The concepts underpinning design acceleration strategies are (i) *design reuse*, and (ii) *raising of the abstraction level*. Both of these themes have been introduced already in the book, and here we focus on the latter in particular, by placing the spotlight on Xilinx *Vivado High Level Synthesis* (or *Vivado HLS*). This is a tool for synthesis of digital hardware directly from a high level description developed in C, C++, or SystemC, i.e. removing the need to manually create a version of the design targeted to hardware, such as a VHDL or Verilog description. Rather, the HLS tool undertakes this task. The defining aspect of HLS is that the designed functionality and its hardware implementation are kept separate — the C-based description does not implicitly fix the hardware architecture, as is inherently true in RTL-level design — and this provides great flexibility.

We will see that working in high level, C-based languages has the potential to significantly reduce design time, due to the raised abstraction level. Additionally, the HLS process provides an integrated mechanism for generating and assessing variations on the hardware implementation, making it easy and convenient to find the best architecture.

This chapter will continue by defining and motivating the need for HLS, while also briefly reviewing the historical context and input languages used. The Vivado HLS tool will be introduced, and the design flow and processes of HLS reviewed from a high level perspective. This will be followed by a more in-depth look at Vivado HLS in Chapter 15.

14.1. High-Level Synthesis Concepts

Before proceeding to a discussion of the Vivado HLS tool and practical design methods in Chapter 15, it is important to establish some of the underpinning concepts. We will start with the most basic question!

14.1.1. What is High-Level Synthesis (HLS)?

It is worth beginning with a clear definition of *high-level synthesis*, and to do this we must first review the concept of abstraction with reference to digital design for FPGAs.

As in many other contexts, abstraction means 'taking away'. This is effectively the hiding of detail — the higher the level of abstraction, the more detail is hidden. There are various models for abstraction presented in the literature, many of which derive from the Gajsku and Kuhn 'Y-chart' for Very Large Scale Integration (VLSI) design [9], published as long ago as 1983! However, here we will focus our discussion on current FPGA design practices, and consider that there are four levels of abstraction, in ascending order: *structural*, *RTL*, *behavioural*, and *high-level*, as depicted in Figure 14.1.

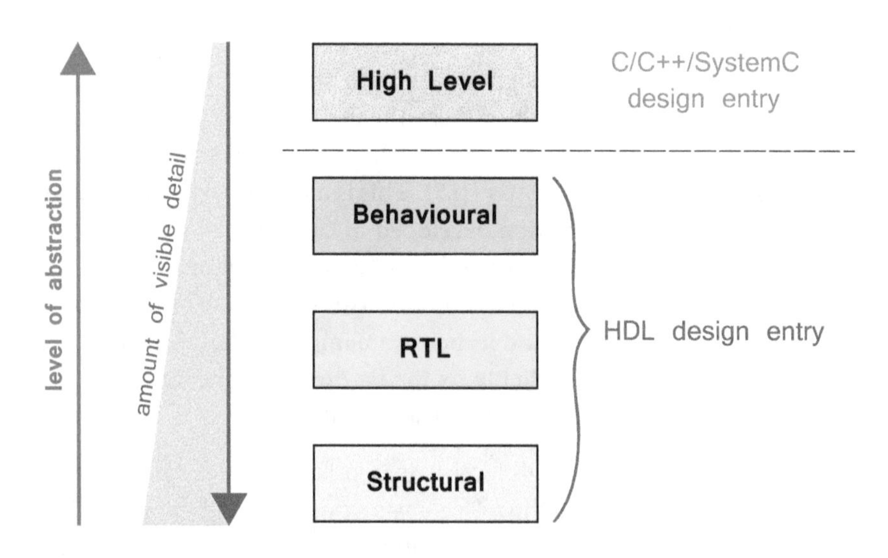

Figure 14.1: Levels of abstraction in FPGA designs

In terms of design entry using HDLs, such as VHDL or Verilog, the lowest level of abstraction (*structural*) involves explicitly instantiating, configuring and connecting each hardware element that comprises the design. This may even extend down to the level of LUTs and FFs. This style represents the designer taking explicit control over all of the design details, which reduces the potential for the synthesis process to optimise the design. More often, designers work at the *'Register Transfer Level' (RTL)* level. This level of abstraction hides the technology-level details, but still represents a description style from which registers, and the operations occurring between registers, can be interpreted. Logic synthesis tools are designed to work at this level of abstraction, i.e. to translate RTL code to hardware. *Behavioural* HDL descriptions are at a higher level of abstraction than RTL, and constitute an algorithmic description of the circuit (i.e. how it 'behaves'), rather than an expression of individual operations with respect to registers. The use of behavioural HDL, therefore, places a greater emphasis on the ability of the synthesis tool to infer hardware from the description, and as such the designer concedes some control over the final implementation, but with the implied benefit that designs can be created more rapidly. Finally, the *high-level* design entry method discussed in this chapter is not actually rooted in HDL, but uses a set of languages suited to expressing designs at an algorithmic level of abstraction, namely C, C++, and SystemC. These languages are, in fact, often used to develop System Level Models (SLMs) as a first, high-level step of modelling a system [22].

In the above discussion, we have used the term 'synthesis' in a slightly different context from HLS. *High-level synthesis* is distinct from other types of synthesis. In FPGA design, 'synthesis' usually refers to logic synthesis, i.e. the process of analysing and interpreting HDL code and forming a corresponding netlist. High-level synthesis actually means synthesising the high-level C, C++ or SystemC code into an HDL description, which would thereafter undergo logic synthesis to obtain a netlist. In other words, high-level synthesis and logic synthesis are both applied (one after the other) when taking a Vivado HLS design towards a hardware implementation, as shown in Figure 14.2.

Furthermore, *physical* synthesis refers to the conversion of HDL to a netlist with explicit knowledge of the target FPGA, and hence the optimisation of the design against the physical characteristics and available resources of the device.

14.1.2. Motivations for High-Level Synthesis

With a high-level representation abstracting low-level detail, the implication is that the description of the circuit becomes simpler. This approach does not actually remove the need to specify aspects such as wordlengths, parallelism and sharing, but to a large extent, the designer *directs* the process and the tools implement the details. The result is that designs can be generated much more rapidly than using more traditional methods, which

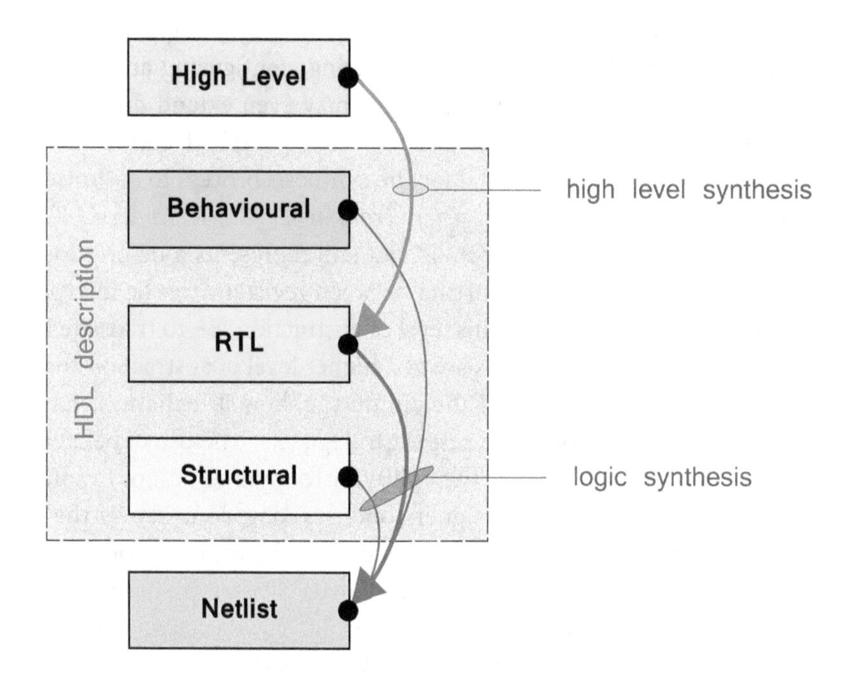

Figure 14.2: High level synthesis and logic synthesis

require the designer to explicitly describe all aspects. Importantly, the separation of functionality and implementation implied in HLS means that the source code does not fix the architecture. Variations on the architecture can be created quickly by applying appropriate directives to the HLS process, rather than having to fundamentally rework the source code, as would be necessary in RTL-level design.

Of course, the implication of giving up full control relies upon the designer trusting the HLS tools to implement lower-level functionality correctly and efficiently, and understanding how to exert influence over this process. There is however a clear motivation to do so, as the potential to accelerate the design process in this way is considerable.

On a more practical level, high-level synthesis from software languages is convenient for many designers, as these languages are widely adopted for developing algorithms and system-level descriptions. It would therefore be useful to leverage the experience of the many engineers comfortable with C, C++ and other programming languages for the purposes of FPGA design. The facility to rapidly convert from software to hardware design languages (crucially, without the need to create and validate an equivalent hardware-targeted design) therefore constitutes a simplification of the design process [22]. The transition from software description to hardware design would normally enforce the distinct design step of developing HDL and validating it against the software-based reference design. With this approach, the two processes can be combined into one. HLS is also beneficial in terms of systems development and software/hardware partitioning, as there is a common language for targeting both elements of the system; this makes it easy to adapt and retarget parts of the design as it is iterated to completion.

As we will see in detail later, the automation present in high-level synthesis tools permits the designer to rapidly produce alternative versions of a given design. For instance, variations on the same design could be produced with different levels of parallelism, allowing the design space to be explored, trade-offs evaluated, and the best solution identified.

In summary, if the motivation for HLS had to be expressed in one word, it would be 'productivity'!

14.1.3. Design Metrics and Hardware Architectures

An important point highlighted earlier in this chapter is that the functional description and the implementation are separate in HLS. Thus, the designer has the opportunity to evaluate possible architectures generated by the HLS process, and optimise according to his or her requirements. This statement does however beg the question, "*What is the designer trying to optimise?*".

In simple terms, hardware design has two fundamental metrics:

- ***Area***, or ***resource cost*** — the amount of hardware required to realise the desired functionality; and
- ***Speed***, or specifically ***throughput*** — the rate at which the circuit can process data.

There are also other factors, and in some cases subtle relationships between them (to be reviewed further in Section 14.4.4), but these are the two most important ones.

The designer is always faced with a trade-off between area (resource cost) and throughput, and this is the key aspect evaluated and optimised during HLS. There may be a certain minimum throughput or maximum area that the system can accommodate, and therefore the task is to decide on the best solution for their particular problem. HLS helps the designer to accomplish this quickly.

This point can be expanded upon using the analogy of decorating a house, and let us assume that the whole house needs decorated. The work could be undertaken by a single painter, and it would take him (or her) 6 weeks. On the other hand, 6 painters could finish the job in 1 week, or 3 painters would require 2 weeks, and so on. Various 'solutions' are possible in the trade-off between *resources* (number of painters) and *throughput* (the reciprocal of the time take to complete the job). A badly organised job might involve 5 painters taking 4 weeks to complete the decoration of the house, and this would of course represent a poor solution!

The realisation of a hardware architecture is similar. At a high level of abstraction, functionality is implemented using units of processing resource. Using more units allows processing to complete more quickly (high cost and high throughput), whereas at the opposite end of the scale, a single processing resource could be deployed and reused over time (low cost and low throughput). Several other points in the trade-off could also be attained. Of course, as with the decorating example, it would also be possible to come up with a poor design that achieves low throughput at high cost.

A simple perspective on the cost/throughput trade-off is provided in Figure 14.3.

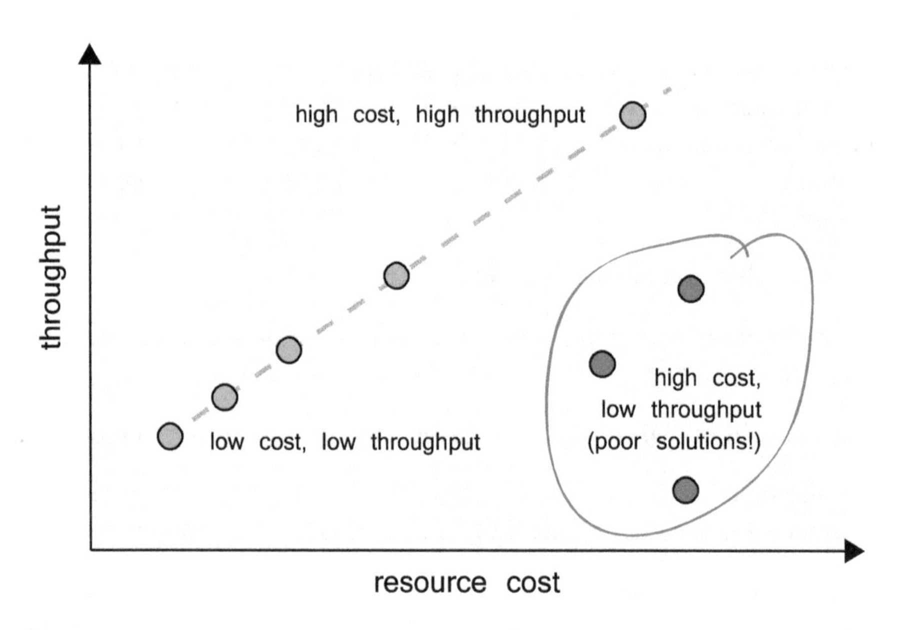

Figure 14.3: A conceptual illustration of the design trade-offs explored via HLS

14.2. Development of HLS Tools

High-level synthesis is not a new concept and, although it has received considerable research interest and found a level of adoption in industry, it has not yet become a widespread favourite among FPGA designers [12], [34]. However, with recent improvements in software design tools for HLS, and bearing in mind the clear motivations for HLS described in earlier sections, it is not surprising that interest in HLS is growing [21].

It is worth noting that contemporary HLS is very similar, in terms of motivations, conceptual framework and terminology, to the early methods reviewed in [20]; in particular, the *scheduling* and *binding* of operations (to be explained in detail later), and the ability to spin out multiple design variations from a single description. At the time the above referenced paper was written (the early 90's), there were still a variety of unresolved issues, including those of verification, human intervention/guidance, handling of complex constraints, and the integration of HLS-generated designs. Design entry methods included programming languages such as Pascal and Ada [29]. The C language subsequently gained prominence, and in fact C-based languages are still the most popular for HLS.

Later, interest grew in object-oriented programming languages for conversion to hardware descriptions, such as C++ and, to a lesser extent, Java [8], [19], [28]. Object-oriented software provides the abstraction mechanisms necessary to represent more sophisticated hardware designs. Further language features specific for hardware design are included in SystemC, namely the ability to model hierarchy, bit accuracy, and the concurrent execution behaviour of hardware [10], [25]. SystemC has gained prominence as a language for modelling complex designs at the system-level.

A number of other languages have enjoyed some level of adoption for C-based design, including *Handel-C*, which originated from the UK-based company, Celoxica, and was later purchased by Mentor Graphics [24]; Cadence's *C-to-Silicon Compiler* [7]; and *Impulse-C*, developed by Impulse Accelerated Technologies [15]. Although all of these represent methods of high-level design entry, the degree to which they fit our definition of HLS (i.e. the separation of function and architecture), varies. All of them represent specialist tools, additional to the algorithm and FPGA system design tools required for FPGA and Zynq development.

Meanwhile, MATLAB-based design entry was the subject of several academic papers in the early 2000's [2], [13], and has since featured in the *AccelDSP* tool (part of the Xilinx ISE suite, now deprecated) which was able to synthesise a subset of high-level MATLAB DSP functions [31]. Now, synthesis from MATLAB code and Simulink models prominently features in MathWorks' own *HDL Coder* product, providing a much more extensive MATLAB-to-RTL capability [23]. The attraction of synthesis from MATLAB code is similar to the motivation for synthesis from C code: high-level design methods can be used for design entry, abstracting many of the implementation details. This is particularly true as, given its status as an underpinning platform for Xilinx System Generator, many FPGA designers may already be familiar with MATLAB programming and its environment.

To bring the discussion back to the topic of this chapter, Xilinx' Vivado HLS tool was originally AutoPilot, developed by AutoESL Design Technologies Inc., a spin-out of the University of California, Los Angeles. Having surveyed the market for the best HLS technology available, Xilinx acquired AutoESL Design Technologies in early 2011 [3], [4], [30]. Xilinx subsequently integrated AutoESL into the ISE Design Suite, and later the Vivado Design Suite, at which time the tool was rebranded *Vivado HLS* [27].

The importance of these developments is significant, as they represent the graduation of HLS from niche tools to mainstream development.

14.3. HLS Source Languages

In the next section we will move on to discuss the Vivado HLS tool in detail, but it is useful to first review source languages for HLS in the general sense. We begin by providing some background on the three languages of Vivado HLS, namely C, C++, and SystemC, and then briefly summarise other languages supported by third party software development tools.

14.3.1. C

The popular, general-purpose C language is a procedural programming language which was initially developed by Bell Labs and has been in use since the 1970s [18], [26]. It was standardised by the American National Standards Institute (ANSI) in 1989, and subsequently adopted by the International Organization for Standardization (ISO) as standard ISO 9899. Since then, several revisions have been made to incorporate new features into the language, the most recent of which was in 2011 [16]. The standardised versions of C may be referred to as 'ANSI-C', 'ISO-C', or simply 'Standard C'.

Historically, C represented a raising in the abstraction level of software programming: it allowed programs to be written not in assembly language, but using higher-level constructs and commands. This had the significant advantages of simplifying the programming task: fewer lines of code were required, with less potential for error, enabling faster debugging and verification processes. It also made the code more portable, because it could be used on different platforms.

Of course, in order for C to be successful, it also had to be efficient enough to justify its use over assembly language, which gave the designer full and explicit control, but with greater design effort. In a sense, the adoption of C is analogous to the concept of HLS for hardware — in both cases designs are developed at a higher level of abstraction, with several benefits including a reduction in development time and effort, ease of verification, and enhanced portability.

Aside from Standard C, which defines the core functionality of the language, C can be extended by including application-specific and target-specific libraries. For instance, when using C to develop software for running on Zynq's ARM, additional libraries are included in the SDK to support floating and fixed point arithmetic, and to target the facilities of the NEON processor.

C has numerous relationships with other programming languages, the most significant of which is with C++.

14.3.2. C++

C++ is an object-oriented language based on C; it extends C with the concepts of classes, templates, polymorphism, and virtual functions, amongst several other features. The level of abstraction of C++ is normally higher than C, because C++ has features that hide detail and permit more sophisticated and flexible code to be developed. On the other hand, the language features and programming styles of C are compatible with C++, and consequently C++ can be considered a superset of C. In summary, C++ represents a higher level language than C, but it also retains support for low-level C operations.

Like C, C++ was also developed at Bell Labs. The standard for C++ was initially published in 1998 as ISO 14882, and has undergone revision to produce subsequent versions, most recently in 2011 [17].

14.3.3. SystemC

Although here we tend to treat SystemC as a language in its own right, strictly speaking it is an extension to C++, i.e. a specific class library which is included in development projects as the header file, 'systemc.h'. SystemC permits C++ style code to be written using many of the hardware-centric principles of HDLs, such as hierarchy, concurrency, and cycle accuracy, which do not form part of Standard C++.

SystemC is commonly used for Transaction Level Modelling (TLM) and Electronic System-Level (ESL) design, in the context of SoC [5]. TLM is the development of very high-level descriptions that model the interactions between system components, while ESL refers to the design of a system at a very high leve of abstraction, i.e. its functionality and basic architecture. As such, both of these represent methods for developing and refining the high-level framework for the SoC, before starting work on the functional implementation of individual components. In TLM, various levels of abstraction are possible in terms of, for example, the implementation target for individual components, and the timing fidelity associated with system transactions [6]. The language can also be used for ESL verification as an integrated part of the development process [11].

Early work on SystemC was supported by the Open SystemC Initiative consortium (which in 2011 merged with the Electronic Design Automation (EDA) standards organisation, Accellera, to form the Accellera Systems Initiative [1]). This represented the initial development phase of SystemC; the SystemC standard was ratified by the IEEE in 2005 as IEEE 1666, and later revised as IEEE 1666-2011 [14]. It is also closely related to *SystemC AMS*, which has analogue and mixed signal support.

Existing users of SystemC for TLM and systems engineering are in a particularly good position to adopt the language for HLS, but even without this background, designers should find that the hardware-oriented features of SystemC make the language very suitable for HLS, particularly for developing complex systems.

The scope of this chapter precludes in-depth introduction of SystemC, and we will use the C language for the majority of our illustrative examples. Readers are referred to [5] and [14] for further information on SystemC.

14.3.4. Other Languages for High-Level Synthesis

As reviewed in Section 14.2, the languages used for HLS are predominantly C-based, including C, C++ and SystemC as supported in Vivado HLS. We also noted that third party HLS design flows include other C-based languages such as HandelC, and also MATLAB code. Research interest in HLS continues, and it is possible that further high-level languages will be supported for FPGA synthesis in the future.

14.4. Introducing *Vivado HLS*

In this section, we will start by defining what Vivado HLS does and the steps involved, before considering its role in the design flow for Zynq. Later, Chapter 15 will cover use of the tool on a practical level, along with further discussions of algorithm and interface synthesis, and the processes involved in creating solutions and evaluating them.

14.4.1. What Does *Vivado HLS* Do?

In short, Vivado HLS transforms a *C*, *C++* or *SystemC* design into an RTL implementation, which can then be synthesised and implemented onto the programmable logic of a Xilinx FPGA or Zynq device [33]. This represents the high-level synthesis step depicted in Figure 14.2 on page 258.

It is important to reiterate that all C-based designs in the context of HLS are for implementation in *programmable logic*; i.e. as distinct from software code intended to run on a processor (whether Zynq's ARM processor, or a soft processor such as MicroBlaze).

In performing HLS, the two primary aspects of the design are analysed:

- The ***interface*** of the design, i.e. its top-level connections, and
- The ***functionality*** of the design, i.e. the algorithm(s) that it implements.

In Vivado HLS design, the functionality is synthesised from the input code via the process of ***Algorithm Synthesis***. The interface is created using one of two alternatives: it can either be (i) manually specified, or (ii) inferred from the code (***Interface Synthesis***). A simple conceptual diagram is provided in Figure 14.4 (note that this depicts only a subset of interface types).

In the specific case of the SystemC input language, interfaces must be manually specified, with two exceptions which will be mentioned later [33].

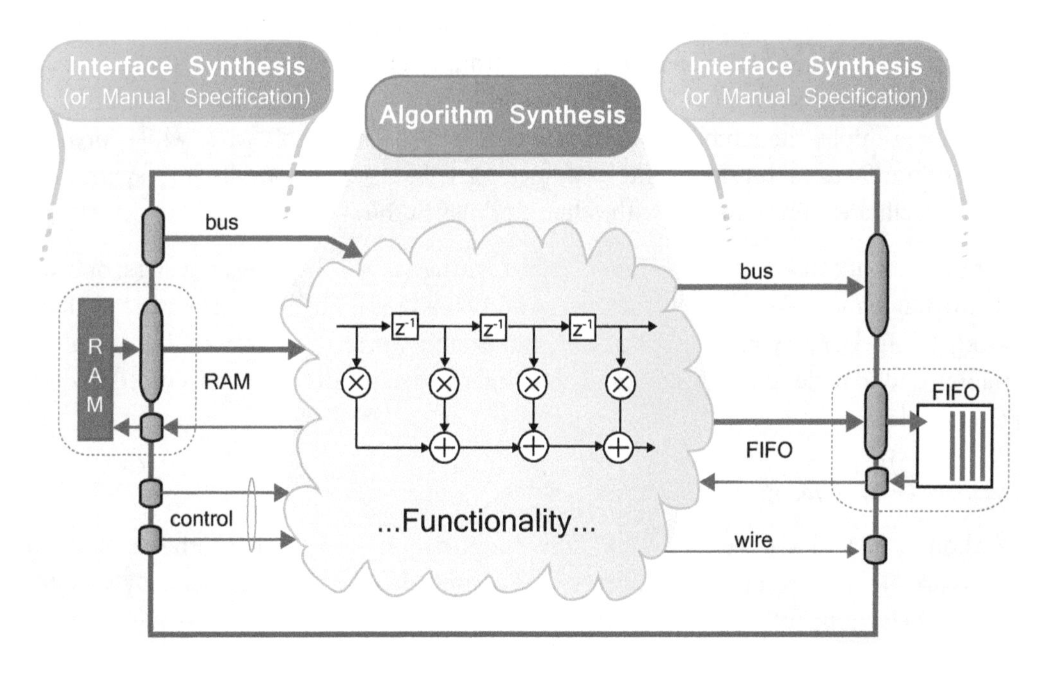

Figure 14.4: Clarification of the algorithm and interface, and showing a subset of interface types

Algorithm Synthesis

Algorithm Synthesis is concerned with the functionality of the design. The desired behaviour is interpreted from the C, C++, or SystemC code that forms the input to the process (the input to the HLS process is a function written in the chosen programming language). Operations are inferred, and these are translated into a set of RTL statements, which are usually executed over several clock cycles.

As will be discussed later, the designer can exert control over the algorithm synthesis process via Vivado HLS *directives*, resulting in variations on the RTL output from HLS being produced. For example, Solution 1 might require 500 slices to implement and 20 clock cycles to execute, whereas Solution 2 might require 1200 slices and 10 clock cycles. The designer is able to experiment and thus achieve a favourable result in terms of their priority implementation metric(s).

Interfaces I: Interface Synthesis

As its name suggests, ***Interface Synthesis*** refers to the interface of the HLS design, and this includes both the *ports* and the *protocols* used. The details of all ports (in terms of their types, dimensions, and directions) are inferred from the top-level function arguments and return values of the source C/C++ file; protocols are inferred from the behaviour of the ports. For example, the simplest interface would be a simple 1-bit wire, while for more complex interfaces, a bus or RAM interface may be used. Naturally the synthesised interface facilitates communicate with other modules in the system.

Interfaces that can be inferred from interface synthesis include: wires, registers, one-way and two-way handshakes, FIFOs, memories, and buses [33]. Further options are available relating to interface synthesis, specifically to permit ports to be inferred from global variables, and to include a global clock enable. Interface synthesis will be discussed in detail in Chapter 15.

Interfaces II: Manual Specification

Although C and C++ designs are fully supported for interface synthesis, SystemC designs are not, and as such, the interfaces of SystemC designs must be manually specified and their behaviour fully described. This corresponds with the hardware-specific features of the SystemC language, and an example will be provided in Chapter 15 which illustrates the similarity between SystemC-based interface specification, and the equivalent in VHDL.

Notably there are two exceptions to the general rule stated above: memory and bus interface types can both be inferred by interface synthesis of SystemC code.

Manual interface specification is also supported for C and C++ designs and may be used if desired; this means that there is the option to explicitly define interfaces, rather than have the interface synthesis process infer them.

14.4.2. Vivado HLS Design Flow

In the previous section, our discussion was limited to the primary processes and outputs, i.e. the execution of HLS algorithms and the production of RTL code. However, the full design flow for HLS comprises further stages, including elements of verification in particular. The full design flow is shown in Figure 14.5, and then summarised.

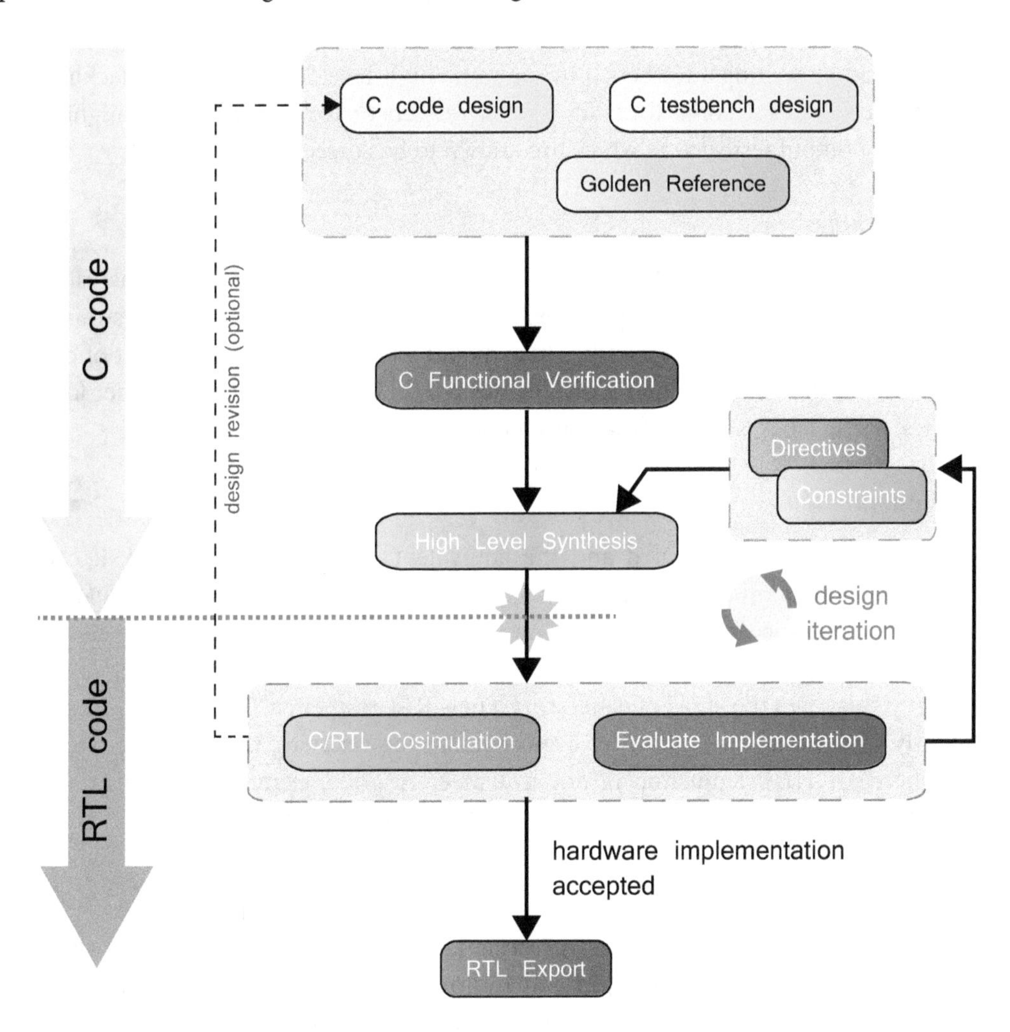

Figure 14.5: An overview of the Vivado HLS design flow

Inputs to the HLS Process

The primary input to the HLS process is a C/C++/SystemC function, along with a C-based testbench which has been developed to exercise the function and verify correct operation. This will involve a 'golden reference' against which to test the outputs produced by the function intended for synthesis; the golden reference may take the form of a prepared set of output values, or it may form part of the testbench itself.

Functional Verification

Firstly, it is necessary to verify the functional integrity of the C/C++/SystemC code that forms the input to HLS, before beginning the process of synthesising it into RTL code. This can be achieved by writing a testbench in the same high-level language, and checking the results produced against some form of 'golden reference'; for instance, this might be a prepared set of output test vectors which are known to be correct.

High-Level Synthesis

The next step is to undertake the HLS process itself, which involves analysis and processing of the C-based code, together with user-supplied directives and constraints, to create an RTL description of the circuit. Once the HLS process is complete, a set of output files is produced, including design files in the desired RTL language. Various other log and report files, testbenches, scripts, etc. are also created.

C/RTL Cosimulation

Once HLS has been performed and the equivalent RTL model produced, it can be checked against the original C/C++/SystemC code via the process of *C/RTL cosimulation* in Vivado HLS. This process re-uses the original, C-based testbench to supply inputs to the RTL version generated by HLS, and check the outputs it produces against expected values. Importantly, this saves the effort of generating a new RTL testbench. The SystemC output is particularly useful here, as it provides a mechanism for verifying the design in environments where an HDL simulator is not available. A good example covering both C functional verification and C/RTL cosimulation is available in [32].

Evaluation of Implementation

Along with verifying the integrity of the design, it is also necessary to evaluate the RTL output in terms of its implementation and performance. For example, the numbers of resources it requires in the PL, the latency of the design, maximum supported clock frequency, and so on. We will outline these metrics in Section 14.4.4, and discuss them in further detail in Chapter 15. For now, it is sufficient to note that, as the designer, we can influence the implementation via the constraints and directives applied to the HLS process.

Design Iterations

Noting the above, as part of the design flow, the implementation of the RTL is evaluated and, as necessary, the constraints and directives are refined; each revision corresponds to a new 'solution' in Vivado HLS terminology (more to follow on *solutions* in Section 14.4.6). It is also possible that evaluation of the design will prompt more fundamental review and refinement of the original algorithm, as designed in C code and input to the HLS process.

Figure 14.5 shows the steps undertaken from C design, to the creation of outputs for RTL synthesis. Note that multiple HLS iterations may be undertaken using modified directives and constraints, in order to find a 'best' solution; this corresponds to the feedback path shown on the right hand side of the diagram.

Should the designer be prompted to change the input C code, a more significant step backwards in the design process is involved, and this is indicated by the arrow at the left hand side of the diagram. Any changes to the C code require functional re-verification, before the subsequent HLS, C/RTL verification, and implementation evaluation processes are again performed and iterated as needed.

RTL Export

Once the design has been validated, and the implementation iterated to the point of achieving the intended design goals, it will be intended for integration into a larger system. This can be achieved directly using the RTL files automatically created by the HLS process (i.e. VHDL or Verilog code), however it may be more convenient to use the facilities of Vivado HLS for packaging IP. Packaging the outputs produced by Vivado HLS means that HLS designs can be easily introduced into other Xilinx tools, namely IP Integrator within Vivado IDE, XPS (for the ISE design flow), and System Generator.

SystemC outputs from the HLS process are produced to enable verification of the hardware produced, but are not themselves synthesisable.

14.4.3. C Functional Verification and C/RTL Cosimulation

Given the importance of verification, it is useful to further detail the *C functional verification* and *C/RTL cosimulation* processes. These are described graphically in Figure 14.6.

Depicted on the left hand side, a C-based testbench has been designed to create and supply input test vectors to the functional C module. The same test vectors are passed through a 'known good' golden reference design, or alternatively read from a prepared file, to give golden reference output test vectors. These are compared with the outputs from the C module, and the testbench reports a pass if the two set of results match, or failure otherwise. The testbench may also be designed to report the total number of errors, or to provide other automated feedback on the results.

As part of the Vivado HLS C/RTL cosimulation process, an equivalent testbench configuration is automatically created by Vivado HLS (shown on the right of Figure 14.6). The testbench verifies the RTL version of the original C module, i.e. the primary output of HLS, against the golden reference, and reports success or failure as before.

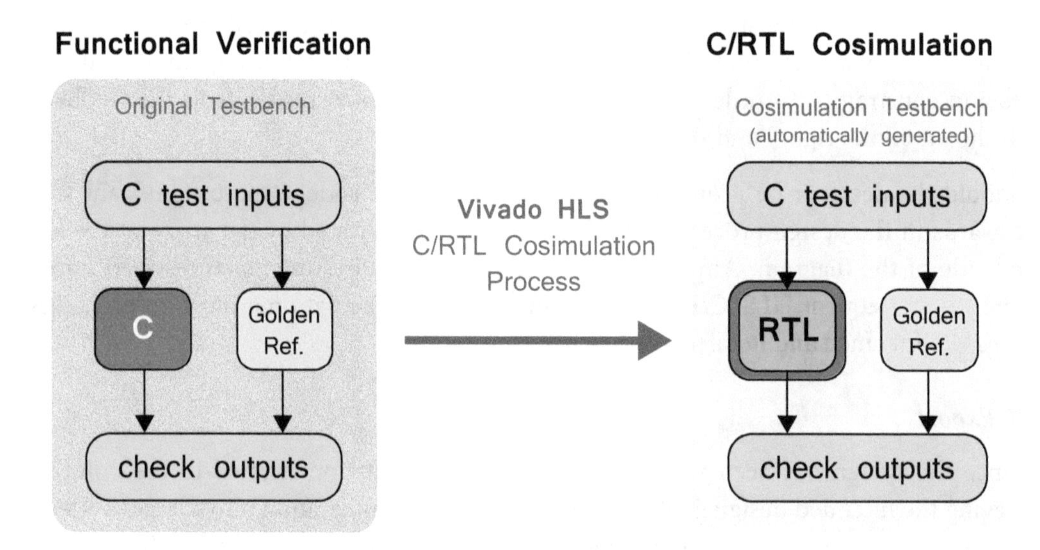

Figure 14.6: C functional verification and C/RTL cosimulation in Vivado HLS

All of the files required for C/RTL cosimulation are created automatically by Vivado HLS, which removes the need for manual RTL testbench creation. The generated testbench includes the necessary translations of data passing between the C-based testbench and the RTL module being tested.

Vivado HLS is also capable of creating a bit-true, cycle accurate System-C model of the generated hardware, and this can be co-simulated in circumstances where an RTL simulation is not available.

Verification is clearly an important aspect of the design process, and the availability of this tool support for RTL-level testing enables increased productivity. In particular, the designer does not need to spend time creating an equivalent testbench for RTL simulation, and the potential for introducing errors by doing so are eliminated. However, it is important to note that RTL simulation is a functional simulation, which does not model realistic timing or bus protocol behaviours. Therefore, it cannot completely verify the correct operation of the module in non-ideal conditions.

14.4.4. Implementation Metrics and Considerations

At this stage, it is useful to define the implementation metrics and related considerations which form part of the design process. We do so here in a simple and informal style; these issues will be returned to and expanded upon later, in Chapter 15.

- ***Resources / area*** — How many resources are needed to implement my design, and how does this compare to the amount available on my target FPGA / Zynq device?
- ***Throughput*** — At what rate can I pass data through the design? Does this meet the needs of my application?
- ***Clock frequency*** — What is the maximum clock frequency that I can run my design at? Is this compatible with the rest of my system?
- ***Latency*** — How many clock cycles does it take for my design to produce an output? Is this delay acceptable in the context of the system in general?
- ***Power consumption*** — How much power does my design consume when it is operating? Is this part of a system sensitive to power consumption?
- ***I/O requirements*** — How complex are the interfaces of my design? Are they compatible with other components of the system?

Any or all of the above factors may be constrained in some way, often with certain factors being prioritised over others. This normally depends on the requirements of the application. For example, a system targeted at a low cost application might prioritise resource minimisation, in order to utilise a smaller device; whereas, on the other hand, a system requiring to adapt quickly to changing inputs may seek to minimise latency and maximise throughput, at the expense of greater resource utilisation.

It is useful to describe briefly the term *clock uncertainty*. The HLS process aims to achieve the target clock period minus the clock uncertainty, which provides a margin for other delays that cannot be modelled at the HLS stage, e.g. RTL synthesis and place and route delays. HLS naturally treats the module in isolation, whereas routing delays are only fully exposed at the system integration stage when the system is populated. The clock uncertainty is user-specified, with a default value of 12.5% [33].

14.4.5. Overview of the High-Level Synthesis Process

Although we have discussed the design flow in general terms, it is useful to consider the process of HLS is more detail. This section will therefore summarise the steps involved in the HLS of C/C++/SystemC design files to achieve an RTL equivalent. For the purpose of providing practical examples, C will be used as the design language.

Recall from Section 14.4.1 that the HLS process performs (i) algorithm synthesis and (ii) interface synthesis (if the interface is not explicitly specified). Here we focus on the former.

Aside from the design files themselves, other inputs to this process are the specification of a particular target device, and the directives and constraints supplied by the designer. As will be discussed in Section 14.4.6, these directly influence the implementation produced by HLS. For the purposes of our current review, we will consider that the target device is fixed, and the constraints and directives applied are constant.

Algorithm synthesis comprises three primary stages, which occur in the following order:

1. Extraction of data path and control;
2. Scheduling and binding; and
3. Optimisations

Each of these will now be briefly explained in turn.

Extraction of Datapath and Control

The first stage of HLS is to analyse the C/C++/SystemC code and interpret the required functionality. This may, for instance, include: logical and arithmetic operations; conditional statements and branching; array operations; and loops.

The implementation will have a *datapath* component, and normally there will also be a *control* component. For clarification, here 'datapath' processing refers to operations performed on the data samples, whereas 'control' is the circuitry required to co-ordinate dataflow processing. The nature of the algorithm fundamentally defines both the datapath and control components but, as we will see in Chapter 15, the designer can take steps during HLS to minimise the complexity of the control component in particular.

Scheduling and Binding

HLS is comprised of two main processes: *scheduling* and *binding*. These are undertaken on an iterative basis, as shown in Figure 14.7, as one affects the other. The operations performed in the two processes are summarised below.

- ***Scheduling*** is the translation of the RTL statements interpreted from the C code into a set of operations, each with an associated duration in terms of clock cycles. The decisions made at this stage are affected by the clock frequency and uncertainty, the target device technology, and any directives applied by the user.
- ***Binding*** is the process of associating the scheduled operations with the physical resources of the target device. The functional and timing characteristics of these resources may affect scheduling, and therefore binding information is fed back into the scheduling process. For instance, the use of DSP48x resources implies a shorter critical path than an equivalent operator built from logic slices.

For example, if the synthesised algorithm requires that a set of arithmetic operations are performed, the HLS process must decide how to *schedule* the operations (how many clock

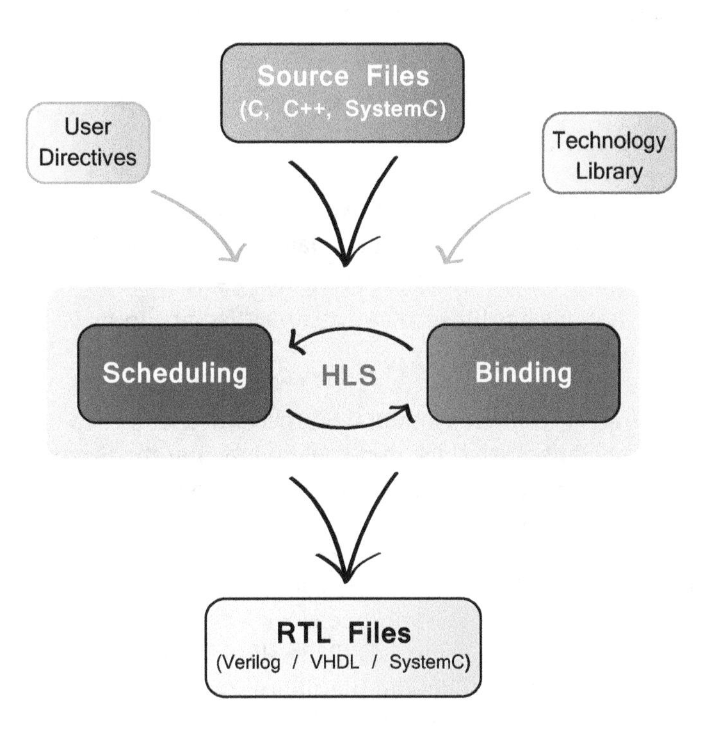

Figure 14.7: Vivado HLS scheduling and binding processes

cycles to allocate to their completion), and how to *bind* the operations (i.e. how to map them to the computational resources on the PL), bearing in mind the target clock frequency and uncertainty. Recall that the hardware architecture is not conveyed or specified by the C source code, but rather different architectural variations can be generated from the source code by applying directives.

The resulting implementation has a set of characteristics, principally in terms of (i) latency, (ii) throughput, and (iii) the resources used.

To illustrate, let us assume that our C algorithm involves calculating the average of an array input, consisting of ten numbers. The implied operations are:

- 9 addition operations to find the total; followed by
- A multiplication by 0.1 to calculate the average.

There are a few different options in terms of scheduling and binding these operations. One possibility is to operate serially over several clock cycles, using a single adder and a single multiplier. Alternatively, the critical path may permit multiple operations to take place within one clock cycle at the targeted frequency, resulting in an implementation with lower latency and higher throughput.

Three variations are depicted in Figure 14.8. Consider in particular the differences in implementation characteristics between them.

1. The first implementation uses the fewest resources (1 adder and 1 multiplier, both constructed from logic fabric), and has a latency of 11 clock cycles. This design has a low throughput — one 11th of the clock rate — because a new operation cannot start until the last one has finished (we assume that pipelining is not used here — to be explained later in Chapter 15).

2. Based on the target technology, the HLS process determines that 3 addition operations can be scheduled per clock cycle, while meeting the timing constraints. This results in an implementation using more resources on the device, but with a shorter latency and higher throughput.

3. Finally, it is determined that, if DSP48x slices are used in place of fabric resources, all operations can take place within one clock cycle. This corresponds to the most costly implementation, requiring 9 DSP48x slices in total (one each for the first 8 additions, with the last addition and multiplication combined into a single DSP48x slice), but it has a much reduced latency (1 clock cycle), and a throughput equal to the clock rate. This style of implementation would of course only be applicable if

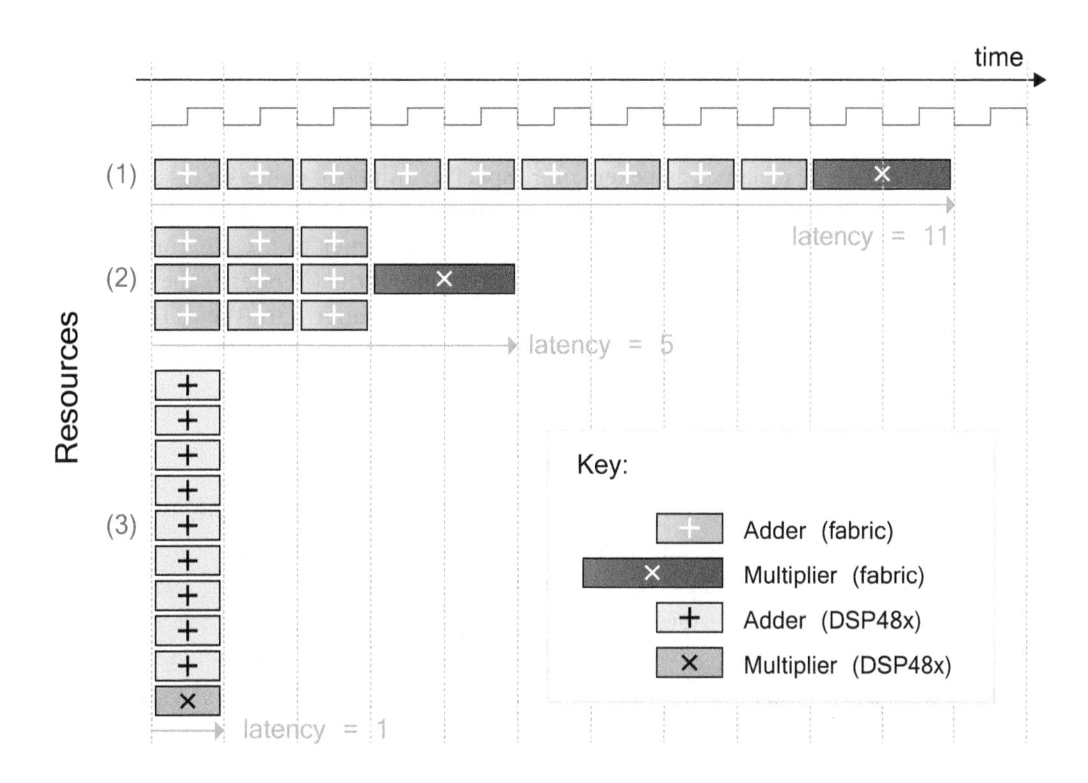

Figure 14.8: Comparison of three possible outcomes from HLS for an example function

timing constraints were met — otherwise Vivado HLS may insert pipelining registers to meet timing.

By default, the HLS process will optimise area, i.e. it will adopt the first strategy outlined above, which consumes the fewest resources. The disadvantages of this implementation are its long latency and low throughput, which may not meet the requirements of the application. However, the designer can exert influence by constraining and directing the HLS processes of scheduling and binding, and thus optimising in a different way.

Optimisations

As mentioned above, the designer has mechanisms available to drive the high-level synthesis process towards his or her implementation goals. There are two methods which can be used to dictate the behaviour of the HLS process, and hence influence the results:

- ***Constraints*** — The designer places a limit on some aspect of the design. For instance, the minimum clock period may be specified. This makes it easy to ensure that the implementation meets the requirements of the system into which it will be integrated. Similarly, the designer may choose to constrain resource utilisation or other criteria, with the aim of optimising the design for the application in hand.

- ***Directives*** — The designer can exert more specific influence over aspects of the RTL implementation via directives. There are various types of directive available which map to certain features of the code, enabling the designer to dictate, for example, how the HLS engine treats loops or arrays identified in the C code, or the latency of particular operations. This can yield significant changes to the RTL output. Therefore, with knowledge of the available directives, the designer can optimise according to application requirements.

A Note on the RTL Output

The user has the option to specify the RTL language for the generated output files, and can choose from VHDL, Verilog, and SystemC. It is, of course, notable that SystemC is also one of the available input languages for HLS, and therefore it might initially seem curious that it is also listed as an output type. However, the SystemC file produced as an output of HLS is an RTL description of the hardware, i.e. a lower-level representation than the input to HLS. In the same way as for the other HLS output languages (VHDL and Verilog), the implementation is based on the chosen target device. This output can be particularly useful in verifying the design where an RTL simulator is not available.

14.4.6. Solutions: Exploring the Design Space

A Vivado HLS project is comprised of a set of design files, a set of testbench files, and project settings, The project can have multiple *solutions* — this is an important term because it relates to the theme of "exploring the design space", i.e. generating a selection of possible implementations which can be compared, and the most suitable version chosen.

It is important to recognise that each solution is a different implementation of *the same* C/C++/SystemC source code. The differences in the synthesised RTL solution are determined by four factors:

- The target technology and part (e.g. Virtex-7, Kintex-7, Zynq-7000...)
- The target clock frequency
- The implementation constraints applied
- User defined synthesis directives

For a given application, the first two of these (target technology and part) are likely to remain constant, and the designer exerts influence over the generated solutions by varying the implementation constraints and synthesis directives, as discussed in the previous section. These will be defined in more detail as they relate to interface and algorithm synthesis in Chapter 15. Once created, each solution contains information about the target, the applied directives and constraints, and the results obtained.

As will be discussed in detail in the next chapter, the results obtained for a set of solutions can differ according to resource cost, throughput, and latency. These can be explored, and further refined, according to the requirements of the task.

14.4.7. Vivado HLS Library Support

It is worth noting that Vivado HLS includes support for arithmetic and mathematical functions, as well as linear algebra, video processing, DSP and others. Full details of library support is available in [33].

14.5. HLS in the Design Flow for Zynq

In Section 14.4.2, the design flow for Vivado HLS was reviewed as an independent process. However, we must also set it in context of the design flow for Zynq.

HLS is a design method for creating functional modules, or IP blocks, for inclusion in a Zynq-based system. IP blocks designed using HLS are for implementation in the PL part of the target Zynq device. There may be multiple such modules in a particular Zynq system design, and part of the task involves appropriately interfacing them to the rest of the design (using for example AXI connections). Referring back to Figure 3.2 on page 53, this corresponds to the stage marked 'Vivado HLS / HDL / System Generator' — modules are created using Vivado HLS for integration into the design as IP blocks.

While Vivado HLS can be used to describe commonly used functional modules such as FIR filters and FFTs from first principles, the designer should be aware that dedicated and optimised support for these operations already exists, and is accessible via the IP Catalog of IP Integrator, or the Xilinx BlockSet of the System Generator tool. The same functionality can be introduced into the Vivado HLS design by calling a function (e.g. an FFT or FIR function) from a supplied library — details of these functions can be obtained from the *Vivado Design Suite User Guide: High-Level Synthesis* [33]. HLS is also a very powerful method for developing custom functionality that does not have a direct equivalent in the existing catalogue of IP.

14.6. Chapter Review

This chapter has reviewed the concept of high-level synthesis, and explained its growing significance as a design method, particularly in the context of Zynq and SoC design. HLS permits an algorithm to be defined at a high level of abstraction using a software language, and it is subsequently converted into an RTL description with the aid of a high-level synthesis tool. This design method has the potential to greatly enhance productivity in terms of both design entry and verification effort.

The Vivado HLS tool was introduced, and its design flow described. It was noted in particular that, in addition to synthesis itself, the Vivado HLS flow also integrates facilities for streamlined verification at a functional level, and also subsequently for validation of the generated RTL code.

Towards the end of the chapter, we reviewed the processes underpinning HLS, and the mechanisms available to the designer to exert influence over the behaviour of these processes, and hence the results obtained. Standard implementation metrics and considerations were outlined, and it was noted that multiple 'solutions' can be generated from the same set of Vivado HLS input files. This permits detailed exploration of the implementation possibilities, evaluation in terms of key metrics, and optimisation with respect to the designer's priority criteria.

In the next chapter, we will build upon this conceptual framework by taking a detailed look at the creation of designs using Vivado HLS.

14.7. References

Note: All URLs last accessed June 2014.

[1] Accellera Systems Initiative website,
available: http://www.accellera.org/

[2] P. Banerjee, "Overview of a compiler for synthesisizing MATLAB programs onto FPGAs", *IEEE Transactions on VLSI Systems*, Vol. 12, Issue 3, March 2004, pp. 312-324.

[3] Berkeley Design Technology, Inc., "An Independent Evaluation of: High-Level Synthesis Tools for Xilinx FPGAs", consultation paper, 2010.
Available: http://www.xilinx.com/technology/dsp/BDTI_techpaper.pdf

[4] J. Bier and J. E. White, "BDTI Study Certifies High-Level Synthesis Flows for DSP-Centric FPGA Design", *Xilinx Xcell Journal*, Issue 71, second quarter 2010, pp. 12 - 17.
Available: http://www.xilinx.com/publications/archives/xcell/Xcell71.pdf

[5] D. C. Black, J. Donovan, B. Bunton and A. Keist, *SystemC: From the Ground Up*, 2nd Edition, Springer, 2009.

[6] M. Burton, J. Aldis, R. Günzel and W. Klingauf, "Transaction Level Modelling: A reflection on what TLM is and how TLMs may be classified", *Forum on Design Languages*, 2007, pp. 92-97.

[7] Cadence, *C-to-Silicon Compiler* webpage,
Available: http://www.cadence.com/products/sd/silicon_compiler/pages/default.aspx

[8] J. M. P. Cardoso and H. C. Neto, "Towards an Automatic Path from Java Bytecodes to Hardware Through High-Level Synthesis", *Proceedings of the IEEE International Conference on Electronics, Circuits and Systems*, 1998, vol. 1, pp. 85-88.

[9] D. Gajski and R. Kuhn, "Guest Editors' Introduction: New VLSI Tools", *Computer*, vol. 16, no.12, pp.11 - 14, December 1983.

[10] D. Gadski, T. Austin and S. Svoboda, "What Input-Language is the Best Choice for High Level Synthesis (HLS)?", *panel session, Proceedings of the 47th ACM/IEEE Design Automation Conference (DAC)*, pp. 857-858, June 2010.

[11] D. Große and R. Drechsler, *Quality-Driven SystemC Design*, Springer, 2010.

[12] R. Gupta and F. Brewer, "High-Level Synthesis: A Retrospective" in *High-Level Synthesis: From Algorithm to Digital Circuit*, edited by P. Coussy and A. Morawiec, Springer, 2008.

[13] M. Haldar, A. Nayak, A. Choudhary and P. Banerjee, "FPGA Hardware Synthesis from MATLAB", *Proceedings of the 14th International Conference on VLSI Design*, January 2001, pp. 299-304.

[14] IEEE Computer Society, "IEEE Standard for Standard SystemC Language Reference Manual", *IEEE Std 1666-2011*, January 2012.

[15] Impulse Accelerated Technologies, *Impulse CoDeveloper C-to-FPGA Tools* product webpage.
Available: http://www.impulseaccelerated.com/products_universal.htm

[16] International Organization for Standardization, "ISO/IEC 9899:2011: Information technology - Programming languages - C", 2011.

[17] International Organization for Standardization, "ISO/IEC 14882:2011: Information technology - Programming languages - C++", 2011.

[18] B. W. Kernighan and D. M. Ritchie, *The C Programming Language*, Prentice Hall, 1978.

[19] T. Kuhn and W. Rosenstiel, "Java Based Object Oriented Hardware Specification and Synthesis", *Proceedings of the Asia and South Pacific Design Automation Conference*, 2000, pp. 579-581.

[20] M. C. McFarland, A. C. Parker, and R. Camposano, "The High-Level Synthesis of Digital Systems", *Proceedings of the IEEE*, vol. 78, no.2, pp 301 - 318, Feb 1990.

[21] G. Martin and G. Smith, "High Level Synthesis: Past, Present and Future", *IEEE Design and Test of Computers*, Vol. 26, Issue 4, July/August 2009, pp. 18 - 24.

[22] A. Mathur, E. Clarke, M Fujita, and R. Urard, "Functional Equivalence Verification Tools in High-Level Synthesis Flows", *IEEE Design & Test of Computers*, July/August 2009, pp. 88 - 95.

[23] Mathworks, *HDL Coder* product webpage.
Available: http://www.mathworks.com/products/hdl-coder/

[24] Mentor Graphics, *Handel-C Synthesis Methodology* webpage.
Available: http://www.mentor.com/products/fpga/handel-c/

[25] M. Meredith and S. Svoboda, "The Next IC Design Methodology Transition is Long Overdue", Open SystemC Initiative, February 2010.
Available: http://www.accellera.org/resources/articles/icdesigntrans/community/articles/icdesigntrans/ic_design_transition_feb2010.pdf

[26] D. M. Ritchie, "The Development of the C Language", *Proceedings of the 2nd History of Programming Languages Conference*, Cambridge, Massachusetts, April 1993.

[27] M. Santarini, "Xilinx Unveils Vivado Design Suite for the Next Decade of 'All Programmable' Devices", *Xilinx Xcell Journal*, Issue 79, second quarter 2012, pp. 8 - 13.
Available: http://www.xilinx.com/publications/archives/xcell/Xcell79.pdf

[28] R. Thomson, V. Chouliaras and D. Mulvaney, "The Hardware Synthesis of a Java Subset", *Proceedings of the Norchip Conference*, 2006, pp. 217-220.

[29] H. Trickey, "Flamel: A High-Level Hardware Compiler", *IEEE Transactions on Computer-Aided Design*, Vol. CAD-6, No. 2, March 1987, pp. 259 - 269.

[30] Xilinx, Inc., Press Release: "Xilinx Acquires AutoESL to Enable Designer Productivity and Innovation with FPGAs and Extensible Processing Platform", 30th January, 2011.
Available: http://press.xilinx.com/2011-01-30-Xilinx-Acquires-AutoESL-to-Enable-Designer-Productivity-and-Innovation-With-FPGAs-and-Extensible-Processing-Platform

[31] Xilinx, Inc., "UG634 - AccelDSP Synthesis Tool User Guide", v11.4, December 2009.
Available: http://www.xilinx.com/support/documentation/sw_manuals/xilinx11/acceldsp_user.pdf

[32] Xilinx, Inc., "UG871 - Vivado Design Suite Tutorial: High Level Synthesis", v2014.1, May 2014.
Available: http://www.xilinx.com/support/documentation/sw_manuals/xilinx2014_1/ug871-vivado-high-level-synthesis-tutorial.pdf

[33] Xilinx, Inc, "UG902 - Vivado Design Suite User Guide: High-Level Synthesis", v2014.1, May 2014.
Available: http://www.xilinx.com/support/documentation/sw_manuals_j/xilinx2014_1/ug902-vivado-high-level-synthesis.pdf

[34] J. Yi and H. Kwon, "Samsung's Viewpoints for High-Level Synthesis" in *High-Level Synthesis: From Algorithm to Digital Circuit*, edited by P. Coussy and A. Morawiec, Springer, 2008.

15

Vivado HLS: A Closer Look

One of the most significant developments in the Xilinx design methodology has been the introduction of a capable tool for high-level synthesis: *Vivado HLS*. Having established the motivation for HLS in the previous chapter, we now take a detailed look at the methods used to design with Vivado HLS.

In doing so, this chapter will cover several topics, including the specification of data types and implications for circuit synthesis, the creation of port and block-level interfaces, and aspects of algorithm synthesis. The use of directives and constraints to influence the solutions produced by HLS will also be demonstrated.

It must be noted that the facilities of HLS are so broad and varied that there is simply not enough space in this chapter to cover everything, so instead we aim for an appealing introduction, and direct the reader to [17], [18], and [19] for more detailed review and tutorial material. This chapter will, however, present a case study focussing on loops, which acts as a good basis to demonstrate a some of Vivado HLS's features and optimisations, and thus to provide the reader with a taste of what can be achieved with HLS.

As with other components of the Vivado Design Suite, there is an emphasis on integration and design reuse, and therefore Vivado HLS includes facilities for packaging IP for convenient integration into system designs. We will briefly review this aspect of the tool towards the end of the chapter.

15.1. Anatomy of a Vivado HLS Project

It is useful to begin by with a high-level model of the Vivado HLS ***synthesis*** process. (Note that simulation and verification are also undertaken using the C testbench (and an automatically generated RTL version of the C testbench), as was previously described in Section 14.4.3 on page 269.)

The inputs to the process are:

- ***C, C++ or SystemC files*** — These contain the functions to be synthesised. In a simple design, there may be a single file containing a single function, or in a more complex case, a hierarchy of sub-functions and multiple files.
- ***C testbench files*** — The C testbench files form the basis for verifying both the C code and the RTL code generated by the HLS process.
- ***Constraints*** — The designer supplies a timing constraint (desired clock period), along with a clock uncertainty figure and details of the target device. Together, these influence the synthesis process along with *directives*.
- ***Directives*** — Directives applied by the designer influence the style of implementation generated from the high-level description (input C code), for instance in terms of pipelining and parallelism.

The outputs produced are as listed below. The designer is able to choose which of these are created.

- ***SystemC model*** — This is an RTL-level model of the output from the HLS process, i.e. a different style of description that an input SystemC file. This SystemC output is not intended for synthesis, but only for RTL simulation.
- ***VHDL or Verilog files*** — The Vivado HLS process generates an RTL-level output in the VHDL or Verilog language, depending on user preference. This is synthesisable code that could be integrated into a project and used to generate a bitstream (*.bit file) for programming an FPGA or Zynq device.
- ***Packaged IP for Vivado, System Generator, or XPS*** — The packaged outputs are convenient for direct inclusion into an IP Integrator project, XPS project, or System Generator design.

The various files outlined above form the basis of a Vivado HLS project. Next, we will move on to introduce the Vivado HLS development environment.

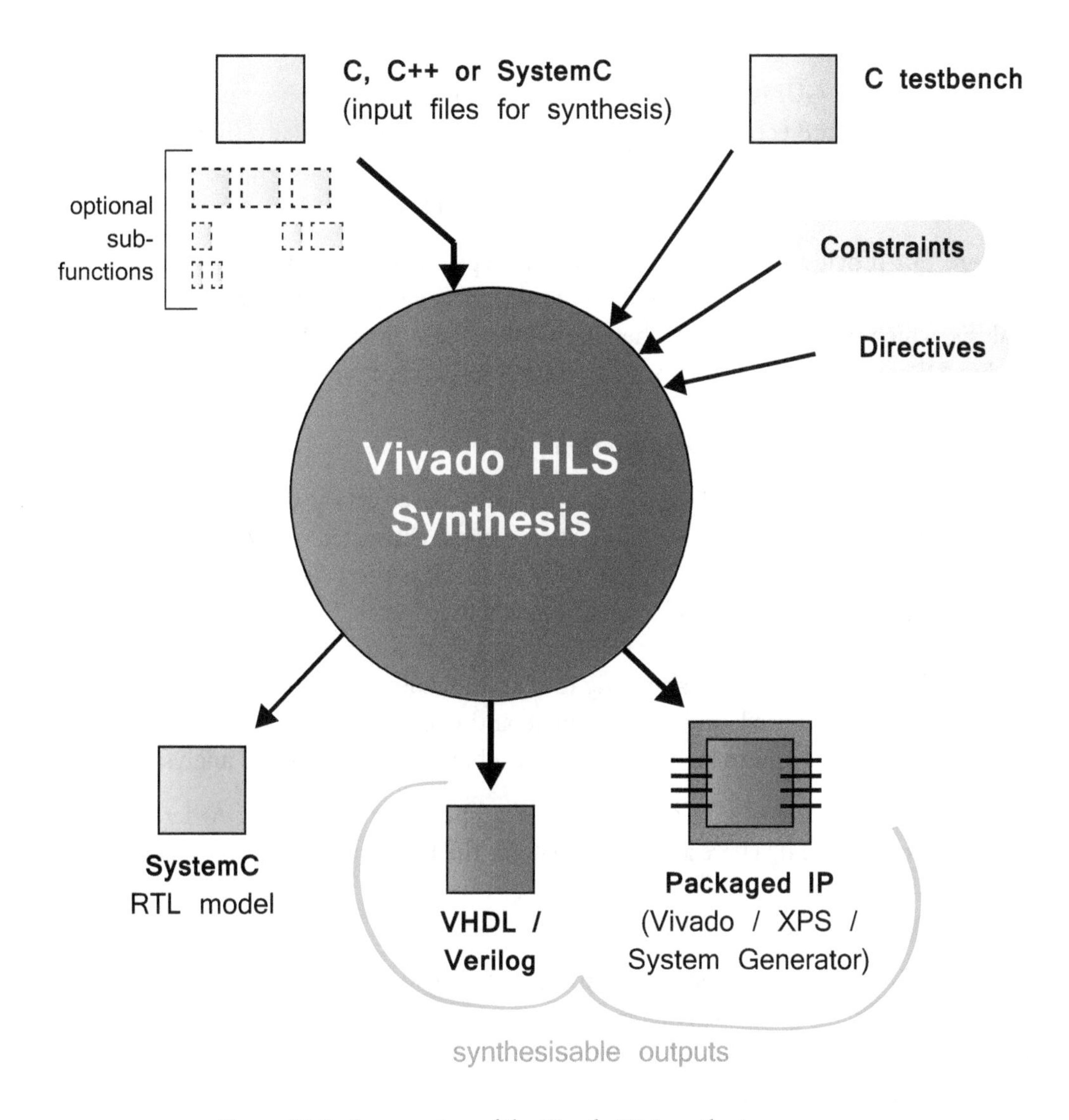

Figure 15.1: An overview of the Vivado HLS synthesis process

15.2. Vivado HLS User Interfaces

The Vivado HLS tool provides both a Graphical User Interface (GUI) and a Command Line Interface (CLI), which may be used separately or in conjunction with each other according to individual preference. Both methods provide access to the same set of functionality, and offer different advantages from a user perspective.

15.2.1. Graphical User Interface

The Vivado HLS GUI is similar to software development environments, providing facilities for managing projects, code editing, and debugging. Additionally, there HLS-specific features for directing the HLS process and evaluating the synthesised hardware. It is useful to highlight these, and the sections following provide a brief overview of each.

Figure 15.2 provides clarification of how these specific components relate to the Vivado HLS GUI. The GUI actually provides three different *perspectives*: Debug, Synthesis and Analysis, which by default lay out the GUI with relevant information panes. The facilities highlighted here relate to the Synthesis and Analysis perspectives.

Synthesis Perspective: Project Organisation

The iterative nature of Vivado HLS development is based on the concept of 'solutions' — alternative implementations of the same source code, generated on the basis of different guidance from the user. The structure of Vivado HLS projects reflects this: there are separate folders for the source code and testbench, together with any included header files, and a set of solution folders. Each solution represents a different implementation, and contains files, reports and results relating to that implementation; the user can generate as many solutions as required, and then evaluate and compare them. At any one time, only one of the solutions is said to be 'active', and the subject of directives and analysis.

The project structure of an example project is shown in Figure 15.2, where the active project is shown in bold. There are also folders for the HLS output files and reports.

Synthesis Perspective: Directives Pane

Vivado HLS also features a pane for setting up and managing directives, which influence how the HLS process behaves. The *Directives* pane reflects the 'active' solution only, and is only visible when the source code is open in the main window.

Note that directives can be either:

- Kept separate from the source code, as TCL commands within a directives file; or
- Integrated into the source file as pragmas.

The choice of inserting directives into the code, or applying them in a separate file, affects how the directives appear in the Directives pane. As indicated in Figure 15.2, a hash (#) symbol indicates a pragma (incorporated into the source code), while a percentage sign (%) indicates a directive located in the dedicated directives file.

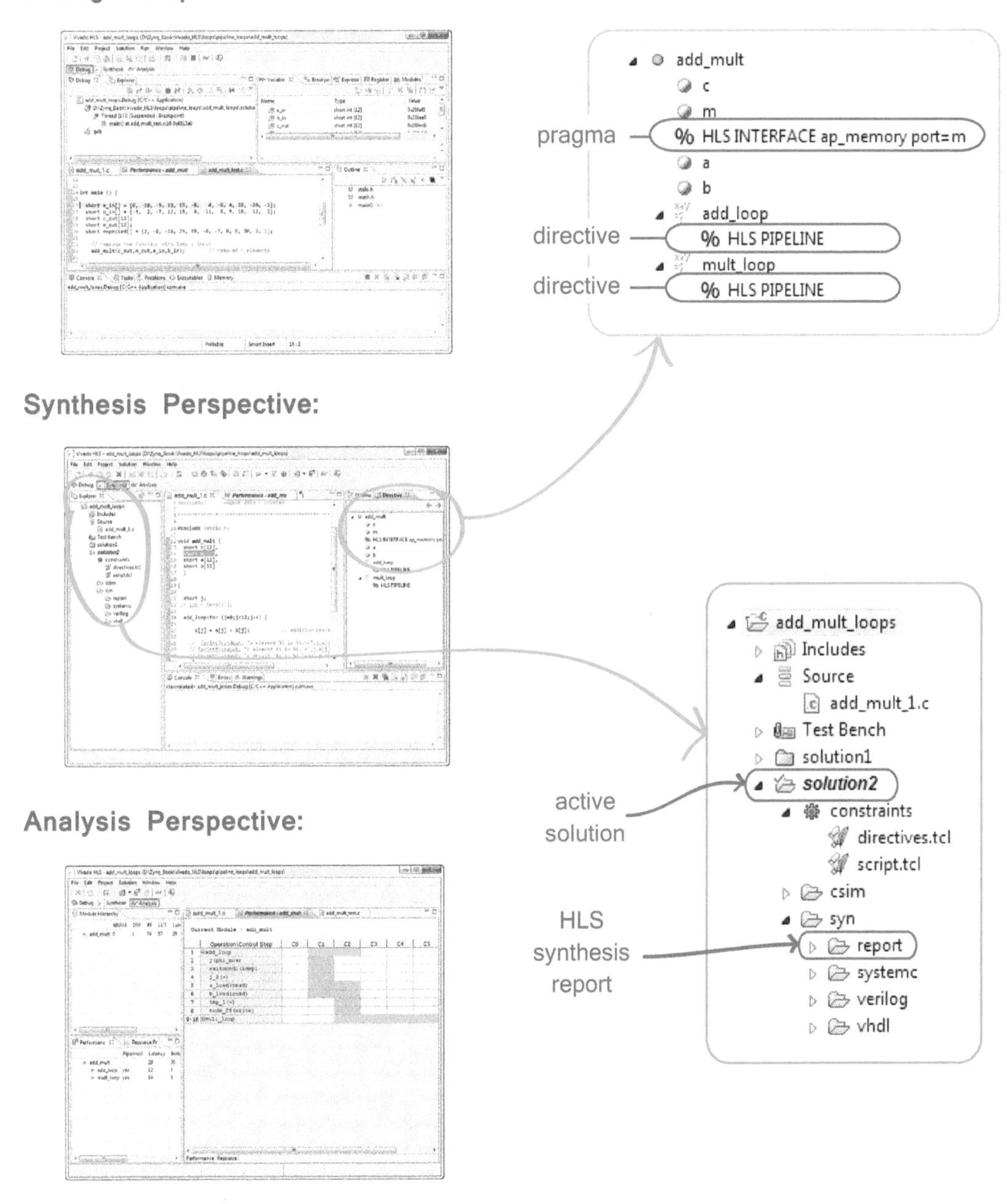

Figure 15.2: Overview of Vivado HLS GUI perspectives

There are advantages to both approaches, and these will be discussed in Section 15.4.6, in the context of interfaces.

Synthesis Perspective: Synthesis Report

Vivado HLS produces a synthesis report for each solution, which represents a consolidated set of statistics relating to that particular implementation. The details include:

- Clock information, and comparison with constraints;
- Latency statistics;
- Details of loops identified within the code (e.g. trip count, latency per loop);
- The estimated cost of the implementation, in terms of different PL resources;
- A list of the synthesised RTL interface ports, including their directions, dimensions and associated protocols.

A separate comparison report can be generated for a specified set of solutions. This permits direct comparison of key implementation metrics across solutions, and it is a very useful facility for iterating towards the optimum solution.

Analysis Perspective

Aside from the synthesis results, there is also an analysis perspective available in the GUI which presents a graphical visualisation of the synthesised design, in terms of operations and controls steps. This information is linked to resources, and also cross-references to the C/C++ code from which it was originally synthesised.

The analysis perspective is useful because it permits the designer a deeper appreciation of how the design has been synthesised, and thus informs further iterations. Viewing the details of the synthesised design helps the designer to identify operations which cause bottlenecks, and which may benefit from further optimisation.

The Analysis perspective will be discussed further in Section 15.6.

15.2.2. Command Line Interface (CLI)

The command line interface / TCL scripting method is well suited when performing repetitive or prescribed tasks, because the required steps can be performed in an automated fashion, thus saving time and ensuring reproducible results.

The method of entering commands is via the TCL language. This is an open source scripting language which is widely used in ASIC and FPGA development. When working with Vivado HLS, TCL can be used to run basic tasks such as setting up projects and running simulations, through to driving extensive sets of tests using defined sets of parameters and directives. In terms of usage, normally it is convenient to prepare a script in advance (e.g. 'my_hls_script.tcl', to include all of the required settings and commands), and execute it once. It is also possible to run commands individually by typing them directly at the prompt.

A comprehensive guide to all TCL commands for Vivado HLS is available, and this is the key resource for developing scripts to drive the tool [18]. Examples of using scripts are also incorporated into the Xilinx tutorials for Vivado HLS [17], and some general guidance is available in a general TCL guide for Vivado Design Suite [16]. You can also gain experience with this method by following one of the tutorials that accompanies this book — see Chapter 16 for more details.

15.3. Data Types

When using traditional design methods for FPGAs, the specification of data type is important as it has a direct effect both on the integrity of the design, and the key metrics of the implementation (namely resource utilisation, timing performance, and power consumption). In the case of numerical data types, underspecifying the wordlength compromises accuracy, while overspecifying can lead to increased resources, inflated power consumption, and a suboptimal maximum clock frequency. It is therefore necessary to specify data types carefully.

This aspect is equally important in Vivado HLS as compared to other methods such as HDL development or block-based design, even if the data types at the point of design entry are different. Understanding the available C, C++ and SystemC data types, and their synthesis, is fundamental to developing effective and efficient designs. To that end, this section will be devoted to reviewing the available types, and explaining how these are translated into an RTL design and hence into hardware. We will start by considering the native data types of the C and C++ languages, followed by arbitrary precision types.

15.3.1. C and C++ Native Data Types

The C and C++ languages have several built-in data types which derive from four basic numeric types: `char`, `int`, `float` and `double`, as summarised in Table 15.1.

In the cases of char, int, and derived types, these are signed by default, but can also be specified as unsigned (or signed to prevent ambiguity). Notably the standard int type, and the short, long, and long long versions of the int type, equate to minimum sizes. Here, typical values are chosen [5], [9].

Table 15.1: Native data types in the C language

Type	Description	Number of Bits[a]	Range[b]
char	Representation of the basic character set.	8	-128 to 127
signed char		8	-128 to 127
unsigned char		8	0 to 255
short int	A reduced precision version of int, requiring less storage.	16	-32,768 to 32,767
unsigned short int		16	0 to 65,535
int	The basic integer data type.	32	-2,147,483,648 to 2,147,483,647
unsigned int		32	0 to 4,294,967,295
long int	In many cases the long int type will be the same length as int, i.e. 32 bits.	32	-2,147,483,648 to 2,147,483,647
unsigned long int		32	0 to 4,294,967,295
long long int	An extended precision integer type.	64	-9,223,372,036,854,775,808 to 9,223,372,036,854,775,807
unsigned long long int		64	0 to 18,446,744,073,709,551,615
float	Single precision floating point (IEEE 754)	32	$-3.403e^{+38}$ to $3.403e^{+38}$
double	Double precision floating point (IEEE 754)	64	$-1.798e^{+308}$ to $1.798e^{+308}$

a. The numbers of bits for the given types are not fixed according the C language definition, but depend on implementation. A set of representative values are shown here.
b. The ranges given are based on the above representative lengths of each type in bits.

There are also some further types of interest, which are briefly summarised below:

- A boolean type, `bool`, is available on inclusion of the header file 'stdbool.h', with the standard values of {true, false}.
- Support for complex numbers is provided via the 'complex.h' library header file. This relates to the floating point types.
- An extended precision floating point type is defined, `long double`, although it may in practice have the same format as the `double` type.

As may be noted from Table 15.1, the native C / C++ data types lie along 8-bit boundaries (8-bit, 16-bit, 32-bit and 64-bit), which arises from the fact that software code is usually targeted at processors of these dimensions. However, such restrictions are not ideal from the perspective of generating efficient hardware architectures.

For optimal hardware implementations, no more bits are used than are necessary, due to the extra implied hardware cost. Arbitrary wordlengths are needed in order to realise circuits requiring arbitrary levels of precision. In fact, the effect of restricting wordlengths to those lying on 8-bit boundaries may be exacerbated when targeting certain dedicated resources of the PL. For example, the multiplication of two 18-bit numbers, A and B, to give the 36-bit result, S, would require A and B to be represented using 32-bits, and S using 64-bits. This would lead to an inefficient multiplier implementation, using four DSP48x slices instead of one — a 300% overhead! (If necessary, refer back to Section 2.2.2 on page 25 for a review of the DSP48x architecture). Furthermore, any registers or other operations on A, B, and S, would be over-sized.

Clearly, then, there is motivation to support arbitrary precision data types in a similar manner to other design entry methods such as HDL and System Generator. Vivado HLS therefore provides specific support for arbitrary precision C / C++ data types. Additionally, SystemC features its own arbitrary precision data types as an integral part of the language, and these are fully supported by Vivado HLS.

15.3.2. Vivado HLS Arbitrary Precision Data Types for C and C++

Having established the need for arbitrary precision arithmetic, i.e. to enable efficient hardware implementations, the direct outcome is an arbitrary precision integer type. However, this does not fully satisfy the requirements of most hardware designers, who normally prefer fixed point arithmetic for certain applications. Therefore, Vivado HLS also provides an arbitrary precision fixed point type, for use in C++ only.

Arbitrary Precision Integer Types

Support for arbitrary precision integer types is achieved using different types and associated libraries for each of the C and C++ input languages, as detailed in Table 15.2. In both cases, wordlengths between 1 bit and 1024 bits are permitted, i.e. $1 \leq N \leq 1024$.

Table 15.2: Arbitrary precision integer data types for use in C and C++ Vivado HLS designs

Language	Integer Data Type	Description	Required Header
C	int*N* (e.g. `int7`)	signed integer of *N* bits precision	`#include "ap_cint.h"`
	uint*N* (e.g. `uint7`)	unsigned integer of *N* bits precision	
C++	ap_int<*N*> (e.g. `ap_int<7>`)	signed integer of *N* bits precision	`#include "ap_int.h"`
	ap_uint<*N*> (e.g. `ap_uint<7>`)	unsigned integer of *N* bits precision	

Note that a different compiler (*apcc* as opposed to *gcc*) must be used when working with arbitrary precision integer types in C, as detailed in [18]. This does not apply to C++.

Figure 15.3 shows equivalent snippets of code demonstrating the use of arbitrary precision integer types in C and C++; notice the slightly different syntax used.

```
// C code example
#include "ap_cint.h"

void top_level_function (..)
{
    // declarations
      int6 small_signed;
      uint10 big_unsigned;
      int22 vbig_signed;
      ...
}
```

```
// C++ code example
#include "ap_int.h"

void top_level_function (..)
{
    // declarations
      ap_int<6> small_signed;
      ap_uint<10> big_unsigned;
      ap_int<22> vbig_signed;
      ...
}
```

Figure 15.3: The use of arbitrary precision integer types in C (left) and C++ (right)

Arbitrary Precision Fixed Point Types

Fixed point arithmetic has the generic word format shown in Figure 15.4, with a specified numbers of integer bits to the left of the binary point, and fractional bits to the right of the binary point. As for integer binary numbers, the MSB has a positive weighting for unsigned numbers, and a negative weighting for signed numbers.

For consistency with Xilinx Vivado HLS documentation [17], [18], in our discussions the overall wordlength is denoted as *W*, the number of integer bits as *I*, and the number of fractional bits as *B*, i.e. $W = I + B$. In the example of Figure 15.4, $I = 5$, $B = 7$, and $W = 12$.

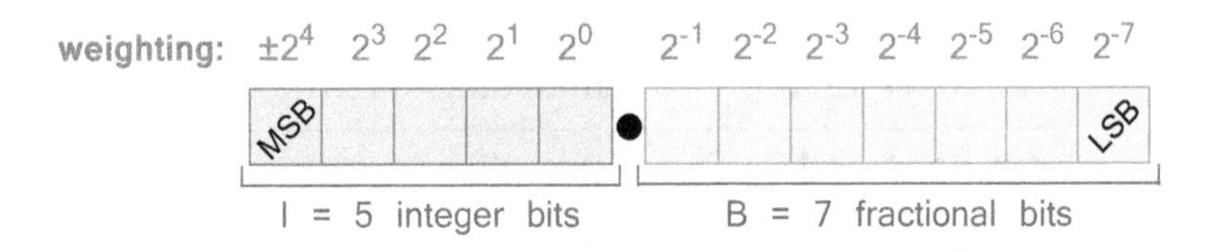

Figure 15.4: An example of the fixed point word format

Vivado HLS fixed point formats for C++ are defined as in Table 15.3; note that the C language is not supported. Here, *W* and *I* are as defined above; *Q* is a string specifying the quantisation mode; *O* gives the overflow mode; and *N* specifies the number of saturation bits in overflow wrap modes (i.e. the *N* most significant bits are set to 1). Details of these options can be found in [18]. The latter three are optional arguments; if unspecified, the quantisation mode, *Q*, defaults to **truncation to zero**, and the overflow mode, *O*, defaults to **wraparound**.

Table 15.3: Arbitrary precision fixed point data types for Vivado HLS designs

Language	Fixed Point Data Type	Description	Required Header
C++	ap_fixed<*W,I,Q,O,N*>	Signed fixed point number of *I* integer bits and *W-I* fractional bits.	`#include "ap_fixed.h"`
	ap_ufixed<*W,I,Q,O,N*>	Unsigned fixed point number of *I* integer bits and *W-I* fractional bits.	

In a similar manner to the previous section on integer data types, it is useful to confirm the syntax for declaring variables using the general type definitions given in Table 15.3. Figure 15.5 provides a code example in which a selection of variables are created, each with different integer and fractional wordlengths. The use of quantisation and overflow modes is also demonstrated.

Note that a set of strings are defined for Q (the quantisation mode) and O (the overflow mode), as listed in Table 15.4 [18], where the defaults are given in bold type.

Table 15.4: Quantisation and overflow modes for the C++ ap_fixed and ap_ufixed types (defaults in bold type)

Parameter	String	Description
Q (quantisation)	AP_RND	Rounding to positive infinity
	AP_RND_ZERO	Rounding to zero
	AP_RND_MIN_INF	Rounding to negative infinity
	AP_RND_INF	Rounding to infinity
	AP_RND_CONV	Convergent rounding
	AP_TRN	Truncation to negative infinity
	AP_TRN_ZERO	**Truncation to zero**
O (overflow)	AP_SAT	Saturation
	AP_SAT_ZERO	Saturation to zero
	AP_SAT_SYM	Symmetrical saturation
	AP_WRAP	**Wraparound**
	AP_WRAP_SM	Sign magnitude wrap around

15.3.3. Arbitrary Precision Types for SystemC

As mentioned earlier, SystemC features in-built support for integer and fixed point types. Use of these types follows a very similar style to C++, as summarised by Table 15.5. Note, however, that SystemC provides two different data types to cater for small (up to 64 bit) and large (up to 512 bit) integer wordlengths, whereas C and C++ use a single data type for up to 1024-bit words.

Code examples for SystemC, and an equivalent set of mode specifiers to Table 15.4, can additionally be found in [18].

```
// C++ code example
#include "ap_fixed.h"

void top_level_function (..)
{
    // declarations
      ap_ufixed<8,3> small_unsigned; // 3 int, 5 fract, defaults
      ap_fixed<10,4,AP_RND> big_signed; // round to + inf.
      ap_ufixed<10,4,AP_RND_ZERO> big_unsigned; // round to zero
      ap_fixed<21,10,AP_TRN,AP_SAT> vbig_signed; // trunc., satur.
      ap_ufixed<21,10,AP_RND_CONV> vbig_unsigned; // conv. round.
      ...
}
```

Figure 15.5: Example C++ code showing declaration of fixed point variables

Table 15.5: Summary of SystemC data types

SystemC Data Type	Description	Required Preamble
sc_int<*W*> sc_bigint<*W*>	signed integer: (up to 64 bits) (up to 512 bits)	`#include "systemc.h"`
sc_uint<*W*> sc_ubigint<*W*>	unsigned integer: (up to 64 bits) (up to 512 bits)	

sc_fixed<*W,I,Q,O,N*>	signed fixed point	`#define SC_INCLUDE_FX`
		`[#define SC_FX_EXCLUDE_OTHER]`
		`#include "systemc.h"`

sc_ufixed<*W,I,Q,O,N*>	unsigned fixed point	

15.3.4. Floating Point Data Types and Operators

Vivado HLS supports the use of floating point data types and operations, in so far as these map to available cores from the Xilinx technology library. For instance, standard arithmetic operations such as addition, subtraction, multiplication and division all have corresponding Xilinx cores, and these can be inferred by Vivado HLS. Further mathematical functions can be also be used on inclusion of the appropriate header files [18]. However, not all floating point operations are supported by HLS, and code should be written with these restrictions in mind.

To provide an example of compatible floating point operations for HLS, the code provided in Figure 15.6, which represents multiplication and addition operations on (single precision) floating point variables, would be successfully synthesised by Vivado HLS using instances of the floating point adder and multiplier cores. This can be confirmed by inspecting the reports produced by Vivado HLS.

15.3.5. Validation of Arbitrary Precision Models

It is possible to compare and validate a function based on arbitrary precision arithmetic against the equivalent function implemented using the native C / C++ data types, the latter usually being the starting point when developing a Vivado HLS design. This is a rapid and useful method of tuning the arithmetic wordlengths, i.e. achieving a level of arithmetic precision that is sufficiently accurate for the purpose, but which does not imply superfluous hardware resources.

Models can be developed with two type specifications for a given variable (or variables): (i) the original, with which the function was initially validated; and (ii) a reduced precision version, i.e. the candidate for synthesis. Only one of these definitions is adopted within a simulation at any given time, with the switch being made via a C macro [18]. This means

```
// C++ code example
void floating_arith (float *s, float a, float b, float d)
{
  float ab;

  ab = a * b;
  *s = ab + d;
}
```

Figure 15.6: Example code demonstrating the use of the 'float' type

that functional simulations can be conveniently executed for each of the variable type definitions, and the results quickly compared, while retaining all parameters from the original reference design, rather than fundamentally removing or changing them.

If the results indicate that the reduced precision version produces the required results, then naturally it is preferred for synthesis due to the reduced hardware cost implied.

15.4. Interface Specification and Synthesis

As mentioned in Section 14.4.1, the mechanism for specifying the interface of an HLS function differs according to the HLS input language: C and C++ support synthesis of the interface from a high-level description, whereas SystemC requires explicit, manual specification (with exceptions). Here we will focus our discussion on synthesis from C and C++.

In Vivado HLS, the input arguments and return value of the designed top-level C/C++ function are synthesised into RTL data ports, each with an associated protocol. Together, the ports and protocol form a port interface. Port interfaces are used to communicate with other subsystems and, if applicable, the processor in the system. In addition to the port interfaces inferred from the C function arguments, block-level protocols and associated ports are used to coordinate the exchanges of data between subsystems.

Over the next few pages, we will first consider the top-level C/C++ function definition from which interfaces are inferred. The synthesis of ports and protocols from this definition will then be explained, the functionality of block-level protocols will be introduced, and finally, the use of directives to influence the interface synthesis process will be reviewed.

15.4.1. C/C++ Function Definition

The functional part of a Vivado HLS design is a C/C++ function, which may contain other sub-functions in a hierarchical manner. The top-level function, i.e. the highest level of hierarchy, forms the basis of the interface synthesis process.

To provide an example, the code listing in Figure 15.7 represents a simple C design with the top level function, `find_average_of_best_X()`. The detailed internal workings of the function are not of consequence, although the reading/writing operations on each argument determines the direction of the synthesised ports, as will be covered in Section 15.4.2.

```
void find_average_of_best_X (int *average, int samples[8], int X)
{
    // body of function (statements, sub-function calls, etc.)
}
```

Figure 15.7: An example top level function for HLS

The function definition contains three arguments, and it should be interpreted that the 8-element array `samples` is an input, as is the integer `X`, while `average` is an output of the function. At a simple level, these three function parameters are therefore converted by HLS into two input interfaces, and one output interface, as shown in Figure 15.8.

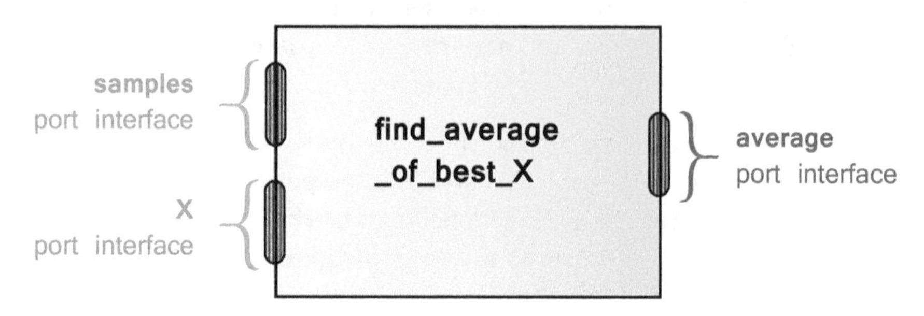

Figure 15.8: Simplified interface diagram for the example function 'find_average_of_best_X()'

Notably, these interfaces may comprise control inputs or outputs in addition to the data port itself, depending on the protocol adopted. We will return to, and update, this interface diagram later in the chapter, after introducing the subject of port protocols.

15.4.2. Synthesis of Port-Level Interfaces

Following our intuitive example of interface synthesis in Section 15.4.1, it is useful to present more formally the conventions adopted when synthesising an RTL port-level interface from C/C++ code. (Note that block-level interfaces will be covered in Section 15.4.5, and these are distinct from port-level interfaces.)

The RTL-level description of a port includes the following aspects:

- The name of the port;
- Its direction (*input*, *output* or *inout*);
- The data type and dimension.

Therefore, when designing using Vivado HLS, all of these properties must be synthesised from the high-level C/C++ code. We will consider each of them during the remainder of this section.

Port Name

The name of the port is drawn from the name of the corresponding function parameter. For example, in the function of Figure 15.7, 'samples' is an array type input parameter of the function, and therefore the name 'samples' would also be used for the array data port. An exception is any port synthesised from a function return statement, which is given the name 'ap_return' (and there is no return statement present in our working example).

In some cases, additional control signals and associated ports are synthesised along with a data port, and this depends on the protocol adopted. We will cover protocols in more detail in Section 15.4.3, but for now, it is useful to note that any control signals associated with a synthesised data port will adopt the same name, with the relevant suffixes for the control types.

Port Direction

The interpretation of port direction follows a set of rules as summarised in Table 15.6. For example, an argument of the C/C++ function that is only ever read from by the function, and never written to, will be synthesised into an RTL *input* port. Likewise, an argument that is always written to, and never read from, will be translated into an *output* port.

Table 15.6: Synthesis of port directions

C/C++ Function Argument	RTL Port Type
An argument which is read from and never written to	in
An argument which is written to and never read from	out
A value output by the function return statement	out
An argument which is both written to and read from	inout (bidirectional)

Data Type and Dimension

The data types and dimensions of ports synthesised from C/C++ function arguments follow the same conventions as for data types in general, as reviewed in Section 15.3. Some interface protocols (to be reviewed in the next section) require additional control ports to be generated, and these are generally 1-bit signals, with some exceptions.

15.4.3. Port Interface Protocol Types

In addition to the port itself, an associated protocol defines the style of interchanges taking place via that port. Vivado HLS specifies a set of available protocols, ranging in sophistication from 'none' (i.e. no explicit protocol) to 'hs' (a hand shaking protocol), 'ack' (an acknowledgement protocol), and even AXI protocols. A full list of the available protocols is included below, where the names used in the Vivado HLS tool are given, along with a brief explanation of each [18].

- ***ap_none*** — This is the simplest protocol type, with no explicit interface protocol, no additional control signals, and no associated hardware overhead. However, there is an implication that timing of input and output operations is independently and correctly handled.

- ***ap_stable*** — This is a similar protocol to *ap_none*, in that it does not involve additional control signals or related hardware. The difference is that *ap_stable* is intended for inputs (only) that change infrequently, i.e. that are generally stable apart from at reset, such as configuration data. The inputs are not constants, but neither do they require to be registered.

- ***ap_ack*** — This protocol behaves differently for input and output ports. For inputs, an output acknowledge port is added, and held high on the same clock cycle as the input is read. For outputs ports, an input acknowledge port is added. After every write to the output port, the design must wait for the input acknowledge to be asserted before it may resume operation.

- ***ap_vld*** — An additional port is provided to validate data. For input ports, a *valid* input control port is added, which qualifies input data as valid. For output ports, a *valid* output port is added, and asserted on clock cycles when output data is valid.

- ***ap_ovld*** — This protocol is the same as *ap_vld*, but can only be implemented on output ports, or the output portion of an inout (bidirectional) port.

- ***ap_hs*** — The _hs suffix of this protocol stands for 'handshaking', and it is a superset of *ap_ack*, *ap_vld*, and *ap_ovld*. The *ap_hs* protocol can be used for both input and output ports, and facilitates a two-way handshaking process between the producer and consumer of data, including both validation and acknowledgement transactions. As such, it requires two control ports and associated overhead. It is, however, a robust method of passing data, with no need to ensure timing externally.

- ***ap_memory*** — This memory-based protocol supports random access transactions with a memory, and can be used for both input, output, and bidirectional ports. The only argument type compatible with this protocol is the array type, which corresponds with the structure of a memory. The *ap_memory* protocol requires control signals for clock and write enables, as well as an address port.

- ***bram*** — The same as *ap_memory*, except that when bundled using IP Integrator, the ports are not shown as individual ports, but grouped together into a single port.

- ***ap_fifo*** — The FIFO protocol is also compatible with array arguments, provided that they are accessed sequentially rather than in random order. It does not require any address information to be generated, and therefore is simpler in implementation than the *ap_memory* interface. The *ap_fifo* protocol can be used for input and output ports, but not bidirectional ports. The associated control ports indicate the fullness or emptiness of the FIFO, depending on the port direction, and ensure that processing is stalled to prevent overrun or underrun.

- ***ap_bus*** — The *ap_bus* protocol is a generic bus interface that is not tied to a specific bus standard, and may be used to communicate with a bus bridge, which can then arbitrate with a system bus. The *ap_bus* protocol supports single read operations, single write operations, and burst transfers, and these are coordinated using a set of control signals. In addition to this generic bus interface, specific support for AXI bus interfaces can be integrated at a later stage, using an interface synthesis directive.

- ***axis*** — This specifies the interface as AXI stream.

- ***s_axilite*** — This specifies the interface as AXI Slave Lite.

- ***m_axi*** — This specifies the interface the AXI Master protocol.

It is beyond the scope of this chapter to explain in detail the mechanics of each of these protocols; however, extensive further information can be found in [18], including detailed timing diagrams, and there are also a number of relevant practical examples in [17].

15.4.4. Synthesis of Port Interface Protocols

Having defined the available protocols, we will now concentrate on the synthesis of specific protocols from the high-level description, and related restrictions.

The designer has the option to choose a protocol for each individual port by supplying an appropriate directive. The set of supported protocols is restricted according to (i) the implied port direction, and (ii) the type of the C/C++ function argument, as shown in Table 15.7 [18]. If the protocol is not explicitly defined via a directive, or if an unsupported protocol is mistakenly selected, then Vivado HLS will apply the default protocol. The defaults are also a function of (i) and (ii) above, as indicated in Table 15.7.

Table 15.7: Protocol synthesis: supported types and defaults (S = supported; D = default) [18]

Argument Type	**Variable**			**Pointer Variable**			**Array**			**Reference Variable**		
	pass-by-value			**pass-by-reference**			**pass-by-reference**			**pass-by-reference**		
Interface Type[a]	**I**	**IO**	**O**	**I**	**IO**	**O**	**I**	**IO**	**O**	**I**	**IO**	**O**
ap_none	**D**	-	-	**D**	S	S	-	-	-	**D**	S	S
ap_stable	S	-	-	S	S	-	-	-	-	S	S	-
ap_ack	S	-	-	S	S	S	-	-	-	S	S	S
ap_vld	S	-	-	S	S	**D**	-	-	-	S	S	**D**
ap_ovld	-	-	-	-	**D**	S	-	-	-	-	**D**	S
ap_hs	S	-	-	S	S	S	S	-	S	S	S	S
ap_memory	-	-	-	-	-	-	**D**	**D**	**D**	-	-	-
bram	-	-	-	-	-	-	S	S	S	-	-	-
ap_fifo	-	-	-	S	-	S	S	-	S	S	-	S
ap_bus	-	-	-	S	S	S	S	S	S	S	S	S
axis	S	-	-	S	-	S	S	-	S	S	-	S
s_axilite	S	-	S	S	S	S	-	-	-	S	S	S
m_axi	-	-	-	S	S	S	S	S	S	S	S	S

a. Reading along the row: **I** = input port; **IO** = inout (bidirectional) port; **O** = output port.

Given the dependencies between protocol, port type and direction, it is important to consider the *type* of the C/C++ function arguments when developing the high-level C/C++ description. As given by the column headings in Table 15.7, values can be passed into and out of C/C++ functions using four different argument types, i.e. as: (i) variables; (ii) pointers; (iii) arrays; and (iv) references. Therefore, a particular argument type corresponds to a limited set of available protocols. For example, passing an array argument as an input restricts the set of available protocols to: *ap_hs*, *ap_memory*, *bram*, *ap_fifo*, *ap_bus*, *axis* and *m_axi*, with *ap_memory* as the default.

To link this to the practical issue of design entry, you may wish to refer back to the function defined in Figure 15.7, and confirm that function has three arguments:

- ***samples*** — an array input;
- ***X*** — in integer (scalar) input, passed by value;
- ***average*** — an output pointer variable.

Then, with reference to Table 15.7, it may be noted that the default protocol for ***samples*** is *ap_memory*, the default for ***X*** is *ap_none*, and the default for ***average*** is *ap_vld*. Taking into account the additional control ports required for these protocols, we can update the synthesised RTL interface originally given in Figure 15.8, to that shown in Figure 15.9. Notably the data ports are 32 bit, as a result of using the C `int` data type.

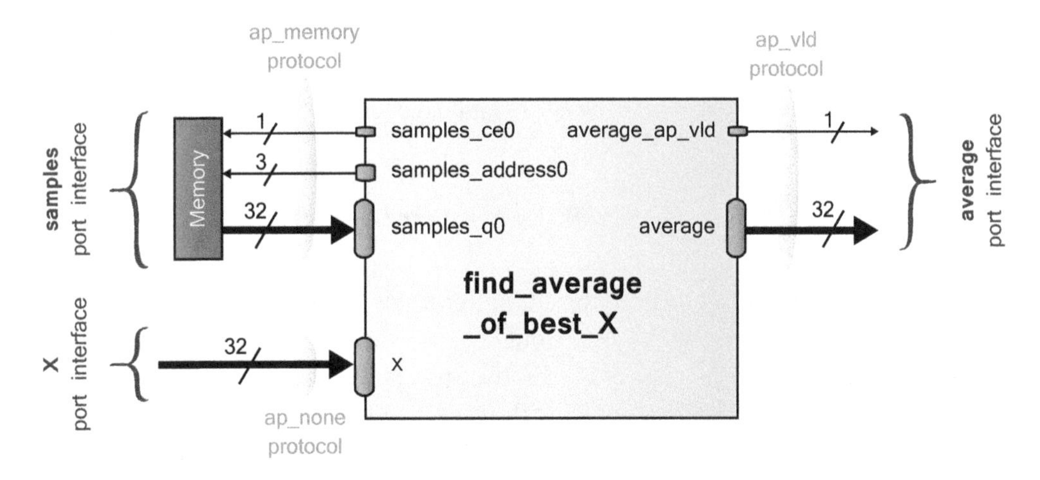

Figure 15.9: RTL interface diagram for the 'find_average_of_best_X()' function, showing default port level ports and protocols

Note also that this diagram shows ports and port-level protocols only — it does not include any block-level protocols, an aspect to be covered in the next section. The clock signal will be added later as part of the block-level protocol.

15.4.5. Block-Level Interface Ports and Protocols

In addition to the logical ports inferred from the arguments of the C/C++ function, and their associated protocols, it is also possible to add block-level protocols to the design (these may also be referred to as function-level protocols). This provides a mechanism to control the execution of the subsystem, and it is useful when integrating one or several Vivado HLS blocks into a system and managing the flow of data between them.

Before defining the protocols, it is worthwhile briefly confirming some terminology, which we will do with the aid of Figure 15.10. This diagram depicts five blocks in cascade, with the direction of data flow indicated; in practice there may also be FIFO buffers between blocks, which are not shown. (Please be aware that, in choosing this example, we cover one possible use model of Vivado HLS blocks — there is no intention to imply that a chain of blocks is the only, or indeed the typical, scenario.)

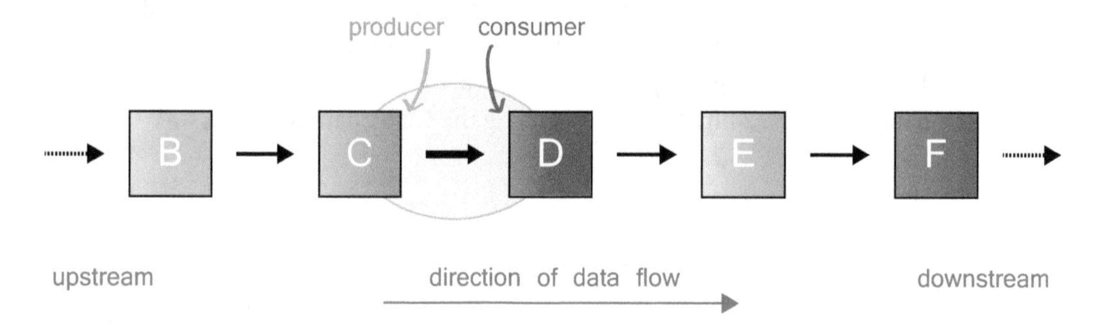

Figure 15.10: Data flow between Vivado HLS blocks

With reference to Block D, blocks to the left of this point (C, B...) would be referred to as *upstream*, while blocks to the right (E, F...) would be considered *downstream*. When considering the interface between any pair of blocks, the upstream block outputs data and passes it to the downstream block. For example, with respect to the interface between Blocks C and D, Block C is the *producer* of data, while Block D is the *consumer* of data.

In some cases it is desirable for a block to exert 'backpressure' on upstream blocks. In other words, the consumer block may wish to prevent the producer block from presenting more data until it is ready to receive it (if necessary, propagating as a ripple effect to further upstream blocks). This is one aspect of operation that can be implemented using block-level control.

There are three types of block-level protocol, and these are listed below according to the terms used in Vivado HLS:

- ***ap_ctrl_none*** — Choosing this option simply means that a block-level protocol is not added. Instead, control is exerted entirely at the port interface level, using port-level protocols.

- ***ap_ctrl_hs*** — A block-level control protocol with handshaking. An *ap_start* control input is asserted to prompt the block to begin operation, and the block produces three output control signals (*ap_ready*, *ap_idle*, and *ap_done*) to indicate its stage of operation. Specifically, the *ap_ready* signal indicates that the block is ready for new inputs, the *ap_idle* indicates when it is processing data, and *ap_done* is asserted when output data is available. To provide a usage example, the *ap_ctrl_hs* protocol is appropriate when a single HLS block is to be interfaced with the controlling processor.

- ***ap_ctrl_chain*** —This protocol is similar to *ap_ctrl_hs*, but has an additional input control signal, *ap_continue*, and is designed for chaining multiple Vivado HLS blocks together. The *ap_continue* input indicates the ability of the downstream block to accept new data, and therefore it can exert backpressure on upstream blocks if necessary. If *ap_continue* is de-asserted, the block will complete its current computation to the stage of presenting the results at the output, but will then stall until *ap_continue* is set high again.

If a block-level protocol is used, it operates independently of any port-level protocols that are adopted on individual ports. There are also two input signals applied to the block that are added regardless of the block-level protocol chosen: *ap_clk* and *ap_rst*. These are required because the internal operation of the block is synchronous, thus it requires a clock signal, and because the block must be capable of being externally reset.

As a general point, it is worth noting that an AXI4-Lite bus interface can be applied to the block-level interface protocol, thus enabling block-level control signals to be passed between the block and a controlling processor. In some circumstances, it is also possible to bundle the block-level control ports with the ports for the port-level interfaces, resulting in a consolidated AXI4-Lite interface [18].

To round off our discussion, let us reconsider our example function, augmented with the *ap_ctrl_hs* block-level control protocol (the default), as shown in Figure 15.11. In this case, six additional ports are added: *ap_clk* and *ap_reset* (as for all Vivado HLS designs); the *ap_start* control input; and the *ap_done*, *ap_ready*, and *ap_idle* control outputs. These new, block-level interface ports are shown towards the top of the diagram.

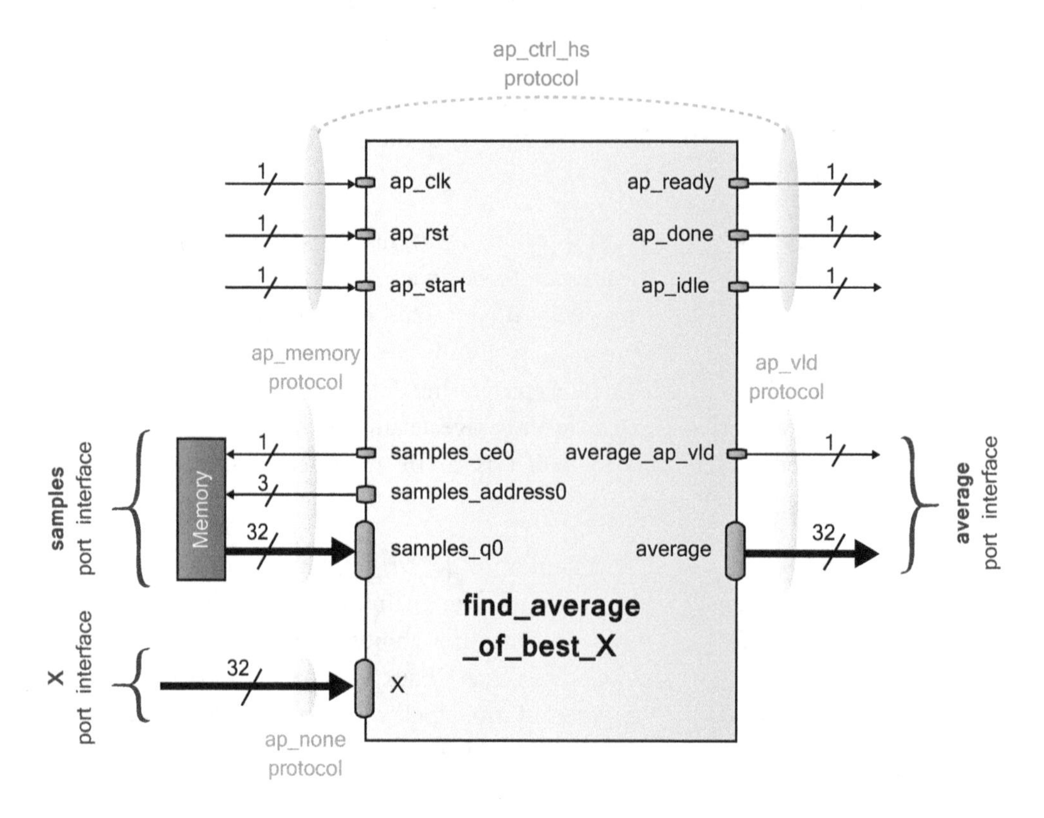

Figure 15.11: RTL interface diagram for the 'find_average_of_best_X()' function, showing both default port level ports and protocols, and default block level ports and protocol

15.4.6. Interface Synthesis Directives

Directives are the mechanism by which the designer can exert high-level control over the implementation of the designed C/C++/SystemC code. Of the HLS directives available, a subset relate specifically to interface synthesis, and these can be used for both port-level directives and block-level directives, reflecting the two types of interface as reviewed over the previous few pages. In addition to explicitly specifying protocols, other aspects of interfaces can also be specified, such as the form of array inputs, and the resources used to implement memories or FIFO buffers.

Directive Types

There are six types of directive which can be applied to both port-level and block-level interfaces, as reviewed below.

- ***Array Map*** — Combines several arrays to form one larger array, with the goal of using fewer FIFO or RAM resources and control ports to implement the interface.
- ***Array Partition*** — Separates array interfaces into several smaller sections, resulting in an expanded set of ports, control signals, and implementation resources, but with increased bandwidth.
- ***Array Reshape*** — In this case, an original array is partitioned into smaller arrays, which are then recombined to form an array with fewer elements, and wider data elements. This implies fewer memory locations and shorter addresses.
- ***Interface*** — This directive can be used to explicitly specify a port-level interface protocol as one of the available options (as listed in Table 15.7 on page 300), or the block-level protocol as *ap_none*, *ap_ctrl_block* or *ap_ctrl_chain*, as covered in Section 15.4.5.
- ***Resource*** — A particular resource can be chosen to implement the interface. For instance, a one or two port RAM can be specified for an *ap_memory* interface, or an *ap_fifo* interface can target a FIFO constructed from a Block RAM or LUTs.
- ***Stream*** — This directive specifies the interface as a streaming port, utilising FIFOs and permitting the depth of the FIFO to be explicitly chosen.

The specification of port-level interface type is made via the corresponding function argument, while the block-level interface is applied to the top-level function.

By applying the required directives, the design example shown in Figure 15.11 can be exported to an IP Integrator block as shown in Figure 15.12. The result of consolidating the port- and block-level interfaces into a single AXI4-Lite interface results is shown in Figure 15.13 (note that the array input `samples` is separate, as it requires a stream interface). This is typically done to enable control by software running on a processor or microcontroller.

Further Options

In addition to the directives detailed above, there is also the option to register individual ports, meaning that an additional clock cycle of delay will be added at the port interface. This brings a potential improvement to the timing performance of the design.

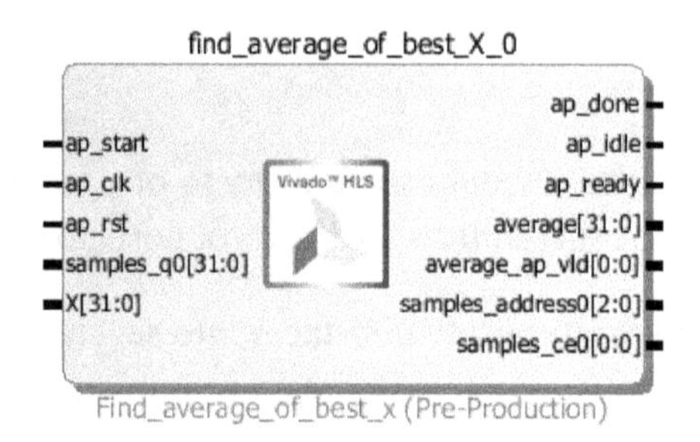

Figure 15.12: An IP Integrator block produced for the "find_average_of_best_X" function, with port- and block-level protocols as specified in Figure 15.11.

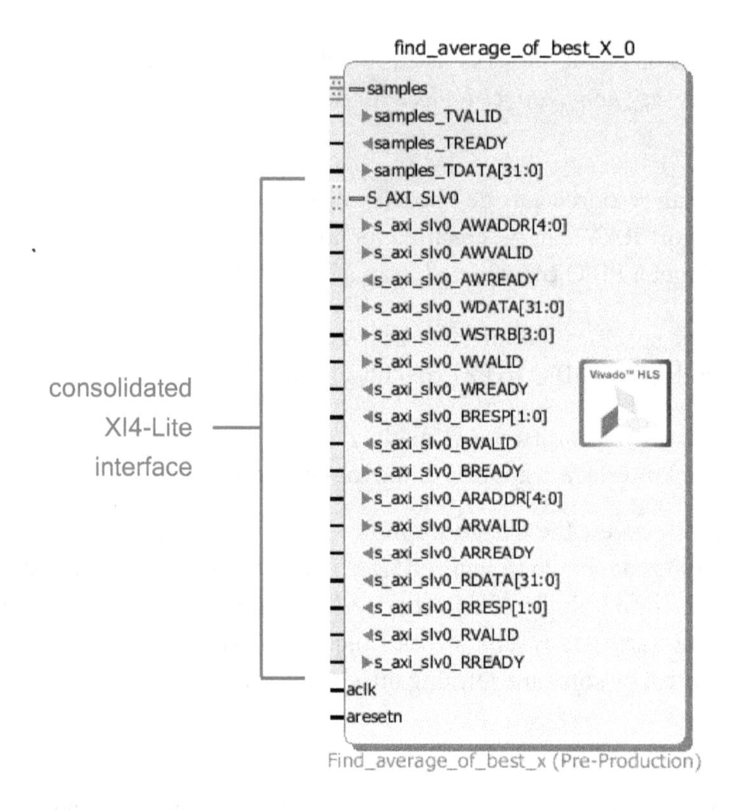

Figure 15.13: An IP Integrator block produced for the "find_average_of_best_X" function, with the block-level protocol ports, and some port-level protocol ports, combined into an AXI4-Lite interface.

Source File or Directive File?

Directives can be kept together in a separate file within the Vivado HLS project, or integrated into the source file as pragmas. An example showing the use of pragmas is given in Figure 15.14. These would be automatically inserted by Vivado HLS, directly into the top of the function body (as indicated in Figure 15.7 on page 296), thus there is no need to type these lines of code manually (although you can if you want to!). In both cases, it is useful to recognise how pragmas are formed.

In the example of Figure 15.14 there are three pragmas: the first two set interface directives, and the third sets a resource directive. Specifically, the first pragma specifies the *ap_memory* protocol for the `samples` input port, and likewise the second line sets the *ap_vld* protocol for the `X` input port. Notice also that a register is inserted on the `X` port. The third directive specifies that the `samples` input port is implemented using a specific hardware resource, in this case a single port ROM implemented with LUTs (i.e. using slice logic as opposed to Block RAM resources).

```
#pragma HLS INTERFACE ap_memory port=samples
#pragma HLS INTERFACE ap_vld register port=X
#pragma HLS RESOURCE variable=samples core=ROM_1P_LUTRAM
```

Figure 15.14: An example of pragmas inserted in the C/C++ source code for interface synthesis

As the interface is often a static aspect of the system, with the designer concentrating optimisation efforts on the functional implementation by creating different *solutions*, interface directives are commonly implemented as pragmas in the source file. This means that the interface settings can be easily ported between Vivado HLS solutions, without the need to re-specify them, or to take particular care when passing directives from one solution to the next. However, the designer may, for good reason, prefer to specify the directives via file, thus keeping the directives and source code separate — this is also a perfectly valid approach.

In a general sense, the insertion of pragmas makes code less portable, and so pragmas best suited to aspects that the designer intends to fix. For instance, it may be desirable to use pragmas to fix the configuration of the interfaces in the design.

Aspects that are changed between 'solutions' as the designer attempts to optimise his or her code, should be specified using a separate directives file, so as not to embed them within the code. Keeping the code and directives separate enables a greater degree of high-level design flexibility.

15.4.7. Manual Interface Specification

Up until this point, the discussion has focussed on interface synthesis from C and C++ functions, but recall that interfaces can also be explicitly specified. This is compulsory when coding in SystemC (with the exception of the *ap_bus* and *ap_memory* interface types, which can be synthesised), and optional when coding in C or C++.

Interface Specification in SystemC

In SystemC, any designed component is represented by a C++ class inherited from the base class, SC_MODULE. This is used to define the interface and functionality of the component [1].

Specification of interfaces in SystemC is similar to an HDL-level description, and includes an explicit specification of the type, direction, and dimension of each port. An example is provided in Figure 15.15 for a simple counter, and this shows the first section of the `my_counter` module (class) definition, where the ports are declared. You may notice the similarities between this code extract and a VHDL entity declaration.

As a result of all port information being manually specified, no synthesis of the interface is required in this case.

```
SC_MODULE (my_counter)  {

     // top level ports
      sc_in<bool> clk;
      sc_in<bool> ce;
      sc_in<bool> reset;
      sc_out<int> count;

     // the rest of the module body definition...
}
```

Figure 15.15: Manual definition of the ports of a counter design using SystemC

Interface Specification in C/C++

In some circumstances it may be desirable to define an interface which has a different set of control ports and associated protocol from those available in Vivado HLS. This may be accomplished in C or C++ by adding an extra block of code to specify the control signals and the required series of transactions on these control ports (recall that these would normally *not* be user-specified, but rather the user would specify data ports, with control ports being included by the HLS process). In a Vivado HLS pragma implementation, the ordering of transactions is specified with the aid of an `ap_wait()` function, which instructs Vivado HLS to insert clock cycles between IO operations within a specifically labelled section of code (a 'protocol region').

A code example demonstrating this manual interface specification technique can be found in [18].

15.5. Algorithm Synthesis

The theme of this section is the synthesis of functional hardware from designed C/C++/SystemC code. The limited scope of this chapter precludes a detailed treatment of all aspects, and rather we focus on a few topics that are of key interest, and representative of the design methods necessary to successfully develop systems using Vivado HLS. In other words, the content that follows in this section should be considered a 'taster' of the processes and possibilities of HLS algorithm synthesis, and not a comprehensive guide. Extensive further information can be found in the Xilinx User Guide 902, "*Vivado Design Suite User Guide: High-Level Synthesis*" [18].

One particular aim of this section is to highlight the control that the designer can exert over the eventual, synthesised hardware implementation through the use of directives, and the opportunities for readily generating and comparing alternatives without the need to significantly change the source code. We accomplish this through a feature case study on the synthesis of loops.

The remainder of this chapter will consider algorithm synthesis from select different styles of C designs, with illustrative examples. However, before doing so, it is important to recognise the metrics used to evaluate and constrain Vivado HLS designs in general.

15.5.1. Implementation Metrics and Constraints

As HLS translates the algorithm expressed by a C/C++/SystemC function into hardware, quantities are required to measure the characteristics of the implementation. This is especially useful when comparing implementation variations, i.e. the set of Vivado HLS 'solutions' generated by applying different directives to the HLS process.

Limits based on implementation metrics can be input by the user, and act as design constraints. These influence the behaviour of Vivado HLS in executing high-level synthesis: the tool attempts to meet the targets set where possible and, if it cannot meet them, it instead produces a 'best effort' RTL design. The rest of this section is devoted to defining the various metrics that may be used to both benchmark and constrain an HLS design.

Area / Resources

The most intuitive measure of an implementation is the hardware cost of building the circuit, in terms of the resources (or equivalently 'area') on the FPGA or PL. With resources inherently fixed for a particular target device, there may be motivation to minimise the cost of a particular Vivado HLS component, depending on the requirements and circumstances of the system as a whole.

By default, Vivado HLS seeks to minimise area, which implies time-sharing of hardware. This generally leads to increased *latency* and reduced *throughput* (both defined below).

Clock Period, Clock Rate, and Clock Uncertainty

The clock period metric expresses the minimum period and inversely the maximum clock frequency that can be supported by a design. It is a function of the physical characteristics of the target device, as well as the *critical path* of the RTL design synthesised from HLS. The critical path is defined as "*the longest combinatorial logic path between two clocked elements*", and it directly limits the maximum clock frequency. Critical path is normally managed in hardware designs via pipelining techniques (i.e. the strategic insertion of registers) and, in similar fashion, can be influenced by the use of pipelining in Vivado HLS.

Vivado HLS prompts the user to specify a target ***Clock Period***, together with a ***Clock Uncertainty***, and together these act as timing constraints. Where possible, the tool creates a design to meet the target minus the uncertainty value, based on Xilinx technology library information. The uncertainty figure is included to cover other factors not known at the HLS stage, in particular RTL synthesis, placement, and routing delays [18].

Latency

In Vivado HLS, the term 'latency' adopts its usual definition as the number of clock cycles between applying an input, and achieving the corresponding output. Latency can be examined at different levels of hierarchy, from the top-level function down to sub-functions, loops or specific sections of code. In the context of loops, latency refers to the completion of all iterations of the loop; the term *iteration latency* is used when referring to a single iteration. The total latency is equal to the iteration latency, multiplied by the number of iterations of the loop (the 'trip count').

Latency can also be specified as a design constraint, i.e. the maximum acceptable latency is defined by the user, and the Vivado HLS tool optimises the design (where possible) to meet the requirement.

Initiation Interval and Throughput

The Iteration Interval (II) is the number of clock cycles that separate the acceptance of inputs to the Vivado HLS design. Without applying directives, the initiation interval and latency may be the same, because the default behaviour of Vivado HLS is to optimise for area, resulting in a serial design. However, the strategic use of pipelining can reduce the iteration interval to much less than the latency of the design. On the other hand, this may increase the area of the design, so there is a trade-off involved.

Initiation interval corresponds directly to throughput, and is analogous to the relationship between clock period and clock frequency. Throughput expresses the rate at which data can be passed through the system. The best possible initiation interval is 1, meaning that new input samples can be accepted on every clock cycle, in which case the throughput is equivalent to the clock rate. Higher levels of throughput can sometimes be achieved through use of partial loop unrolling, or by replicating a synthesised function.

15.5.2. Data Types

As for interface synthesis, the choice of data types in a high-level algorithm has a fundamental impact on the synthesised hardware. The use of wordlengths that are longer than required can lead to a needlessly expensive implementation, in terms of the amount of PL resources required to create the design. It also has the potential to impact on other implementation metrics, such as latency, maximum clock frequency, and initiation interval.

The use of arbitrary precision data types, as reviewed in Section 15.3, is an effective mechanism for specifying appropriate wordlengths and hence contributing to an efficient overall implementation.

15.5.3. Pipelining

The term 'pipelining' is widely used in the context of hardware design to refer to the insertion of registers into a circuit, usually for the purpose of minimising the critical path (i.e. the longest combinatorial logic path between clocked elements), and hence maximising the achievable clock frequency. Consequently, we can think of data samples moving along the processing path in a regular, synchronous fashion, with storage of intermediate samples in the pipeline registers. Alternatively, it could be said that the samples are "in a pipeline".

When the term is used to refer to processor operation, it means that a task is broken down into a regular structure of subtasks that can be performed simultaneously by different 'pipeline stages' of the processor. The number of pipeline stages, and the operations performed by each of them, are fixed features of the processor architecture. For instance, a 5-stage pipeline might allow the processor to simultaneously: (i) read an instruction; (ii) decode an instruction; (iii) read data; (iv) execute; and (v) write a result, all on the same clock cycle, using a dedicated pipeline stage for each. Individual pipeline stages process consecutive samples, meaning that a new output can be computed on each clock cycle.

In HLS, pipelining has a meaning which is related to both of the above, but distinct. In particular, we can define the concept of pipelining as referring to the partitioning into sub-stages of an arbitrary set of dependent operations, whether these are combinatorial or clocked.

To define HLS pipelining further, we must abstract the detail of low-level operations and consider that pipelining relates to the segmentation of *logical processing stages*. Unlike the hardware case, these are not necessarily physical, combinatorial processing elements (although they could be). Pipelining is incorporated to permit the overlapping of operations, similar to the concept of pipelining in processors. However, unlike processors with fixed architectures, there are no restrictions on the nature of the operations themselves, because the FPGA fabric or Zynq PL provides a blank canvas with which to implement any arbitrary functionality.

The motivation for pipelining is to provide an opportunity for parallel processing, and thus to increase the throughput supported by the design. Pipelining can be applied as a directive in Vivado HLS, at the level of functions and loops. Over the next few pages, we will consider the pipelining of operations, while the pipelining of loops will be covered in Section 15.5.5.

Algorithm Execution and Data Dependency

It may be assumed that any algorithm expressed in software includes a set of functional steps, or operations. Each step depends on a certain set of data samples being available as inputs, and in some cases these data samples may not be available until a previous step has been completed. Consequently, there is an implied data dependency between the steps forming the algorithm, and a required order of execution.

A direct synthesis of the algorithm may therefore lead to a set of operations which must logically occur at the same time, due to the data dependency between them. In other words, all operations belong to the same processing *stage*, which must be entirely completed before new inputs can be processed. Whether the execution of the stage corresponds to a single clock cycle or multiple clock cycles, the important point is that there is no opportunity to begin computing the next output until the previous one has been completed.

To give a simple example, consider a function given in Figure 15.16, which comprises three processing steps, Op1, Op2, and Op3. The second step, Op2, relies on the outputs of the first step, Op1, while Op3 depends on the outputs from Op2. As such, and with no memory to store intermediate results, there is a dependency between the operations: the final output is generated only after all operations have been completed (in order). Therefore we can designate the group of Op1, Op2, and Op3 as a processing *stage*.

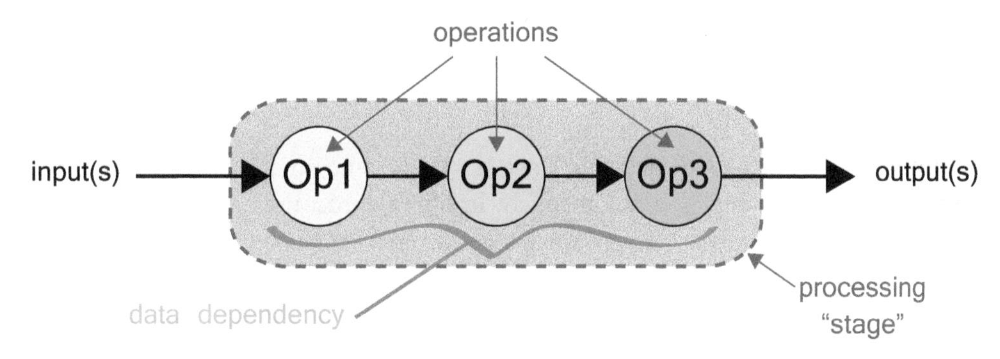

Figure 15.16: Example function with data dependencies between operations

In terms of processing data, the data dependency within the stage means that only one sample output can be processed at a time; the three operations must run sequentially on each set of input samples, and the next set of inputs cannot be accepted until the current output is ready. Thus the *data sample period* is equivalent to the total processing time of Op1, Op2, and Op3, and this directly governs the data throughput supported by the design. The other important metric, input-to-output *latency*, is equivalent to the individual latencies of the three operations. Both of these are indicated on the waveforms shown in Figure 15.17.

It may be noted that a clock signal is not drawn in Figure 15.17, and this is deliberate. At an abstract level, Op1, Op2, and Op3 may represent logic in a combinatorial path *or* a set of clocked operations. In fact, as pipelining has the effect of separating the stages, it will result in increased throughput irrespective of whether the circuit is clocked or combinatorial.

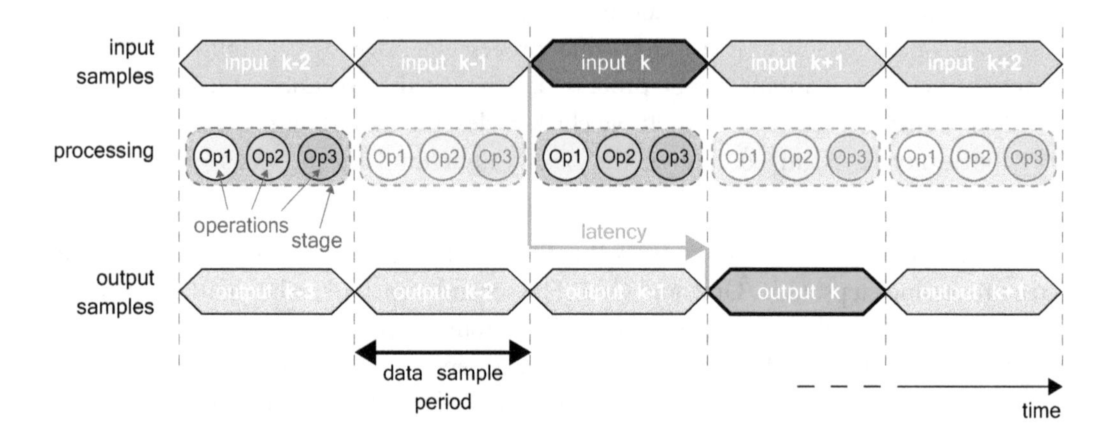

Figure 15.17: Waveform showing throughput and latency, before pipelining

Pipelining an Algorithm

Pipelining means that the processing stage is split up into smaller stages that can each process different data samples simultaneously. In other words, the data dependencies that caused a set of operations to be grouped together are broken up, and this allows the operators to execute in parallel.

In hardware terms, the method used to achieve this separation of stages is to insert registers between the new, smaller stages, which allows data samples to be held in memory. A direct consequence is that the overall latency from input to output increases with respect to the sample period (a disadvantage), however the sample period can be reduced because the stages are shorter. The latter has a direct impact on throughput, which is normally considered the more important performance metric. A further benefit is that, where the original stages represent a combinatorial logic path, the insertion of pipeline registers may also cause an increase in the maximum supported clock frequency.

From a hardware-oriented perspective, the pipelined function is equivalent to the signal flow graph shown in Figure 15.18. As before, the final outputs depend on signals that have rippled through Op1, Op2 and Op3, but now there is memory to hold the intermediate results. The outcome is that the three operators are now free to operate simultaneously, each acting upon consecutive input data samples as shown in Figure 15.19. This removes the data dependency between operations, and they can now be considered as each representing a separate processing stage.

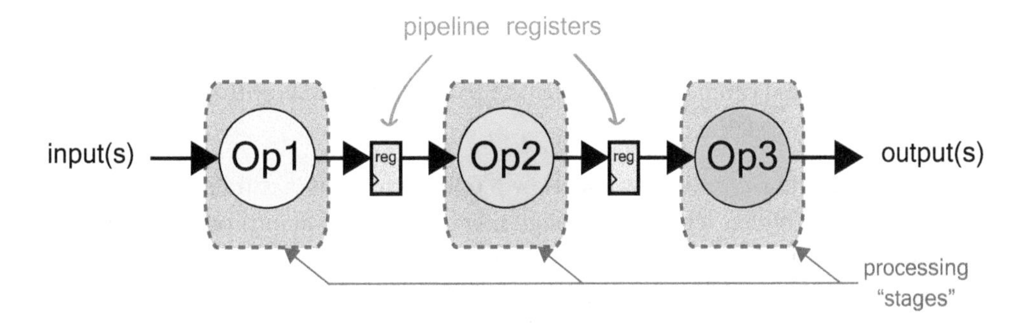

Figure 15.18: Partitioning of operations into separate stages via pipelining

The updated waveform diagram reflects the smaller stages (each effectively a single operation) and confirms a corresponding shortening of the data sample period. This means that the throughput has increased by a factor of three. It is also apparent that the operators can simultaneously compute results for consecutive samples; for instance, Op1 can process sample *k*+1 as soon as it has finished with sample *k*, and passed it to Op2. The latency is now greater with respect to the duration of a single stage, as the inserted pipeline registers add to the delay between input and output, but this is an acceptable overhead.

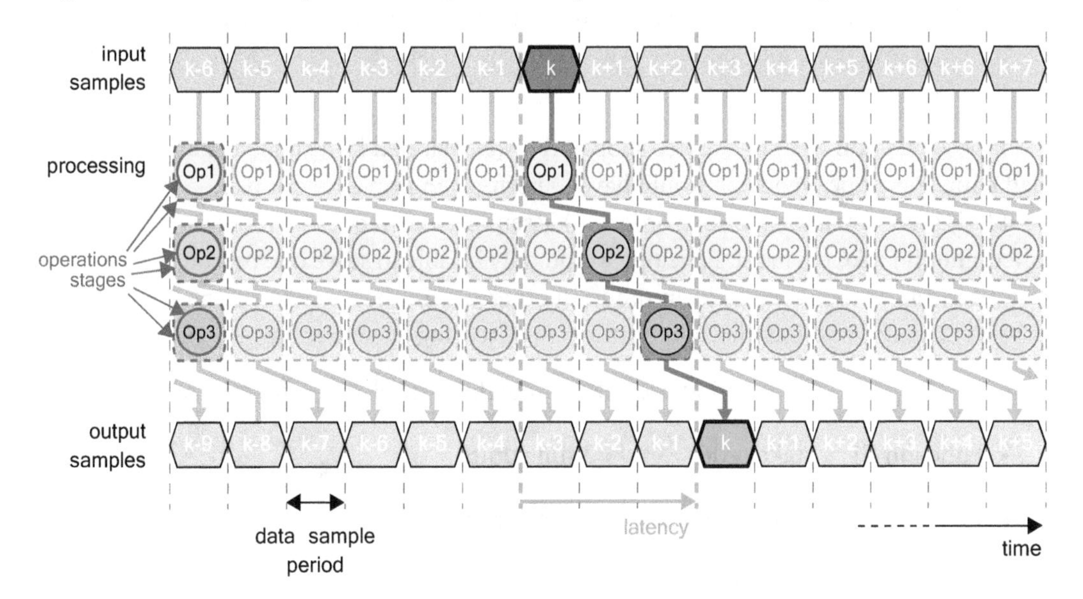

Figure 15.19: Waveform showing throughput and latency, after pipelining

15.5.4. Dataflow

As discussed in the previous section, pipelining is a method of increasing the concurrency of the hardware produced from a software description, and thus improving the throughput. Pipelining can be performed at the level of operations within a function, or on loops.

Dataflow pipelining (or simply 'dataflow') optimisation is a similar concept, but is actually applied at a higher level in the design hierarchy: it acts to improve the concurrency between functions. For example, a top-level design may include four functions, F1, F2, F3 and F4, with data dependencies between them. A direct synthesis of the top-level function would result in the sub-functions executing in a sequential, non-overlapping manner, analogous to the grouping together of operations depicted in Figure 15.16 (which is labelled a processing *stage*). With dataflow pipelining, the equivalent optimisation to pipelining is performed, in the sense that registers are inserted between functions to partition them into separate stages that can execute concurrently.

Given that functions are more complex than simple operations, (which was discussed in the context of pipelining), dataflow optimisation can actually go a step further, by analysing the content of functions and the dependencies between them. This might allow a further degree of function 'overlapping', with the effect of shortening overall execution latency as well as increasing throughput. For instance, it might be possible to start function F2 once F1 is 50% complete, rather than waiting until it has entirely finished. This is most easily illustrated by example, and here we will attempt to relate dataflow optimisation it to an everyday scenario.

Suppose that you went to a coffee shop and it was manned by a single person, whom we shall name Penelope. Penelope would have to perform multiple functions to serve you. She would: (i) greet you and take your order; (ii) assemble your requested food; (iii) make your coffee; and (iv) take payment. Penelope would only be able to serve one person at a time, by performing all of the above tasks in order: in other words, the latency (i.e. the total time to serve a customer) and the throughput (the rate of serving customers) would be limited.

For the purposes of illustration, we will assume that the following functions are defined:

- Function F1: take order 3 time units
- Function F2: prepare food 2 time units
- Function F3: prepare coffee 4 time units
- Function F4: accept payment 3 time units

Therefore, it will take Penelope a total of 12 time units to serve each customer. She can only serve one customer at once, so the customer throughput is 1 every 12 time units, and each customer must wait for 12 time units to complete their purchase.

Matters could be improved by adding more staff. Suppose that Penelope now has the assistance of Cameron, Hamish, and Isla. Cameron will greet customers and take their orders, Hamish will assemble food orders, Isla will take payments, and Penelope will make the coffees. As well as increasing customer throughput, latency can be reduced, because knowledge of the operations that comprise the functions allows them to be overlapped. For example, Cameron can ask for drinks orders first (to enable Penelope to start serving them) and then food orders (to enable Hamish to start assembling them), such that Penelope's function can begin before Cameron's has completed. Likewise, Isla can commence her function of taking payment at the till as soon as the complete order is known, and while Penelope and Hamish are still completing their serving functions.

These two variations on the coffee shop scenario are depicted in Figures 15.20 and 15.21. Originally, as shown in Figure 15.20, only one customer could be served every 12 time units (analogous to clock cycles), and customers also had to wait for 12 time units to complete their transactions. This is because Penelope had to do everything by herself, and could only complete one task (or 'operation') at a time.

Notice that, as a result of adding more staff (as in Figure 15.21), one customer can be served every four time units, rather than one in every 12. There is also a reduction in the time each customer has to wait, from 12 time units to 4. The improvement in throughput arises from the increased concurrency (different operations can now take place at the same

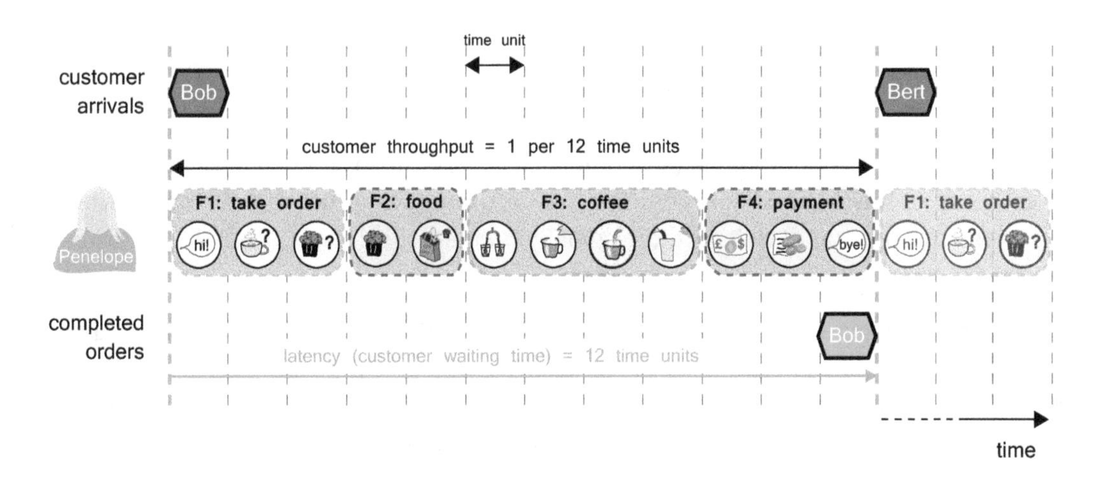

*Figure 15.20: Coffee shop example **without** dataflow optimisation*

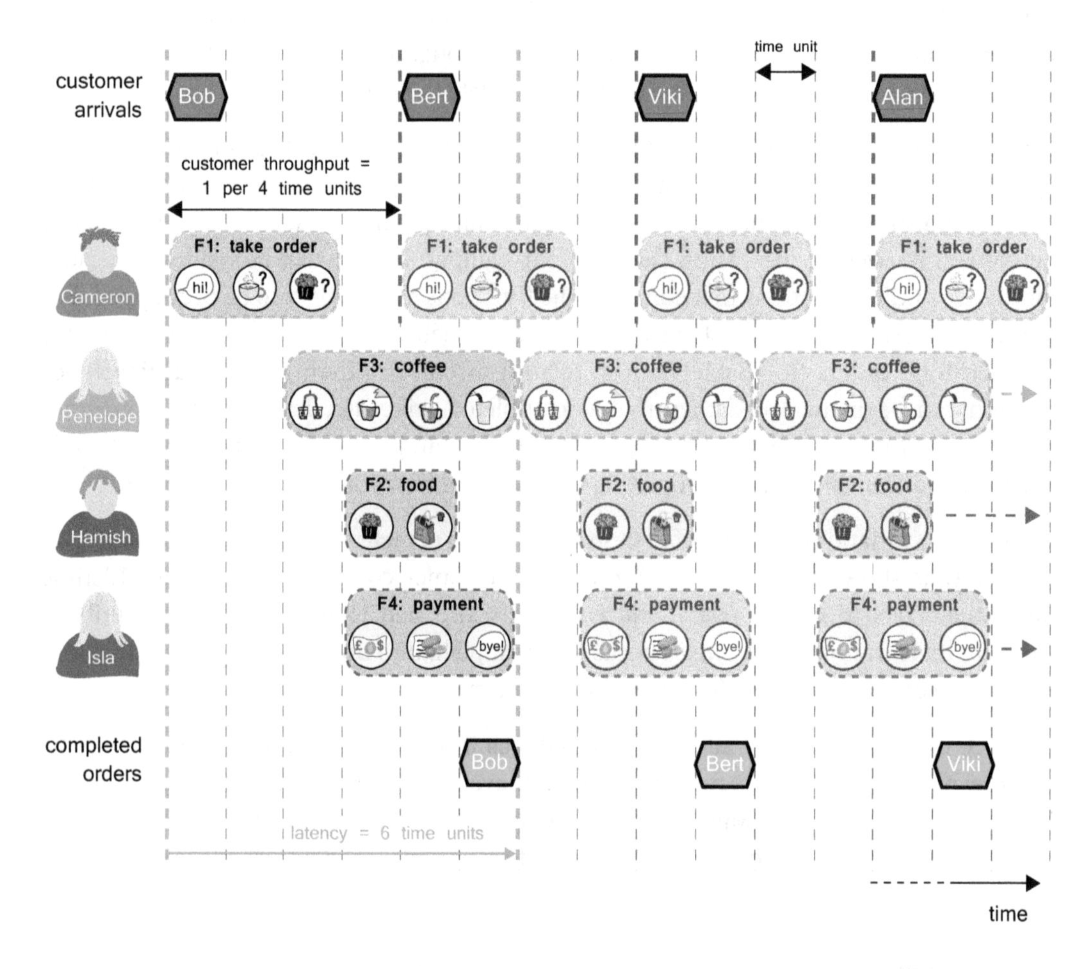

*Figure 15.21: Coffee shop example **with** dataflow optimisation*

time), while the reduction of latency (time a customer has to wait) is attributable to the overlapping of functions. Simply adding more staff, but having each of them undertake all functions in order (as Penelope did originally), would still leave Bob, Bert and friends waiting 12 time units to receive their orders!

We note in Figure 15.21 that the throughput is limited by Penelope's coffee making function, which takes 4 time units, whereas all of the others require less time to complete. If it were possible to pipeline *within* this function, e.g. to use one or more additional baristas to prepare the espresso, froth the milk, etc. concurrently, then further improvements could be achieved.

Similarly to the design of a hardware circuit, adding more concurrency (i.e. employing more staff) has the effect of improving performance. This is not 'free' as obviously the staff must be paid(!), but it may be desirable in order to obtain performance improvements.

The use of dataflow optimisation can be specified by directive in a similar fashion to pipelining. As we have seen through this example, the motivation for dataflow optimisation is to increase the obtainable throughput as a result of concurrent processing, and also to reduce latency by overlapping functions.

15.5.5. Algorithm Case Study: Loops

Loops are used extensively in software programming, and constitute a very succinct and natural method of expressing operations that are repetitive in some way. They are also used in a similar manner in HDLs, for example to iteratively instantiate and connect circuit components. However, an important difference is that the designer can prompt the loop(s) to be synthesised in different ways via the mechanism of directives in Vivado HLS. This contrasts with the use of loops in HDL, where code expressing loops is directly converted into hardware, usually resulting in prescribed and fixed architectures.

As an important software construct, loops are very well supported by Vivado HLS for hardware synthesis. Several loop optimisations can be made using directives, allowing the architecture of the resulting implementation to be altered with few or no changes required to the software code. During the remainder of this section, we will consider the default synthesis of loops, and architectural variations that can be achieved through the use of directives. Simple code examples are chosen in order to clearly make the necessary HLS-related points.

Default Loop Synthesis

By default, Vivado HLS seeks to optimise area, and therefore unless the designer directs otherwise, loops are automatically 'rolled', meaning that they time-share a minimal set of hardware. This means that the repetitive operations described by the loop are realised by a single piece of hardware implementing the body of the loop. To take a simple illustrative example, if a loop was designed to add the individual elements of two, 12 element arrays, then conceptually the implementation would involve a single adder (the body of the loop), shared 12 times according the number of loop iterations.

There is some latency associated with each iteration of the loop, and in this case the latency is affected by interactions with the memory interfaces at the inputs and output of the function. Memory interfaces are inferred due to the use of array arguments, according to the default interface protocols as discussed in Sections 15.4.3 and 15.4.4. Additional clock cycles are also required for entering and exiting the loop.

Code for this example is provided in Figure 15.22. Analysis of HLS synthesis with default settings shows that the overall latency is 26 clock cycles: 2 cycles each for 12 iterations (including reading the inputs from memory, adding, and writing the output to memory), and a further two clock cycles for entering and exiting the loop. Execution is illustrated in Figure 15.23.

```
void add_array (short c[12], short a[12], short b[12])
{
   short j;                              // loop variable

   add_loop: for (j=0;j<12;j++) {        // loop through elements (x12)
           c[j] = a[j] + b[j];          // addition operation
       }
}
```

Figure 15.22: Example code for a loop adding the elements of two arrays

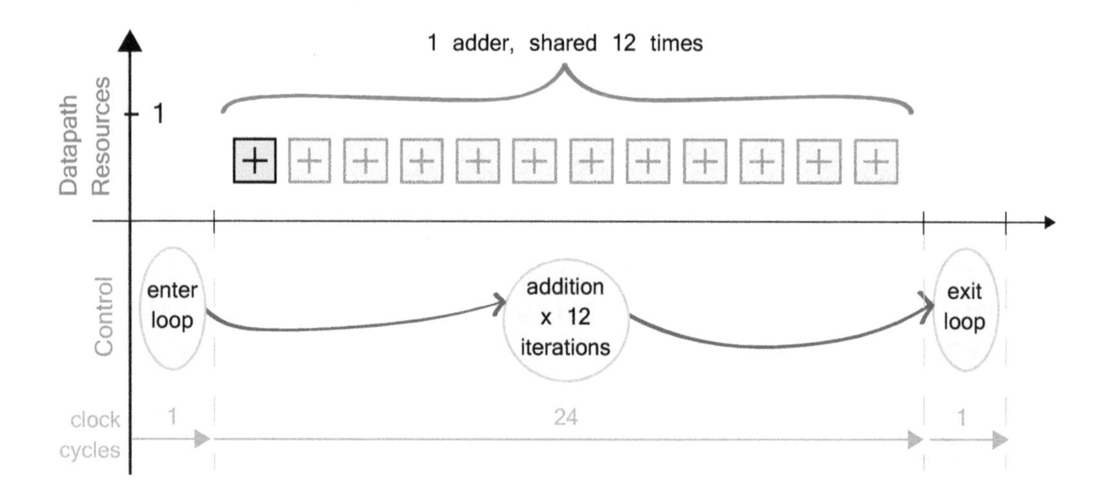

Figure 15.23: Extraction of addition loop into datapath and control logic

Simple Loop Architecture Variations

The default, rolled loop implementation may not always be desirable, but there are alternatives. Directives can be used to specify the architecture as summarised below.

- ***Unrolled*** — In a rolled implementation, a single instance of the hardware is inferred from the loop body and shared to the maximum extent. *Unrolling* a loop means that instead the hardware inferred from the loop body is created up to *N* times, where *N* is the number of loop iterations. In practice, the number of instances may be lower than *N* if other limiting factors are identified in the design, such as memory operations. The clear disadvantage of the unrolled version is that it consumes much larger area on the device than the rolled design, but the advantage is increased throughput.
- ***Partially unrolled*** — This constitutes a compromise between rolled and unrolled, and is typically used when a rolled implementation does not provide high enough throughput. If a rolled architecture represents minimal hardware cost but maximal time sharing (lowest throughput), and an unrolled architecture represents maximal hardware cost but minimal sharing (highest throughput), then we may try to find a different balance between the two. Control is exerted by the use of directives, and a number of different positions in the trade-off may be possible.

With reference to the upper section of Figure 15.23, which depicts a rolled architecture, fully or partially unrolling the loop would cause the number of datapath resources (adders) to increase, but to be shared to a lesser extent. Meanwhile, in the lower section of the diagram, the large, central state constituting the addition of array elements would require fewer clock cycles to complete. When fully unrolled, the implementation effectively does not contain a loop, and as a result the loop entry and exit clock cycles are saved, too.

With these observations in mind, the decision to select a rolled, unrolled, or partially unrolled implementation for the loop will be based on the specific requirements of the application, particularly in terms of the target throughput and any constraint on area utilisation that may apply.

Optimisation: Merging Loops

In some cases there might be two loops occurring one after the other in the code. For instance, the addition loop in the example of Figure 15.22 might be followed by a similar loop which multiplies the elements of the two arrays. Assuming that the loops are both rolled (according to the default mode), a possible optimisation in this case would be to merge the two loops, such that there is only one loop, with both the addition and multiplication operations being conducted within the single loop body.

The advantage of merging loops may not be immediately obvious, but in fact it relates to the control aspect of the design (recall from Section 14.4.5 that the C source code is analysed and decomposed into datapath and control components as part of the HLS process). Control is realised in the form of a Finite State Machine (FSM), with each loop corresponding to at least one state; thus the FSM can be simplified due to the merging of loops, as this results in fewer loops overall, and thus fewer FSM states. This is demonstrated by the code examples in Figures 15.24 and 15.25. The first example shows two separate loops, one each for addition and multiplication of the arrays, while the second shows the effect of merging the loops to create a single loop.

The `add_loop` represents 12 iterations of an addition operation (which takes 2 clock cycles), while the `mult_loop` represents 12 iterations of a multiplication operation (which takes 4 clock cycles). Therefore, the overall latencies of the two loops are 24 and 48 clock cycles, respectively. Merging the loops has the effect that the latency of the new, combined loop reduces to the longer of the original two loops, i.e. 48 clock cycles. One further clock cycle is saved due to the removed loop transition, i.e. the 'exit/enter' state in Figure 15.24.

The merging of loops can be controlled automatically using a Vivado HLS directive, and consequently there is no need to make an explicit change to the source code. The code provided in Figure 15.25 is for illustrative purposes only, i.e. to show the effect of applying the 'merge' directive.

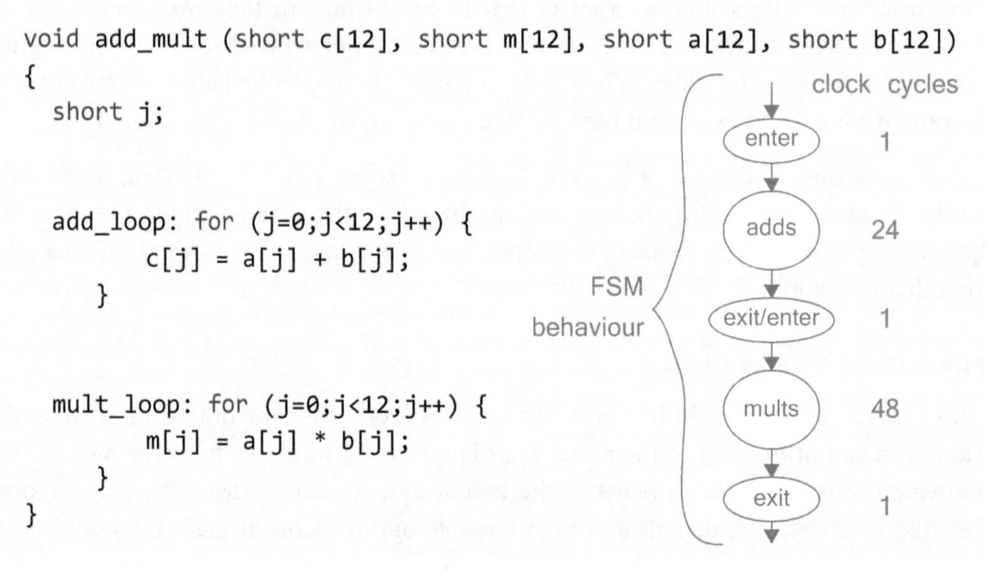

Figure 15.24: Consecutive loops for addition and multiplication within a function

```
void add_mult (short c[12], short m[12], short a[12], short b[12])
{
  short j;

  add_mult_loop: for (j=0;j<12;j++) {
        c[j] = a[j] + b[j];
        m[j] = a[j] * b[j];
     }
}
```

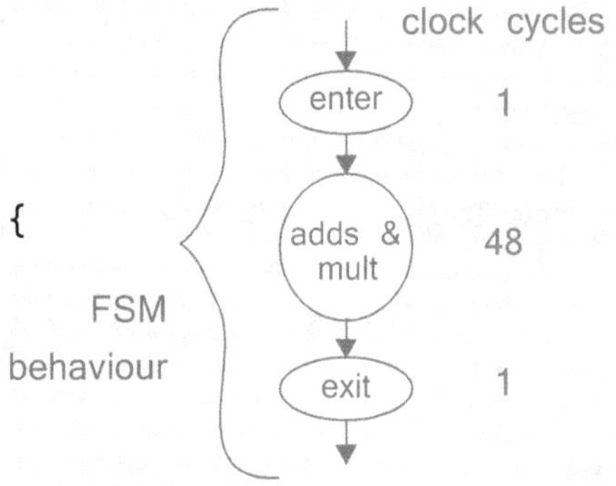

Figure 15.25: Merged addition and multiplication loops

Note that there are some practical restrictions on loop merging, with respect to compatibility and the limits of the loops to be merged. In our simple example, the limits were equal, but this may not always be the case; guidelines on this issue can be found in [18].

Nested Loops

Another common configuration is to nest loops, i.e. place one loop inside another. There may even be multiple levels of nesting. To give an example of a 2-level nested loop, suppose we extend our array addition example from linear arrays to 2-dimensional arrays. Mathematically, this is equivalent to adding two matrices, as shown in Equation (1).

$$\begin{bmatrix} f_{00} & f_{01} & f_{02} & f_{03} \\ f_{10} & f_{11} & f_{12} & f_{13} \\ f_{20} & f_{21} & f_{22} & f_{23} \end{bmatrix} = \begin{bmatrix} d_{00} & d_{01} & d_{02} & d_{03} \\ d_{10} & d_{11} & d_{12} & d_{13} \\ d_{20} & d_{21} & d_{22} & d_{23} \end{bmatrix} + \begin{bmatrix} e_{00} & e_{01} & e_{02} & e_{03} \\ e_{10} & e_{11} & e_{12} & e_{13} \\ e_{20} & e_{21} & e_{22} & e_{23} \end{bmatrix} \quad (1)$$

Now, in order to add the arrays, we must iterate through the rows, and for each row, iterate through the columns, adding together the two values for each array element. Coding the matrix addition operation requires an outer and an inner loop to cycle through the rows and columns, respectively, as shown by the code example in Figure 15.26. According to (1), there are 3 rows and 4 columns, and this determines the limits of the nested loops (note that indexing starts as zero, as per the usual convention).

Extending this idea, it would be possible to work with three dimensional arrays, or even higher dimensions, by increasing the levels of nesting in the loop structure.

```
void add_matrix (short f[3][4], short c[3][4], short d[3][4])
{
  short j,k;                                   // loop variable

  row_loop: for (j=0;j<3;j++) {                // iterate through rows
       column_loop : for (k=0;k<4;k++) {  // iterate through columns
           f[j][k] = c[j][k] + d[j][k];       // addition operation
      }
  }
}
```

Figure 15.26: Example code for nested loops adding the elements of two dimensional arrays

Optimisation: Flattening Loops

In the case of nested loops, we have the option to perform 'flattening'. This means that the loop hierarchy is effectively removed during high-level synthesis, while preserving the algorithm, i.e. all of the operations performed by the loop(s). The advantage of flattening is similar to merging: the additional clock cycles associated with transitioning into or out of a loop are avoided, meaning that the overall duration of algorithm execution reduces, thus improving the achievable throughput.

In order to explain flattening in further detail, it is necessary to clarify the terms *loop* and *loop body*. By *loop*, we refer to an entire code structure of a set of statements repeated a defined number of times. The statements inside the loop, i.e. the statements that are repeated, are the *loop body*. For instance, `column_loop` is a *loop*, and the statements contained within `column_loop` correspond to the *loop body*.

When loops are nested, and again taking the example of a 2-level nested structure, the outer loop body contains another loop, i.e. the inner loop. The outer loop body (including the inner loop) is executed a certain number of times; for instance, `row_loop` has 3 repetitions in the example of Figure 15.26, and hence there are 3 executions of the inner loop, `column_loop`. Each execution of the inner loop involves repeating the inner loop body a specified number of times, as well: in our example, a statement to calculate the matrix element `f[j][k]` is executed 4 times, where `j` is the row index and `k` is the column index.

The overhead of loop transitioning means that two additional clock cycles are required each time the inner loop is executed, i.e. one to enter the inner loop, and one to exit from it.

To clarify this point, a diagram depicting the control flow for our matrix addition example, and associated clock cycles, is provided in Figure 15.27. This represents the original loop structure. The process of flattening 'unrolls' the inner loop, and as a consequence, reduces the number of clock cycles associated with transitioning into and out of loops; specifically, the 'enter_inner' and 'exit_inner' states in Figure 15.27 are removed. These would have been repeated 3 times, hence 6 clock cycles are saved in total in this case.

In our simple 3 x 4 matrix addition example, the saving equates to 6 clock cycles, but in other examples this could be considerably higher (in particular where the outer loop iterates many times, or there are several layers of nesting), and thus there is a clear motivation to flatten loops. Similar to merging of loops, flattening can be achieved by a directive and does not involve explicit unrolling of the loop by manually changing the code. However, depending on its initial form, some manual restructuring may also be required in order to achieve a loop structure optimal for flattening [18].

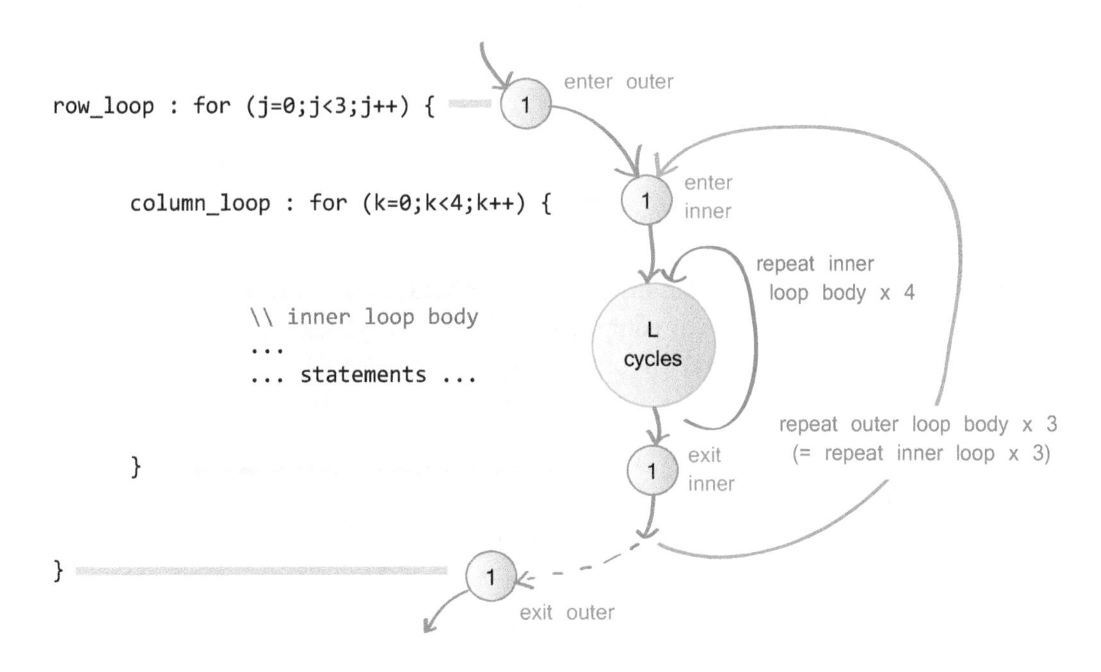

Figure 15.27: Control flow through the matrix addition, without flattening (clock cycles in circles)

Pipelining Loops

The direct interpretation of a loop written in C code is that executions of the loop occur consecutively, i.e. each loop iteration cannot begin until the previous one has completed. In hardware terms, this translates to a single set of hardware (as inferred from the loop body) that is capable of executing the operations of only one loop iteration at any particular instant, and which is shared over time according to the number of loop iterations.

This relates to our previous discussion on pipelining, i.e. as covered in Section 15.5.3. We observed that throughput is limited when a set of operations are grouped together into a processing stage. In the case of loops, the loop body (i.e. the set of repeated operations) forms such a stage, and without pipelining this would result in all stages operating in a sequential manner, and within them, all operations executing in a sequential manner. In effect, all operations from all iterations of the loop body would occur one after the other, similar to Figure 15.17 on page 314. The total number of clock cycles to complete the execution of a loop, N_{loop}, would therefore be:

$$N_{loop} = (J \times N_{body}) + N_{control} \tag{2}$$

where J is the number of loop iterations, N_{body} is the number of clock cycles to execute all operations in the loop body, and $N_{control}$ represents the overhead of transitioning into and out of the loop.

The insertion of pipelining into the loop means that registers separate the implemented operators within the loop body. Given that the loop body is repeated several times, this carries the implication that operations within loop iteration j+1 can commence before those of the loop iteration j have completed. In fact, at any instant, operations corresponding to several different iterations of the loop may be active.

As a consequence of pipelining the loop, the hardware required to implement the loop body is more fully utilised, and loop performance is improved in terms of both throughput and latency. The effect of adding a pipelining directive can therefore be considerable, especially where there are multiple operations within the loop body, and many iterations of the loop are performed.

When working with nested loops, it is useful to consider at which level of hierarchy pipelining should be applied. Pipelining at a certain level of hierarchy will cause all levels below (i.e. all nested loops) to be unrolled, which may produce a more costly implementation than intended. Therefore, a good balance between performance and resource utilisation may be achieved by pipelining only at the level of the innermost loop (for instance, `column_loop` in Figure 15.27).

15.5.6. Arrays

In Vivado HLS, array types normally represent storage, and hence arrays are usually synthesised into memories.

The memories inferred during the HLS process are mapped to physical resources on the PL — whether Block RAM, or distributed RAM constructed from logic slices — and therefore it is important to be aware of the size and shape of synthesised memories, and how these map to the resources available on the device. In fact, often it is desirable to exert influence over the synthesised memory to enhance its mapping to physical memory resources, and designers may achieve this through the use of directives.

A number of array optimisations are available, and these can be specified via directive as detailed below. Note that these are similar to the interface directives of the same name, but in this case they relate to elements within the circuit.

- ***Resource*** — The designer can choose to map an array in the C-based HLS source code to a specific memory resource.

- ***Array Map*** — Allows several small arrays to be combined into a single, larger array. This can bring the advantage of reducing the total amount of memory resources required (e.g. it may be possible to use a single Block RAM to implement all memories combined, rather than one Block RAM each for four discrete memories). Mapping can be either *horizontal* (arrays are concatenated to form an array with more elements), or *vertical* (array elements are combined, resulting in an array with longer words).

- ***Array Partition*** — This directive could be considered the opposite to the Array Map directive, in the sense that it allows the designer to determine the subdivision of a large array into a set of smaller ones. The motivation for partitioning is generally to improve the aggregate rate at which memory transactions can take place, e.g. a large dual port RAM can accommodate two transactions per clock cycle, whereas four smaller dual port RAMs can accommodate a total of eight transactions per clock cycle (two each). At the extreme, array partitioning can prompt the subdivision of an array into individual register elements.

- ***Array Reshape*** — This directive allows an array with many elements, each with short words, to be reshaped into an array with fewer, longer words. The motivation for applying this directive is to reduce the number of required memory accesses.

- ***Stream*** — Applying the stream directive prompts an array to be synthesised into a FIFO, rather than a RAM.

Where combination of memories is based on arrays with differing dimensions, this is accommodated by the tool as described in [18].

As is evident from the selection of array directives reviewed here, the designer can mould arrays in various different ways, according to his or her needs. In some circumstances, it is desirable to combine arrays in order to optimise resource usage. On other occasions, it may be more important to optimise memory bandwidth, and this can be accomplished by separating arrays into smaller memories, such that there are more memory access ports available.

Array manipulation can therefore be considered a flexible and powerful technique available to the HLS designer, whether the principal implementation aim is to minimum resource utilisation, or maximise performance.

15.6. Design Evaluation and Optimisation

As mentioned extensively throughout this chapter, designs developed in Vivado HLS are based on a structure wherein the source code is constant, and a series of variations ('solutions') are generated by supplying different constraints and directives. The designer can explore possibilities and iterate towards the optimum solution by tuning these parameters.

In this section, we consolidate some of these points, and consider in particular the steps a designer might take to refine a design via the mechanisms provided in Vivado HLS.

15.6.1. Design Constraints

Certain constraints can be applied within the design process to limit some aspect of the produced solution. The most commonly applied type of constraint is a timing constraint, which usually places an upper limit on the clock period (although other aspects of timing can also be defined). Another possibility is to constrain the latency of a design, meaning that the number of clock cycles between applying an input, and observing the corresponding output, is given an upper or lower limit. Aside from aspects of timing, the designer can also constrain the resources used to implement the desired functionality.

The HLS process may produce different results depending on the applied constraints. For example, if a maximum latency is prescribed, then the generated design may use more resources to implement the required algorithm; on the other hand, if resource utilisation is limited, then the resulting implementation is likely to employ time sharing, and thus to exhibit higher latency.

15.6.2. Synthesis Directives

We have seen through the discussion in this chapter that there are several types of directive available, which permit the designer to influence certain aspects of the synthesised hardware. For instance, we noted that interface constraints can be applied to prompt a particular type of protocol to be used, and that pipelining directives influence concurrency, latency and throughput.

Directives can be applied as TCL commands and consolidated in dedicated file, or inserted inline in the C/C++/SystemC source code, as pragmas. These methods may each be preferred for different reasons. To provide a common example, often the interface of a design is defined first, and fixed; therefore applying interface directives as pragmas means that these choices are consistent across all solutions. On the other hand, while actively exploring the algorithm synthesis design space, it is appropriate to keep the directives separate from the source code, as this makes it easier to manage the application of directives, and create 'fresh' solutions where desired.

15.6.3. Statistics and Reports

With each solution produced by Vivado HLS, an accompanying report will be generated to encapsulate its various statistics, including estimates of timing (clock) performance, latency, and resource utilisation. (Note that these are estimates, bearing in mind that full detail is not available until the more time consuming stages of RTL synthesis and implementation are undertaken.) Full details of the synthesised interface are also provided. Where applicable, the report will also contain details of individual loops within the design, in terms of their trip count (number of iterations), latency and initiation interval.

A further option is to generate a report consolidating the statistics from a set of solutions. Vivado HLS prompts the user to make a selection from all solutions in the project, and a report is prepared accordingly. This summary is a helpful method of comparing statistics across a set of solutions, and thus identifying the solution providing the best fit with requirements, or the trend associated with applying a certain directive.

15.6.4. Design Iterations and Optimisation

It should be apparent from previous discussion that the designer can exert considerable control over the results of HLS via the use of directives and constraints. The reports produced for individual solutions will help to identify issues such as memory bottlenecks, excessive loop latency, and over-utilisation of resources. He or she can then refine the existing directives, or augment them to better direct the synthesis process towards the optimum solution.

15.7. Exporting from Vivado HLS

Designs can be exported from Vivado HLS to form a selection of different outputs. These are provided to permit easy integration of Vivado HLS IP with other development tools in the Vivado and ISE design suites.

At the stage of exporting, there is an opportunity to 'Evaluate' the design, meaning that the stages of RTL synthesis and implementation are performed, and this produces a further report confirming actual values for resource utilisation and timing performance. This can be undertaken using VHDL or Verilog as the RTL language.

15.7.1. Vivado IP Catalog (IP-XACT Format)

The primary option is to export from HLS to the IP-XACT format, which allows the module to be integrated into a Vivado IP Integrator design. An option is presented allowing the designer to label the IP package, and to insert author and version information. The result is a zip folder residing in the 'impl\ip' subfolder of the relevant solution, and this represents the IP Catalog package.

Once in IP-XACT format, the IP produced from Vivado HLS can be easily shared and distributed.

15.7.2. System Generator for DSP

A further option is to export the HLS design as an IP block for use in System Generator. When doing so, the user is prompted to specify either ISE or Vivado, depending on their tool of choice for performing logic synthesis and implementation. That is to say, if the final System Generator system (including the IP originating from HLS) is intended to be synthesised using ISE, then ISE should be chosen when exporting from Vivado HLS.

15.7.3. Pcore for XPS

Users working with the XPS tool for embedded systems design have the option to export Vivado HLS IP as a *pcore*, which can be easily integrated into an XPS-based system.

15.8. Chapter Review

This chapter has provided a detailed look at the Vivado HLS development tool, which provides the facility to rapidly develop hardware designs from C-based software descriptions.

Although there is insufficient scope in a single chapter to review all aspects of the tool, we have covered the Vivado HLS environment, the use of data types (including the facilities for working with arbitrary precision formats), and various aspects of interface and algorithm synthesis. Several conceptual and code-based examples have been presented to illustrate the points made.

A running theme throughout has been the ability of the designer to influence the synthesis from input C code, via the use of directives and constraints, and thus to produce different 'solutions'. As part of this discussion, key performance and implementation metrics were reviewed, and examples were presented to illustrate the designer's ability to control these using directives.

Finally, it was noted that designs produced in Vivado HLS can be readily exported for integration into larger system projects, whether in IP Integrator, XPS, or System Generator.

15.9. References

Note: All URLs last accessed June 2014.

[1] D. C. Black, J. Donovan, B. Bunton and A. Keist, *SystemC: From the Ground Up*, 2nd Edition, Springer, 2009.

[2] M. Burton, J. Aldis, R. Günzel and W. Klingauf, "Transaction Level Modelling: A reflection on what TLM is and how TLMs may be classified", *Forum on Design Languages*, 2007, pp. 92-97.

[3] D. Gajski and R. Kuhn, "Guest Editors' Introduction: New VLSI Tools", *Computer*, vol. 16, no.12, pp.11 - 14, December 1983.

[4] D. Gadski, T. Austin and S. Svoboda, "What Input-Language is the Best Choice for High Level Synthesis (HLS)?", *panel session, Proceedings of the 47th ACM/IEEE Design Automation Conference (DAC)*, pp. 857-858, June 2010.

[5] GNU, *The GNU C Reference Manual.*
Available: http://www.gnu.org/software/gnu-c-manual/gnu-c-manual.html

[6] D. Große and R. Drechsler, *Quality-Driven SystemC Design*, Springer, 2010.

[7] IEEE Computer Society, "IEEE Standard for Standard SystemC Language Reference Manual", *IEEE Std 1666-2011*, January 2012.

[8] B. W. Kernighan and D. M. Ritchie, *The C Programming Language*, Prentice Hall, 1978.

[9] S. G. Kochan, *Programming in C: A Complete Introduction to the C Programming Language*, 3rd Edition, Sams Publishing, 2005.

[10] M. C. McFarland, A. C. Parker, and R. Camposano, "The High-Level Synthesis of Digital Systems", *Proceedings of the IEEE*, vol. 78, no.2, pp 301 - 318, Feb 1990.

[11] G. Martin and G. Smith, "High Level Synthesis: Past, Present and Future", *IEEE Design and Test of Computers*, Vol. 26, Issue 4, July/August 2009, pp. 18 - 24.

[12] A. Mathur, E. Clarke, M Fujita, and R. Urard, "Functional Equivalence Verification Tools in High-Level Synthesis Flows", *IEEE Design & Test of Computers*, July/August 2009, pp. 88 - 95.

[13] M. Meredith and S. Svoboda, "The Next IC Design Methodology Transition is Long Overdue", Open SystemC Initiative, February 2010.
Available: http://www.accellera.org/resources/articles/icdesigntrans/community/articles/icdesigntrans/ic_design_transition_feb2010.pdf

[14] D. M. Ritchie, "The Development of the C Language", *Proceedings of the 2nd History of Programming Languages Conference*, Cambridge, Massachusetts, April 1993.

[15] Xilinx, Inc., "UG634 - AccelDSP Synthesis Tool User Guide", v11.4, December 2009.
Available: http://www.xilinx.com/support/documentation/sw_manuals/xilinx11/acceldsp_user.pdf

[16] Xilinx, Inc., "UG835 - Vivado Design Suite Tcl Command Reference Guide", v2014.1, April 2014.
Available: http://www.xilinx.com/support/documentation/sw_manuals/xilinx2014_1/ug835-vivado-tcl-commands.pdf

[17] Xilinx, Inc., "UG871 - Vivado Design Suite Tutorial: High Level Synthesis", v2014.1, May 2014.
Available: http://www.xilinx.com/support/documentation/sw_manuals/xilinx2014_1/ug871-vivado-high-level-synthesis-tutorial.pdf

[18] Xilinx, Inc, "UG902 - Vivado Design Suite User Guide: High-Level Synthesis", v2014.1, May 2014.
Available: http://www.xilinx.com/support/documentation/sw_manuals/xilinx2014_1/ug902-vivado-high-level-synthesis.pdf

[19] Xilinx, Inc., "UG998 - Introduction to FPGA Design with Vivado High-Level Synthesis", v1.0, July, 2013.
Available: http://www.xilinx.com/support/documentation/sw_manuals/ug998-vivado-intro-fpga-design-hls.pdf

16

Designing With Vivado High Level Synthesis

This practical chapter and its corresponding tutorial will provide an introduction to High Level Synthesis for Zynq using the Vivado HLS tool.

16.1. Prerequisites

Before beginning this exercise it is recommended you have read up to and including Chapter 15, entitled *Vivado HLS: A Closer Look*, as this chapter provides a detailed account of the various techniques we will be exploring in the following tutorial.

16.2. Aims and Outcomes

The aim of this practical exercise is to provide a taster of the use of Vivado HLS in the design and implementation of systems for Zynq.

After completion of this tutorial you will be able to:

- Create Vivado HLS projects using both the graphical user interface and Tcl scripting.
- Synthesise interfaces through use of various HLS directives.

- Optimise a Vivado HLS design to meet various constraints, exploring multiple HLS solutions with differing directives.

The main aim of this tutorial is to give an appreciation of the potential of HLS in the development of Zynq systems. As such, we will not be catering to an in depth study of HLS and those seeking to learn more can continue to do so through **Xilinx UG871** - *Vivado Design Suite Tutorial: High-Level Synthesis.*

16.3. Overview of Exercise 3A

The first practical exercise introduces the Vivado HLS tool and the generation of HLS projects using it. We will demonstrate the creation of projects utilising the provided GUI, and also using Tcl scripting to facilitate rapid creation of projects whilst minimising repetitive tasks.

Exercise 2A is available on the website: www.zynqbook.com

16.4. Overview of Exercise 3B

This exercise will utilise the project created previously to explore the Vivado HLS toolflow with regards to design optimisation. In particular, this exercise concerns a matrix multiplier system which will be synthesised, analysed, and optimised through use of various directives. Completion of this exercise will allow you to

1. Create, analyse and compare multiple solutions for an HLS design.
2. Implement directives to meet critical performance requirements in a design.
3. Negotiate a trade-off between required performance and hardware utilisation.

Exercise 2B is available on the website: www.zynqbook.com

16.5. Overview of Exercise 3C

This final exercise will return to the matrix multiplier system and discuss how interfaces are synthesised in Vivado HLS from the source code, through use of directives and block level protocols.

Exercise 2C is available on the website: www.zynqbook.com

16.6. Possible Extensions

For a comprehensive introduction to Vivado HLS it is recommended you view Xilinx UG871 as mentioned in Section 16.2.

16.7. What Next?

The next practical chapter will cover the creation of IP for use in a Zynq system, using a variety of tools including HDL, Xilinx System Generator, MathWorks HDL Coder and Vivado HLS.

The concept of IP integration will be covered in the remainder of this section of the book.

17

IP Creation

This practical chapter focuses on the creation of custom IP modules for implementation on the Zynq-7000 platform. The IP creation methods used in the practical exercises presented in this chapter will coincide with those detailed in Chapter 13 - IP Block Design.

17.1. Aims and Outcomes

The overall aim of these practical exercises is to introduce the IP creation flow of the various languages and tools available for the creation of IP. All of the individual IP modules created during this practical tutorial will be brought together to form part of a DSP based system which will be implemented in a future practical in Chapter 20 — Adventures with IP Integrator.

After completion of this tutorial you will be able to:

- Create IP blocks in the following languages/tools:
  - HDL
  - MathWorks HDL Coder
  - Xilinx Vivado HLS
- Be familiar with the Vivado IP Packager flow

17.2. Overview of Exercise 4A

In this exercise a simple IP module will be created in HDL (VHDL) which will allow the LEDS on the ZedBoard to be controlled by software running on the PS. This will utilise the *Create and Package IP Wizard* in Vivado to create an AXI-Lite interface wrapper which will allow the custom IP to be connected as an AXI-Lite slave to the Zynq processor. The IP packaging process will be introduced, detailing the automatic AXI-Lite interconnect detection and the inference of memory mapped registers. The external LED pins will be mapped to the LED interface in the IP Integrator design by creating a new XDC file.

The steps involved in this Exercise are:

1. Create a new project in Vivado IDE.
2. Invoke the *Create and Package IP Wizard* and create a new AXI-Lite slave IP.
3. Add custom functionality to the generated IP template.
4. Package the newly customised IP with IP Packager.
5. Add the packaged IP to the IP Catalog.
6. Create a block diagram and connect the LED IP to the Zynq processor via the AXI interconnect.
7. Generate and export the hardware design to the SDK.
8. Create a simple software application to test that the custom IP functions correctly.

Exercise 4A is available on the website: www.zynqbook.com

17.3. Overview of Exercise 4B

An adaptive Least Mean Squares (LMS) filter will be created and tested in MathWorks Simulink. HDL code for the LMS subsystem will then be generated through the use of HDL Coder, and packaged as an IP core for implementation in Vivado. This will provide a good introduction to the HDL Coder flow, which is capable of creating IP Integrator compatible IP by automatically generate an AXI-Lite interface that adheres to the correct signal naming conventions. The generated IP will then be imported into a Vivado IDE project and packaged with IP Packager.

An adaptive filter is a form of self-learning filter, capable of adapting to a channel, or a particular set of signals, rather than being designed with a single filter characteristic in mind. The LMS algorithm is one such approach to adaptive filter design, using an iterative weight update algorithm to update to coefficients of an FIR filter to best construct a desired signal from one corrupted with noise. A block diagram of an LMS filter is provided in Figure 17.1.

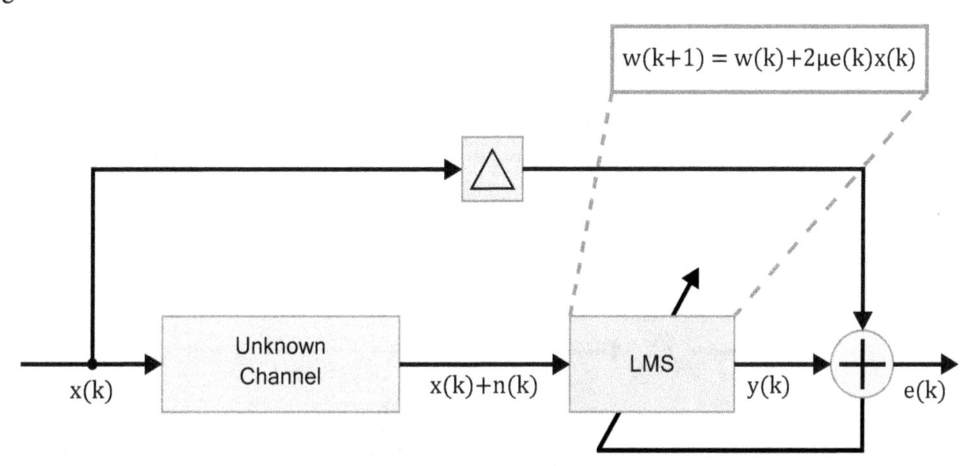

Figure 17.1: Block diagram of LMS filtering of noise created by an unknown channel

The steps involved in this Exercise are:

1. Invoke Simulink and create a LMS system.
2. Explore the fixed point signal types required for HDL generation.
3. Test the LMS to ensure that it is working correctly.
4. Invoke the HDL Coder workflow and set the required parameters to generate a Xilinx compliant IP core.
5. Import the HDL Coder generated IP into the Vivado IP Catalog.

Exercise 4B is available on the website: www.zynqbook.com

17.4. Overview of Exercise 4C

In this exercise, Vivado HLS will be used to create an NCO from an existing C-code implementation. This will build upon the practical examples that were covered in Chapter 16 — Designing With Vivado High Level Synthesis, which introduced the features of Vivado HLS.

The NCO implementation will be run through Vivado HLS to generate HDL code which can be packaged as an IP core for inclusion in Vivado IP Integrator projects.

A common method of implementing NCOs is to use a sine Lookup Table (LUT). A sine waveform can be generated by stepping through the entries of the LUT. When the end of the table is reached we simply wrap around to the start of the table, thereby creating a periodic waveform. This scheme is depicted in Figure 17.2. The ramp function is generated by a fixed point accumulator. The input to the accumulator is added to its latest value on each step. The magnitude of this input value therefore controls the frequency of the oscillator.

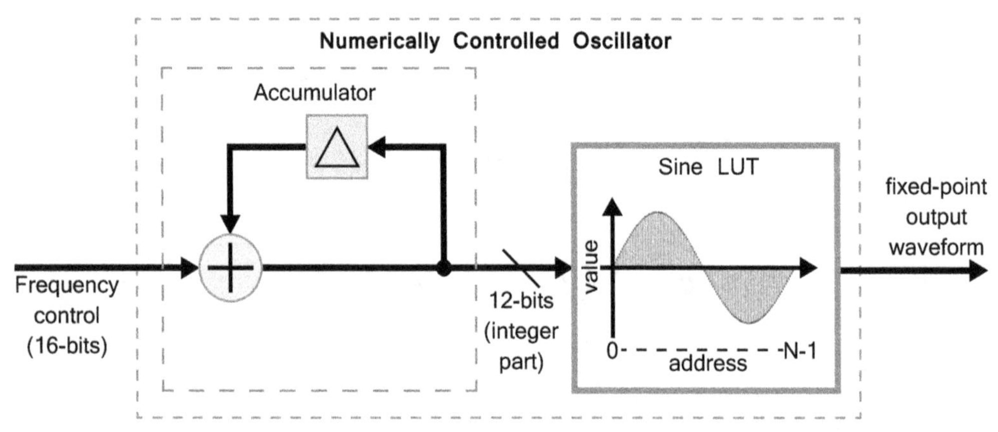

Figure 17.2: Architecture of a Numerically Controller Oscillator (NCO)

The steps involved in this Exercise are:

1. Invoke Vivado HLS and import the existing NCO C-code algorithm implementation.

2. Simulate the NCO C-code algorithm using the provided C-code testbench file.

3. Export the NCO algorithm as a HDL IP core.

Exercise 4C is available on the website: www.zynqbook.com

17.5. Possible Extensions

Having completed Exercise 4C, there are some possible variations you can introduce to personalise the developed system. For example, you could:

- Build upon the IP core created in Exercise 4A, and use HDL to create an IP core which reads the values of the DIP switches and/or push buttons on the ZedBoard.
- Create further IP blocks for the system, using the method of your choice, which introduce audio effects such as echo or reverb.

17.6. What Next?

This set of practical exercises rounds off the topic of IP creation.

Next, we move onto the topic of IP reuse and integration, which focuses on the IP-centric system design approach, and the various IP libraries that are available. The AXI interconnect will be looked at in greater detail, and further practical exercises will follow, with the main focus being on using IP Integrator for overall Zynq embedded system design.

18

IP Reuse and Integration

The Vivado Design Suite provides a design flow that revolves around the IP blocks and the ability to include them in your design from a number of different design sources [6].

In this chapter we will explore the IP-centric system design approach that is presented by the Vivado Design Suite, taking a closer look at some of the IP libraries that are available. The IP integration tools, such as IP Packager and IP Integrator will also be introduced.

18.1. Overview

An overview of the IP-centric design flow that is enabled by Vivado Design Suite is provided in Figure 18.1, and revolves around a central IP repository called the ***IP Catalog***, which amalgamates IP from a variety of sources [6]:

- Vivado Design Suite IP
- Modules from external Xilinx design tools such as Vivado HLS and System Generator.
- Third-party IP modules

The various forms of IP can be instantiated and configured, as well as interactively connected, via IP Integrator which provides a block based environment in which to create entire embedded system designs [5].

The Vivado Design Suite IP-centric design flow revolves around the philosophy of IP design, reuse and integration, allowing you to create your own IP modules (see Chapter 13), but also to package your IP designs for repeat use and easily integrate third-party IP. It is the process of IP reuse and integration which we will be exploring in this chapter.

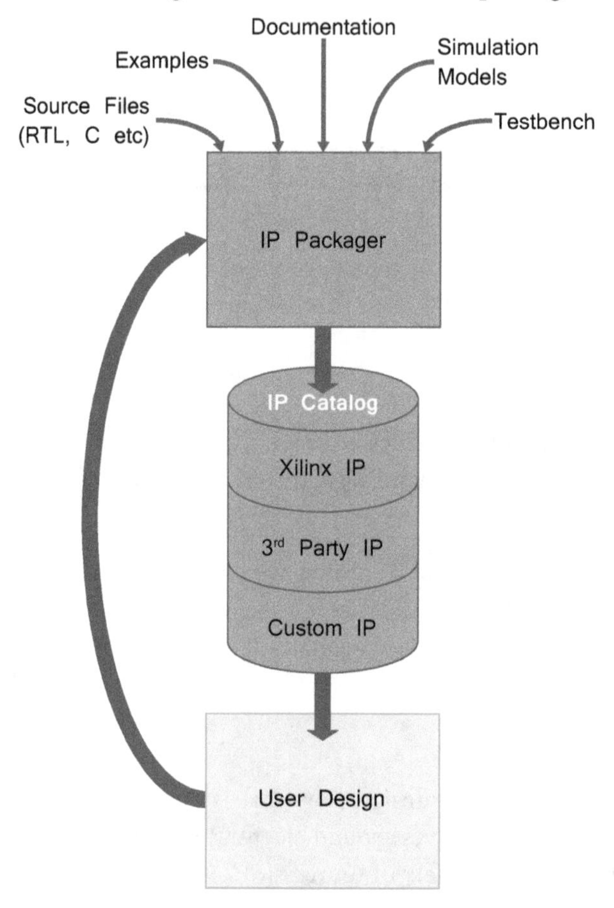

Figure 18.1: Vivado IP-centric flow

18.2. System Design — A System-Level Approach

The Vivado Design Suite provides an environment that easily allows you to configure, implement, verify and integrate IP. The form of the IP can be anything from logic designs, soft processor cores, C-based algorithm designs and DSP modules. All IP can be configured and verified as an individual module or as part of a larger, system-level design. Custom IP is packaged for use in the Vivado Design Suite according to the IP-XACT standard before being available in the IP Catalog. A large selection of Xilinx IP supports the AXI4 interconnect standard to enable easy system-level integration. The AXI interconnect is covered in Chapter 19.

An overview of the Vivado Design Suite system-level design flow is provided in Figure 18.2.

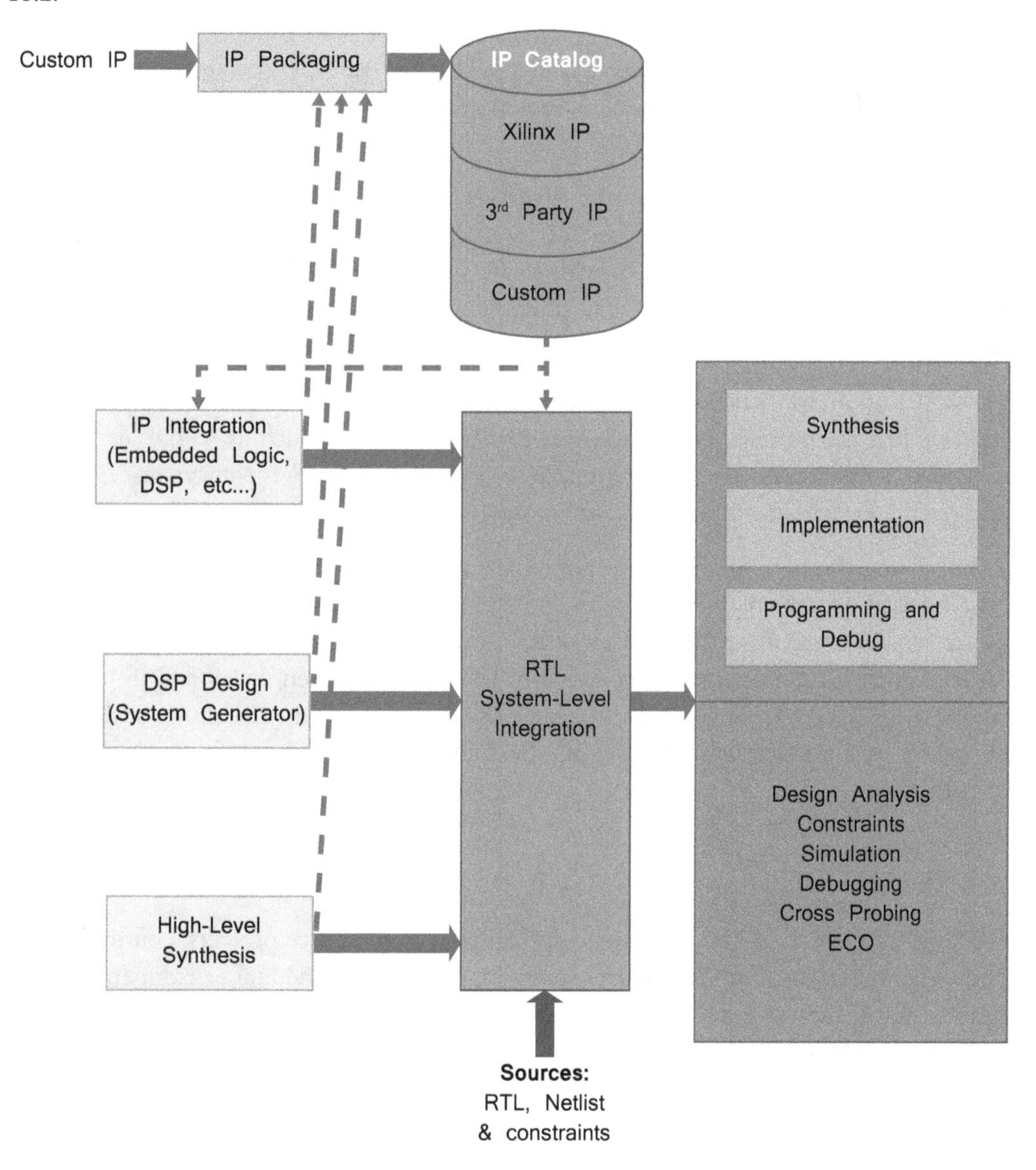

Figure 18.2: Vivado Design Suite system-level design flow

18.3. IP-XACT

The IP-XACT standard is an eXtensible Markup Language (XML) schema for documenting IP using metadata, which is both human readable and machine accessible [2], along with an Application Programming Interface (API) which allows software tools access to the stored meta-data [1]. The tools that are compatible with the IP-XACT standard are able to interpret, configure, implement and change IP blocks that are compliant with the IP meta-data description [1]. This provides all users of the standardised IP - whether it be the IP providers, EDA vendors or systems designers — with a consistent IP interface to facilitate IP reuse.

The standard was developed by the SPIRIT Consortium — a group of IP provider, electronic design, EDA and semiconductor companies — and is now managed by Accellera Systems Initiative. IP-XACT also provides software views, file lists, protocol standardisation and the ability to further extend the schema for additional design and flow information [1].

IP-XACT in no way describes the hardware functionality of IP, but instead describes the interface. It is therefore not a replacement for RTL languages such as VHDL or Verilog, embedded software or documentation, but is instead complementary: an IP-XACT component provides key information about the IP.

The standardisation of IP-XACT addresses a number of issues with IP reuse and provides ease of system integration and integration, as well as the ability to automate connections between IP. All of this amounts to the ability to drastically reduce time-to-market, which is an ever increasing concern. IP-XACT has been developed to provide a standardised data exchange format which is robust enough to allow design flow automation and verification as well as the flexibility to be used by multiple IP designers, vendors, EDA vendors and end users.

18.4. IP Libraries

A large number of IP libraries exist, varying from open source projects, commercial IP vendors as well as in-house IP libraries solely for use by an individual or company. In this section we will further explore the various sources of IP, covering any limitations which exist and provide some specific examples.

18.4.1. Vivado IP Catalog

The Vivado IP Catalog presents a centralised repository for all of your IP modules, including Xilinx, third-party and custom IP which can be shared between multiple end-users to enable effective design reuse. The key features of the Vivado IP Catalog include [6]:

- All available IP is based on the IP-XACT standard, providing an open and consistent IP interface.
- A consistent interface by which to access all Xilinx IP from a common, easy access repository.
- Support for multiple repository locations, including network storage locations, allowing a consistent IP deployment environment for external or custom developed IP.
- Integrated IP design examples which provide the ability to evaluate IP as an instantiated source in a Vivado project.
- Global RTL synthesis of IP with the ability simulate using behavioural simulation models or synthesisable RTL.
- The ability to output simulation models, HDL example designs and instantiation templates on-demand.
- Instantly accessible options to customise and generate IP using the Vivado IDE or Tcl automatic scripting.

18.4.2. Third-Party

In recent years there has been a shift to companies designing and licensing IP cores, rather than manufacturing and selling physical silicon products. A good example of such a company is ARM Holdings who develop and license the ARM processors which form part of the Zynq platform. ARM Holdings do not manufacture any physical processor chips, and instead licenses its designs as IP modules.

Third-party IP vendors can range from an individual who licenses a single design all the way up to a large, multi-billion dollar corporations with thousands of individual IP designs available. Some companies licence IP cores that are optimised for Xilinx devices, with many such companies being members of the Xilinx Alliance Program — a worldwide ecosystem of collaborating companies which work with Xilinx to further the development of All Programmable technologies [7]. Examples of Xilinx-specific IP that is available from Alliance members include memory controllers, image and video codecs, motor controllers and graphics accelerators. As most IP available from Alliance members is optimised for use on Xilinx devices, the use of their IP can help to maximise performance while minimising resource utilisation. Furthermore, most of the available IP provides support for the Xilinx design flow, allowing it to be easily integrated in your design.

There is also a large range of IP vendors who develop and license generic IP modules which are designed to be non-vendor-specific. Such generic IP cores are generally packaged as synthesisable RTL, although in some cases a gate-level netlist is provided. When provided as RTL, the end-user has the ability to modify the IP design to a certain degree, although vendors offer little support for modified designs. A gate-level netlist can be compared to an assembly-code listing in computer programming, and gives the IP vendor a certain degree of security against the reverse engineering of their designs.

More recently, the open source community has begun developing IP cores and providing them free-of-charge. Generally, open source IP cores are distributed under one of two licenses: the GNU Lesser General Public License (LGPL) or the modified-BSD license. In terms of delivery, open source IP is usually provided as generic synthesisable RTL; it is rare to find an open source IP core that is truly vendor-specific. The most notable open source IP provider is OpenCores which is the world's largest community for the development of open source IP cores [3].

There are a number of advantages and disadvantages to using both vendor-specific and non-vendor-specific IP cores. These are summarised in Table 18.1.

Table 18.1: Advantages and disadvantages of IP types

	Vendor-Specific	**Non-Vendor-Specific**
Advantages	Guaranteed to work straight out of the box.	Can be found for free (open source).
	Optimised for a specific device, in terms of power, performance and resource efficiency.	One IP core can be used across multiple devices.
Disadvantages	Not portable (will only work with a small amount of devices).	Will require modifications to comply with tool chain naming conventions.
	Can be expensive.	Not optimised for a specific device.

18.4.3. Custom IP

Custom IP which is created by any of the methods described in Chapter 13 can be easily integrated into the Vivado IP Catalog to add to the comprehensive collection of Xilinx IP. Depending on the design method of choice, the IP may be able to be imported to the IP Catalog directly. IP created using System Generator or Vivado HLS, for example, are packaged in a format that can be imported into the IP Catalog. For other design methods such as HDL design and HDL coder, the custom IP will have to be processed by IP Packager in order to be incorporated in the IP Catalog.

For more information on IP Packager, please refer to Section 18.5.2.

18.5. IP Integration

The Vivado Design Suite provides you with all the tools necessary to integrate IP, both Xilinx and third-party, into your Zynq system designs. Two of those features, IP Integrator and IP Packager will be reviewed in the following section.

18.5.1. IP Integrator

Vivado IP Integrator provides a graphical and Tcl based IP and system-centric design development environment that offers a "correct-by-construction", automated development flow [4]. The functionality, built in to the Vivado Design Suite, provides a platform and device aware environment which can auto-connect fundamental IP interfaces, as well as supporting one-click IP subsystem generation, real-time DRCs and powerful debugging capabilities [4].

IP integrator allows designers to work at interface level to accelerate the creation of complex systems in Vivado, ensuring that designs and IP are correctly configured. The automatic generation of device drivers and address generation, coupled with the automatic interfacing of separate IP modules, streamlines the assembly of a design, making the whole process quicker and easier than ever [4].

18.5.2. IP Packager

Vivado IP Packager enables third-party IP developers to quickly prepare IP for integration in the Vivado IP Catalog, allowing them to include any custom IP into a Vivado Design Suite design. The IP Packager flow ensures that the end IP user always has a consistent experience when using IP from the Vivado IP Catalog, whether it be Xilinx, third-party or custom developed IP.

Using the IP packaging and usage flow in Figure 18.3 as an example, existing IP is packaged using IP Packager which packages the IP source and data files into a ZIP file. The ZIP file can then be integrated into the Vivado Design Suite IP Catalog. Once the IP is selected in a Vivado project, the IP is treated like any other IP module from the IP Catalog allowing the user to customise the IP through parameter selections which are then used generate an instance of the IP.

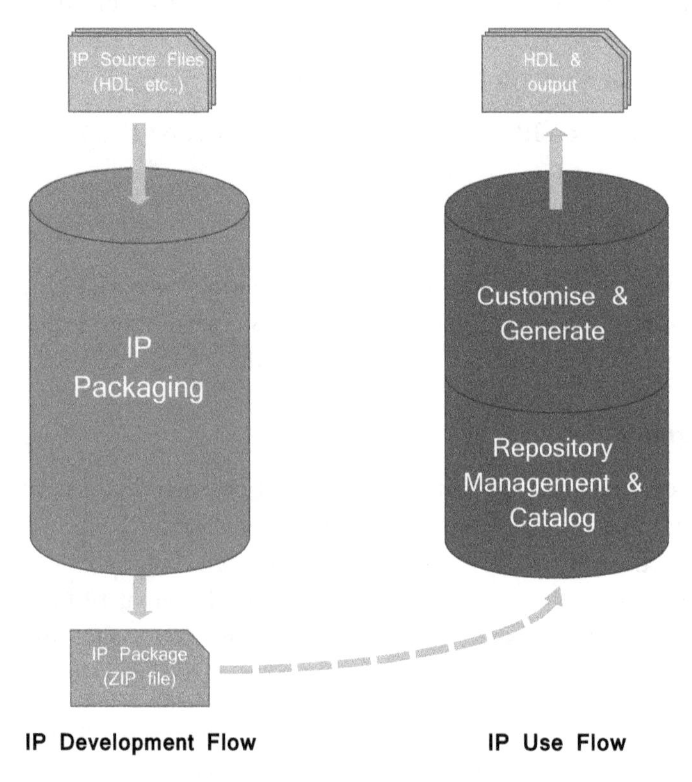

Figure 18.3: IP packaging and use flow

The primary output of IP Packager is an IP-XACT component file that can include default GUI files incorporated in the ZIP file, along with regeneration and report files [6].

One of the key benefits of using IP Packager is that custom IP can be packaged using the IP-XACT standard, without the need to fully understand the details of the IP-XACT coding schema. Some background information on the IP-XACT standard has been provided in this chapter, but it should be noted that no prior knowledge of IP-XACT is required when using IP Packager. The purpose of IP Packager is to take care of the underlying IP-XACT XML coding, and to allow the user to focus solely on creating the IP functionality.

18.6. Chapter Review

This chapter has introduced the concept of IP reuse and integration, providing an overview of the Vivado IP-centric design flow and the system-level approach to design. A discussion of the various ways in which IP can be obtained, including vendor-specific, generic and open source IP was provided, along with a discussion of the various advantages and disadvantages of each.

The tools available in the Vivado Design Suite for the reuse and integration of IP have also been introduced, with specific focus on the IP-centric design flow. IP Integrator and IP Packager features have been discussed as well as an overview of the IP-XACT IP meta-data documentation standard which is widely adopted throughout the IP design, semiconductor and EDA system design industry.

18.7. References

NOTE: All URLs last accessed June 2014.

[1] "IEEE Standard for IP-XACT, Standard Structure for Packaging, Integrating, and Reusing IP within Tool Flows", IEEE Standard 1685-2009, February 2010.

[2] M.v. Hintum and P. Williams, "The Value of High Quality IP-XACT XML". Available: http://www.design-reuse.com/articles/19895/ip-xact-xml.html

[3] OpenCores, webpage. Available: http://opencores.org/

[4] Xilinx, Inc, "Vivado Integration - Vivado IP Integrator", webpage. Available: http://www.xilinx.com/products/design-tools/vivado/integration/index.htm

[5] Xilinx, Inc, "Vivado Design Suite User Guide: Design Flows Overview", UG892, v2014.2, April 2014. Available: http://www.xilinx.com/support/documentation/sw_manuals/xilinx2014_2/ug892-vivado-design-flows-overview.pdf

[6] Xilinx, Inc, "Vivado Design Suite User Guide: Designing with IP", UG896, v2014.1, May 2014. Available: http://www.xilinx.com/support/documentation/sw_manuals/xilinx2014_2/ug896-vivado-ip.pdf

[7] Xilinx, Inc, "Xilinx Alliance Program", webpage. Available: http://www.xilinx.com/alliance/

19

AXI Interfacing

This chapter will introduce the AMBA AXI Protocol used with IP in a Zynq system. The various interfaces provided by the AXI4 protocol will be discussed with reference to how they vary in operation and which applications they are best suited for. Finally, implementing IP with AXI support in Xilinx Vivado IP Integrator will be presented.

19.1. Development of AXI

AXI is a protocol belonging to the ARM AMBA family of microcontroller buses. The AMBA protocol is an open standard on-chip interconnect specification, allowing the connection and management of many controllers and peripherals in a multi-master design. Much like the requirements for the rest of the AMBA family it is targeted at high-performance, high-frequency system designs. The AXI protocol is optimised for FPGA implementation through coordinated development with Xilinx and is used as a means of communication between IP cores of an FPGA design.

The AXI protocol in particular exhibits the following key features[1]:

- address/control phases are separate from data phases
- byte strobes enable unaligned data transfers
- burst-based transactions possible with only start address issued
- read and write data channels are separate allowing low-cost Direct Memory Access (DMA)

- multiple outstanding addresses can be issued
- transactions can be completed out-of-order
- register stages are easily added for timing closure

Whilst AMBA was first introduced in 1996, it wasn't until the release of AMBA 3.0 in 2003 that the first version of AXI was included. AMBA 4.0, released in 2010, includes the latest version of AXI known as AXI4[2].

19.2. Variations of AXI4

There exists three types of AXI4 interface, each suited to a different nature of application[4].

- **AXI4** — The high-performance interface, suited for memory mapped communication allowing bursts of up to 256 data transfer cycles per address phase.
- **AXI4-Lite** — A light-weight variant of the interface, used for memory mapped single transactions. This variant has the benefit of a smaller logic footprint with a simplified interface. This variation does not support burst data and so only provides a single data transfer per transaction.
- **AXI4-Stream** — No address phase means this is not memory mapped and allows for an unlimited data burst size. A single channel is defined for the transmission of streaming data, modelled after the Write Data Channel in Figure 19.1 but allowing bursts of an unlimited amount of data. Connection is from master to slave only, so if bidirectional transfers are required both peripherals must be of type master/slave.

19.3. AXI Architecture

The AXI protocol features burst-based transactions, with each containing address and control information on the address channel. Several AXI masters can be connected to several AXI slaves through an AXI interconnect. An AXI master transfers data to an AXI slave through the AXI interconnect using a write data channel (or a read data channel from slave to master). The write transactions in particular have an additional write response channel, as all data flows from master to slave and as such this is used for the slave to signal completion of a write transaction. The following figures demonstrate communication between an AXI master and slave. Note that the AXI interconnect is omitted in this demonstration.

Figure 19.1 shows the write channel architecture, where address and control data is passed from master to slave before a burst of data is transmitted, and a write response signalled following completion.

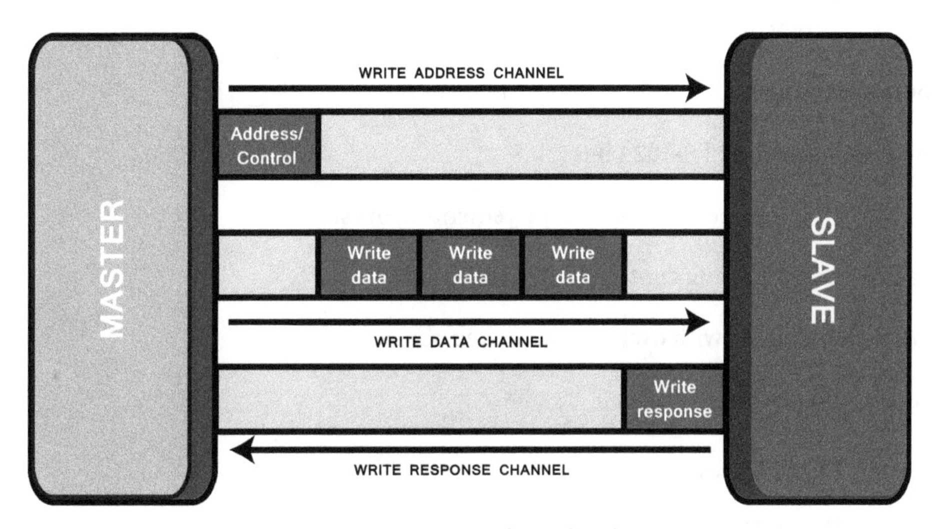

Figure 19.1: AXI4 write channel architecture

Figure 19.2 on the other hand shows a read transaction, with address and control data transmitted to the slave before a burst of read data is transmitted to the master.

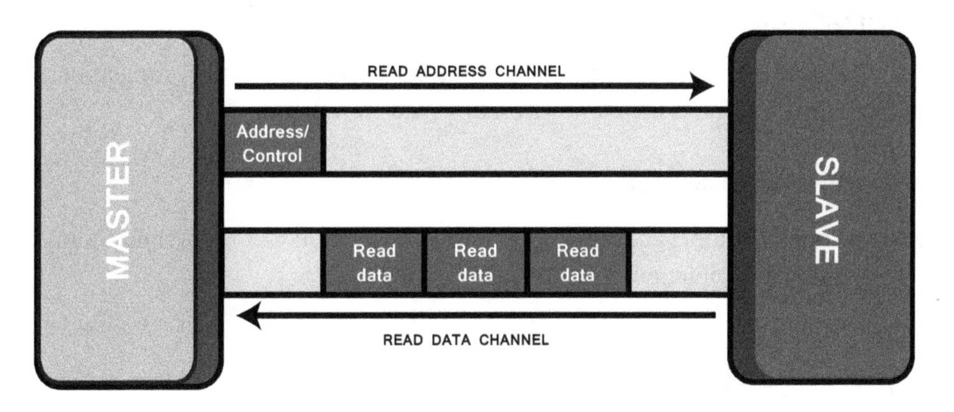

Figure 19.2: AXI4 read channel architecture

The AMBA AXI Protocol Specifications defines each of the channels in the protocol[1].

19.3.1. Address Channels

Read and write address channels are separate, and provide all the address and control information regarding a transaction. This information dictates the operation of the following AXI protocol mechanisms:

- bursts varying from 1 to 16 data transfers per burst
- burst transfer size of 8-1024 bits
- wrapping, incrementing and non-incrementing bursts
- caching and buffering control at the system-level
- atomic operations with exclusive or locked accesses
- secure and privileged access

19.3.2. Write Data Channel

The write data channel includes:

- a data bus ranging from 8-1024 bits wide
- a byte lane strobe for every eight bits of data, used for identifying valid bytes in the data bus

19.3.3. Read Data Channel

Similarly the read data channel includes a data bus with an identical range but also a read response which indicates the completion of a read transaction.

19.3.4. Write Response Channel

The write response channel allows signalling of a completion to a write transaction from the slave by sending a completion signal once for each burst.

19.4. Examples of Applications

Table 19.1 details some of the available Xilinx IP that utilises each of the various AXI4 interfaces.

As previously mentioned, the standard AXI4 interface is the high-performance version of the interface, best suited for more demanding IP requiring consistent, high-speed performance.

AXI4-Lite features a smaller hardware footprint than its counterparts and so ideal for lower-performance IP requiring minimal hardware utilisation.

As the name implies, AXI4-Stream is best suited for applications requiring a constant stream of data.

Table 19.1: Example applications of various AXI interfaces[1]

Interface	Industry	Example Applications
AXI4	Audio and Video/ Image Processing	Video Direct Memory Access
	Communications/ Networking	Ethernet VOIP Receiver, 3GPP LTE Channel Decoder
	Embedded Processing	AXI/PLBV46 Bridges (for backwards compatibility), ChipScope AXI Monitor (for debugging/ embedded system diagnostics)
AXI4-Lite	Audio and Video/ Image Processing	Deinterlacer, Gamma Correction, Image Edge Enhancement
	Automotive	Controller Area Network (CAN)
	Communications/ Networking	10 Gigabit Ethernet Media Access Controller, Digital Predistortion (DPD), Crest Factor Reduction (CFR)
	Embedded Processing	Hardware ICAP, BRAM Interface Controller, External Peripheral Controller

Interface	Industry	Example Applications
AXI4-Stream	Audio and Video/ Image Processing	Streaming video input/output, image noise reduction
	Communications/ Networking	Encoders/decoders, interleavers/deinterleavers
	DSP	CORDIC, FFT, FIR Compiler
	Embedded Processing	Streaming FIFO, Ethernet peripheral,

19.5. AXI Transactions

19.5.1. AXI Write-Burst Transaction

Figure 19.3 shows a *simplified* write-burst transaction using AXI4 for data to be written at an address, ***A***. The master drives the slave and transaction begins with the sending of address and control information via the signal *AWADDR*. Following confirmation of a valid address with *AWVALID*, a signal is sent to confirm that the system is ready for the transaction on *AWREADY*. The master then sends the data blocks ***DATA(A0)-DATA(A2)*** to the slave on the *WDATA* signal, with the final data item being indicated through the *WLAST* signal going high and confirming the completion of the transaction[3]. The master also sends various control signals regarding data bursts but these have been omitted from the diagram for reasons of clarifying basic operation only.

19.5.2. AXI Read-Burst Transaction

Figure 19.4 demonstrates a *simplified* read-burst transaction using AXI4 for data being read from a specific address, ***A***. As with the write-burst, the slave is driven by the master through transmission of address and control information in signal *ARADDR*. *ARVALID* goes high signalling a valid address and the system is confirmed ready for transmission with the signal *ARREADY*. Data blocks are read from address ***A*** via the signal *RDATA*, and as before the final data block is indicated via signal *RLAST*[3]. Note the *RVALID* signal, kept low by the slave until read data is available.

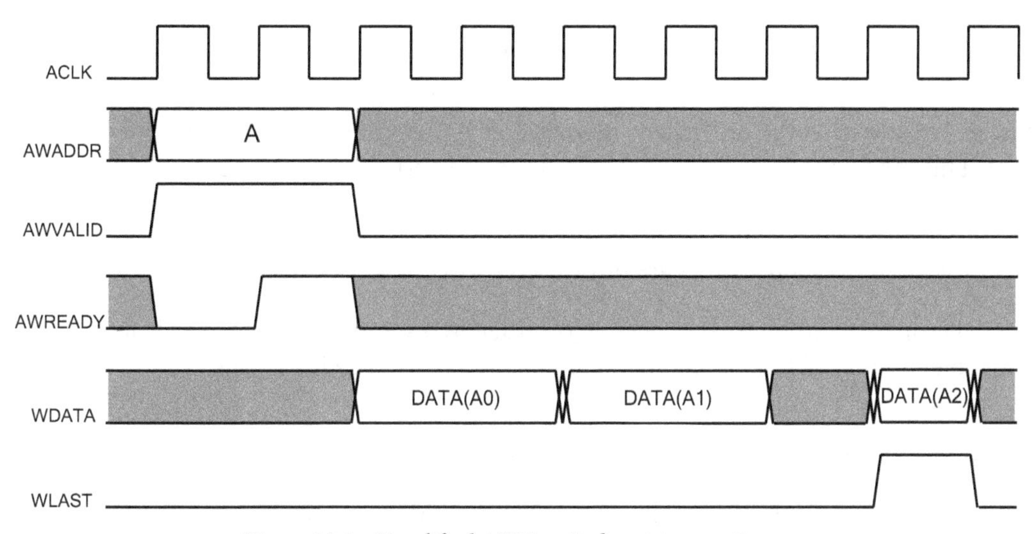

Figure 19.3: Simplified AXI4 write burst transaction

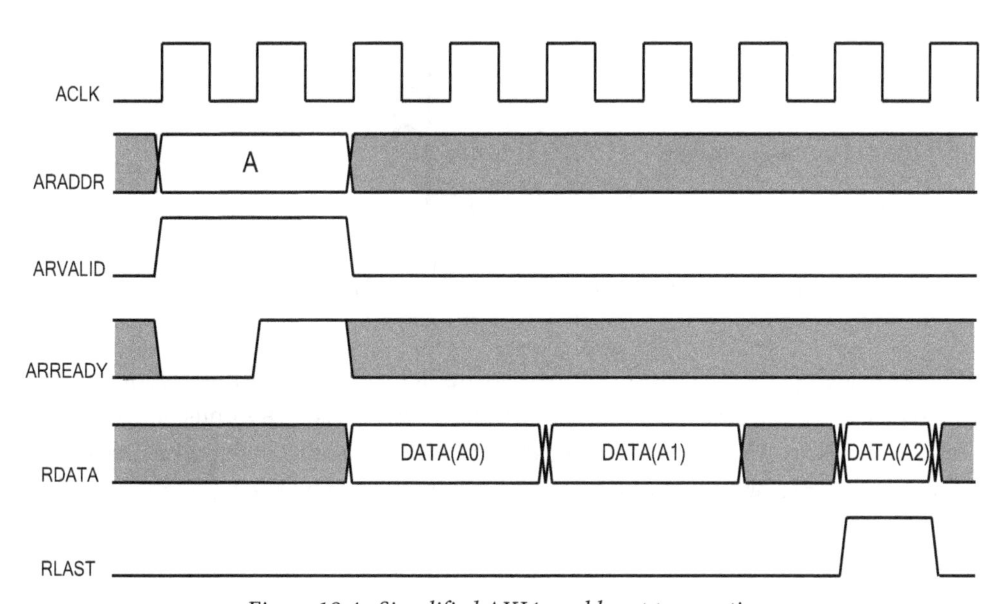

Figure 19.4: Simplified AXI4 read burst transaction

As stated in Section 19.1, the AXI protocol allows out-of-order transactions. Each transaction is given an identifying tag issued by the interconnect that governs whether a transaction must be completed in the order they are issued or if they can in fact be transmitted out-of-order. This is beneficial in the instance where system performance can be improved as[3]:

- Slaves with faster-response times are prioritised above those that are slower.
- Read data can be returned from slaves out-of-order, for example if data required later is available in the buffer before more immediately required data is available.

19.6. AXI in the Xilinx Toolflow

The process of implementing IP incorporating AXI interfaces on a Zynq chip is easily approached through use of the Vivado IP Integrator.

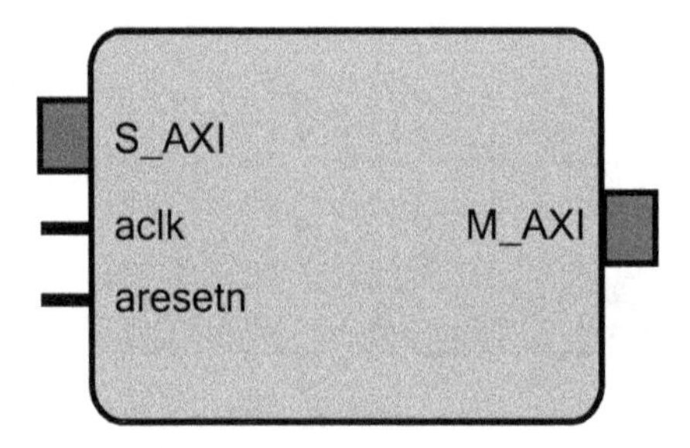

Figure 19.5: Representative AXI4 data FIFO in Vivado IP Integrator

Figure 19.5 shows a single AXI Data FIFO part as implemented using the Vivado IP Integrator. The block features a slave AXI bus, *S_AXI*, as well as a master AXI bus, *M_AXI*, both based on the standard AXI4 interface. Expanding these buses by clicking on the + reveals the signals contained within. In Figure 19.6, the slave bus has been expanded to reveal all the corresponding signals.

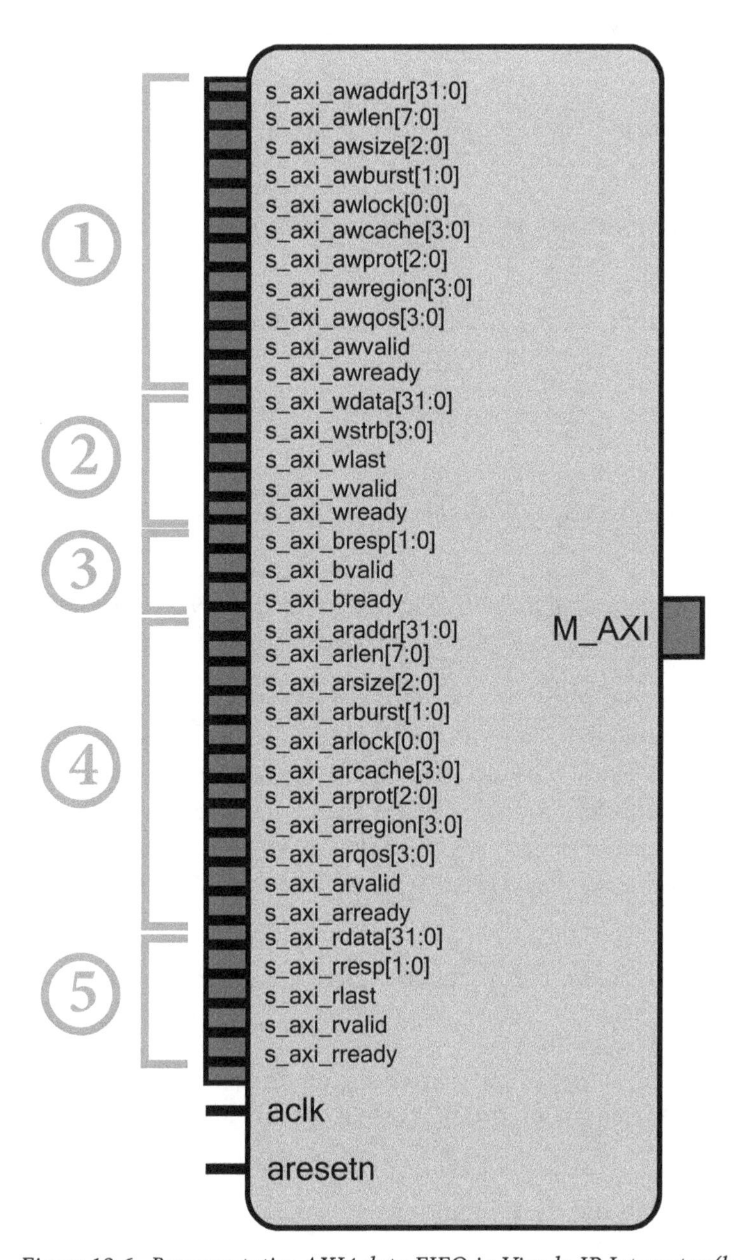

Figure 19.6: Representative AXI4 data FIFO in Vivado IP Integrator (bus expanded)

For a full breakdown of each of the listed signals you should refer to the Xilinx AXI

Reference Guide[4], however we can relate some of the signals in Figure 19.6 to the channels highlighted in Section 19.4.

①
- **Write Address Channel** — the signals contained within this channel are named in the format *s_axi_aw...*

②
- **Write Data Channel** — the signals contained within this channel are named in the format *s_axi_w...*

③
- **Write Response Channel** — the signals contained within this channel are named in the format *s_axi_b...*

④
- **Read Address Channel** — the signals contained within this channel are named in the format *s_axi_ar...*

⑤
- **Read Data Channel** — the signals contained within this channel are named in the format *s_axi_r...*

You should note the presence of signals highlighted in Section 19.5 regarding the read and write operations, as well as a wealth of different control signals.

Figure 19.7 gives an example of the connections between AXI devices and the Zynq processing system as configured in Xilinx Vivado IP Integrator. The key to operation is the use of the AXI Interconnect block (instance named *axi_intercon_1* in this system). This interconnect is a slave to the master processing system, and is a master routing signals from the processor to various instances of slave AXI devices. In this system, the interconnect is connected to two slave devices; AXI GPIO (*gpio_1)* and an AXI BRAM controller (*bram_ctrl_1).*

Focusing on the AXI Interconnect in particular, there are several different signals to consider.

- There is one slave input, *S00_AXI* and corresponding clock input *S00_ACLK* coming from the master AXI output of the processing system and the processor clock respectively.
- This clock also supplies the clock input for the two AXI devices and their master inputs on the AXI Interconnect.
- All reset signals come from a processor reset block not included in this view for simplicity.

- The outputs of the AXI Interconnect are both master channels containing the signals detailed in Figure 19.6, each driving a respective slave device.
- Further detail on the utilisation of Xilinx IP Integrator with reference to connecting blocks is discussed in Chapter 18.

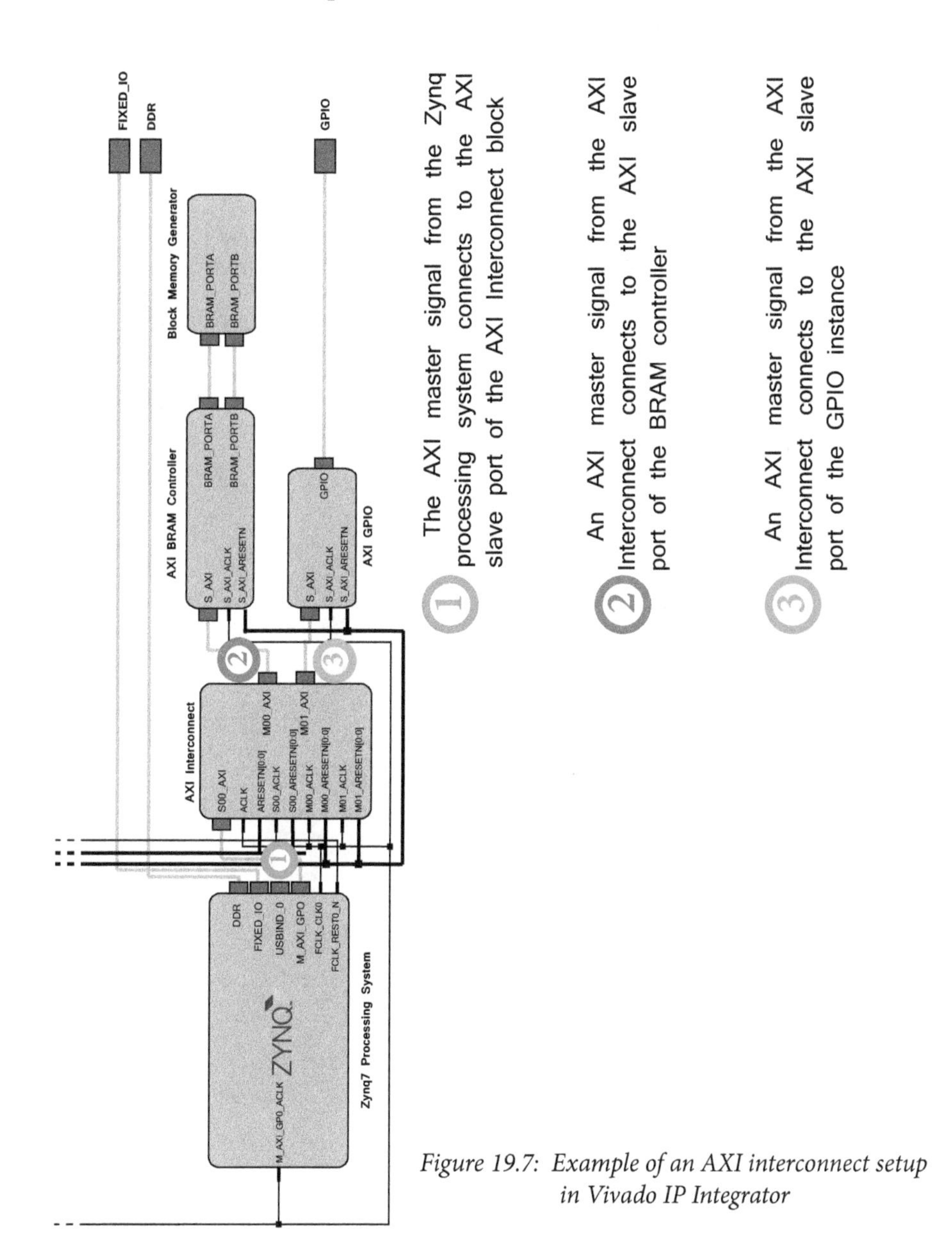

Figure 19.7: Example of an AXI interconnect setup in Vivado IP Integrator

19.7. Summary

In this chapter the AMBA AXI4 interface for IP integrated on a Zynq chip has been introduced. The variations of AXI4 have been presented, with some example IP applications from the Xilinx IP catalogue. Utilisation of the AXI interface with Xilinx IP via Vivado IP Integrator has also been highlighted.

19.8. References

Note: all URLs last accessed June 2014.

[1] ARM, "Introduction — Channel Definition" in *AMBA AXI Protocol Specification*, v1.0, June 2003..

[2] Xilinx, Inc, "AXI4 IP Catalogue", http://www.xilinx.com/products/intellectual-property/axi_interconnect.htm

[3] ARM, "Introduction — Basic Transactions" in *AMBA AXI Protocol Specification*, v1.0, June 2003.

[4] Xilinx, Inc, "AXI Reference Guide", UG761, v14.3, November 2012.

20

Adventures with IP Integrator

In this practical chapter, the IP blocks that were created in Chapter 13 - IP Block Design will be brought together to form a DSP system which will run on the ZedBoard. The features of Vivado IP Integrator block design will explored, and existing IP from the IP Catalog will also be used. Further, the Xilinx Design Constraints (XDC) format will be introduced in order to map the external ports from the IP Integrator design to individual pins on the ZedBoard.

A prepackaged I2S controller for the Analog Devices ADAU1761 audio codec on the ZedBoard will also be used in order to allow input and output of audio signals to the design, and the necessary changes and connection to the Zynq processor will be highlighted.

The complete Zynq embedded system will take audio input from the provided audio codec and add a tonal noise component which will be generated from the NCO. The corrupted audio signal will be passed as the input to the LMS filter, and the pure NCO tonal signal will be used as the reference signal. The LMS filter will then be used to remove the tonal noise from the audio signal, resulting in a clear audio signal being produced at the output. An echo effect will also be introduced at the output of the LMS filter, with a user definable delay length. The delay length will be displayed on the ZedBoard LEDs.

An overview of the complete DSP system is provided in Figure 20.1.

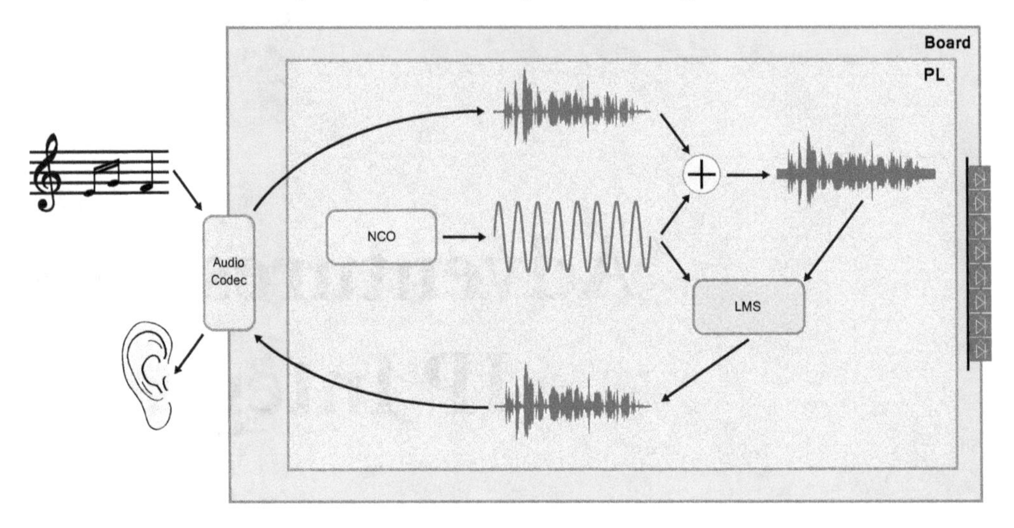

Figure 20.1: Overview of complete DSP system

20.1. Aims and Outcomes

The main aim of this set of practical exercises is to use Vivado IP Integrator to bring together the custom IP modules that were created in the previous exercises, as well as IP from the IP Catalog, to create a DSP system for implementation on the ZedBoard.

After completing this tutorial you will be able to:

- Add a IP to the IP Catalog from a wide-array of sources.
- Combine and connect IP from various sources to create a full IP Integrator block design.
- Fully utilise the IP Integrator Designer Assistance and Connection Automation tools.
- Customise the Zynq Processor system.
- Map external ports to device pins using the XDC file format.
- Create software applications to fully control custom IP modules in the Zynq PL.

20.2. Exercise 4A

In this exercise, the IP modules which were created in the previous practical exercises will be imported into the IP Catalog by adding the required IP repositories. A ZedBoard IP Integrator design will be created, with all the required connections made.

The steps involved in this exercise are:

1. Add all required custom IP repositories to the IP Catalog and import the IP.
2. Create initial IP Integrator block diagram.
3. Explore the individual custom IP blocks and their available block parameters.

Exercise 4A is available on the website: www.zynqbook.com

20.3. Exercise 4B

Building upon the IP Integrator design that was created in Exercise 3B, prepackaged IP that interacts with the audio codec on the ZedBoard will be introduced, imported into IP Integrator and required modifications to the Zynq processor block made. External port and interface connections will be made in the design, and XDC constraints file created to map all external ports to their corresponding pins on the ZedBoard.

The steps involved in this exercise are:

1. Add audio codec controller IP to the IP Catalog.
2. Include the audio codec controller in the existing IP Integrator block diagram.
3. Make required changes to the Zynq Processor to add an addition fabric clock and enable an I^2C interface to communicate with the audio codec conroller.
4. Create required external interfaces for communication with the audio codec.
5. Map the external interfaces of the design to individual pins on the ZedBoard by adding entries to the XDC constraints file.
6. Generate hardware for the finalised design.

Exercise 4B is available on the website: www.zynqbook.com

20.4. Exercise 4C

The finalised design from Exercise 3B will be exported to the SDK for software development. The software application which controls the interactions between the various IP modules in the DSP system will be created, and the various software driver files introduced.

The final step to test the DSP system will be carried out by running the newly created software application on the ZedBoard to ensure that all parts are working together correctly.

The steps involved in this exercise are:

1. Export the finalised hardware design to the SDK.
2. Create a new application project.
3. Import the required custom IP driver files.
4. Create a software application which will communicate, and control interactions, with the custom IP blocks.
5. Run the test application and ensure that everything functions correctly.

Exercise 4C is available on the website: www.zynqbook.com

20.5. Possible Extensions

Having completed Exercise 4C, there are some possible variations you can introduce to personalise the developed system. For example:

- As the development of this system has focussed solely on audio, you could develop a system which performs some video processing using the VGA or HDMI connections.

20.6. What Next?

This set of practical exercises concludes Part B of the book, "*Zynq SoC & Hardware Design*".

Next, we move onto Part C, which looks at the idea of operating systems on Zynq, with particular emphasis on the Linux OS.

PART C

Operating Systems & System Integration

21

Introduction to Operating Systems on Zynq

One of the main questions that a designer is faced with when developing an embedded system is whether or not to include an embedded OS. The information presented in this chapter aims to help answer this question and will detail the scenarios where an embedded OS should be used, the categories of OS that exist and the numerous different OSs that are available. The options that are available in terms of making the most of the dual-core processor available on the Zynq platform will also be explored.

21.1. Why Use an Embedded Operating System?

Although an embedded OS is not necessary for all embedded systems applications, there are a number of advantages to their use; this section highlights many of these advantages and will provide some examples where appropriate.

21.1.1. Reducing Time to Market

When developing an embedded system there are a number of key areas in which an embedded OS can reduce the development time.

OS vendors provide support for a wide array of architectures and platforms; this is advantageous if there comes a point during product development that it is deemed necessary to move to a new processing platform. As the software that has already been developed is targeted at an OS, rather than a specific device, the process of moving to a new architecture or device should not be problematic.

For example, there is a large degree of similarity between embedded Linux and the traditional Desktop version. If a designer has a familiarity with developing applications for the Desktop variety of Linux, the move to developing for embedded Linux will be straightforward and the learning curve will be relatively shallow. This will substantially reduce, if not remove, the time required for a designer to become familiar with a new development environment. It is also worth noting that most embedded OS vendors include, or have third-party support for, the Eclipse IDE.

Improved portability can be achieved through the use of an OS with a standardised system interface, such as the Portable Operating System Interface (POSIX), whereby the user command line and scripting interface of the OS is constant. POSIX was jointly developed, and continues to be maintained, by a working group consisting of members of the IEEE Portable Applications and Standards Committee, The Open Group, and members of the ISO/IEC Joint Technical Committee 1 [6]. This group is collectively known as the Austin Group [6].

By using POSIX standardised OSs, the impact of moving development from one OS to another is reduced as the high-level calls do not need to be ported to the new OS due to the use of the POSIX API. Generally speaking, the use of POSIX standardised OSs provides a greater flexibility in OS-level portability.

21.1.2. Make Use of Existing Features

Embedded OSs offer support for many features which would otherwise have to be developed by the system designer; a number of these features are highlighted below.

When incorporating a display into an embedded platform — whether it be a low resolution, internal LCD panel or an external display output via HDMI — the system must be able to support it. The degree of support that is necessary depends on the specific application but it can generally be split into two levels: *driver support* and *graphical interface support*. By choosing to use an established OS you will generally have access to both driver-level and interface-level support.

Driver-level support provides the low-level software that makes the connection between the embedded processor and the display, allowing the system to support the specific interface used. Certain aspects that are controlled at the driver- level include signal timing and synchronisation, screen resolution, buffer formats and screen refresh rates. Many OSs come with support for a large range of video drivers; this removes the need to develop a driver from scratch, which can be a complex and time consuming process.

Graphical interface-level support deals with the high-level graphical content that is to be displayed. This can range from a single line text display, to a fully featured GUI. OSs provide high-level APIs for graphical development, as well as providing a functioning GUI.

21.1.3. Reduce Maintenance and Development Costs

By making use of an embedded OS you reduce the amount of custom code that needs to be developed and tested. In reality, this amounts to far fewer potential bugs being introduced to the software, which in turn reduces the time and cost of testing the system to find and remove them. An OS will provide a stable platform, which has been extensively tested, so that you can concentrate on developing the custom features of your system and not debugging low-level code. By making use of the tool suite provided by the OS vendor you will have access to OS-aware profiling and debugging tools which will speed up development and aid in finding any performance bottlenecks. The tool suite will also be essential if developing software for multi-threaded processing systems.

Another area in which OSs can reduce maintenance costs is through the provision of support outside of the in-house development team. Most commercially available OS vendors will have support teams which will be able to help with OS platform problems, whether it be over the phone or via email. While open source based OSs may not have dedicated support teams available, they do have extensive online knowledge bases and dedicated users who can offer support via online forums. In some cases, companies may produce a commercial version of an open source OS which includes proprietary features. If this is the case, support is usually provided by the OS vendor, who can also help with open source licensing issues.

OS vendors will periodically release updates and improvements to the OS meaning that you do not have to worry about improving the underlying code, and can instead concentrate on custom functionality. By focusing your time on custom development you will stay ahead of the competition.

21.2. Choosing the Right Type of Operating System

There are a number of possibilities when determining the type of OS to use on an embedded system. Such examples include a simple standalone OS, a RTOS or specialised embedded OS such as the numerous variations of embedded Linux.

Before that choice can be made, however, we should consider the type of embedded OSs that are available.

21.2.1. Standalone Operating Systems

A standalone OS, also known as a bare metal OS, is a simple OS that aims to provides a very low-level of software modules that the system can use to access processor-specific functions.

Regarding the Zynq platform specifically, Xilinx provides a standalone OS platform that provides functions such as configuring caches, setting up interrupts and exceptions and other hardware related functions. The standalone platform sits directly below the OS layer and is used whenever an application requires to access processor features directly [8].

A standalone OS enables close control over code execution but is fairly limited in terms of functionality. It should only be used for applications where the software functions are straightforward and repetitive. The number of tasks being carried out by a standalone OS should be relatively small, as adding further tasks can increase the task management required by the standalone rapidly.

21.2.2. Real-Time Operating Systems (RTOS)

The defining feature of a RTOS is the degree of determinism that is guaranteed by the scheduler; the purpose of a RTOS is not to achieve a high throughput, but instead to respond both quickly and predictably for a given task.

The function of many embedded systems demand that the software responds to events within a short, defined response time. Given this requirement, real-time systems can be categorised as one of three types: *soft real-time*, *hard real-time* or *firm real-time* [3].

A soft real-time system is one in which the meeting of a response deadline is preferred but not critical. A failure to meet the specified response time will not destroy the performance of the system, but may degrade it.

A hard real-time system, however, is one in which the missing of a response time is unacceptable and could lead to the overall failure of the system.

Firm real-time systems are a middle ground between hard or soft systems; a small number of missed response deadlines will not lead to the overall failure of the system, but a larger number of missed deadlines may result in total failure of the system [3].

Most modern RTOS systems include a set of high-level functions that complement the real-time kernel. Such functions can include a GUI, communications protocol stacks and a certain degree of peripheral device management. In an embedded system, the RTOS controls the device and is responsible for providing the required level of responsiveness. Software tasks are controlled by the RTOS which schedules the CPU time allocated to each of the tasks accordingly[1].

21.2.3. Other Embedded Operating Systems

While a RTOS is suitable for the management of real-time applications on embedded systems, they do not generally offer the highest system throughput or performance. For applications that require high system performance, another type of OS is usually required.

Traditionally, the preferred option would be an embedded Linux solution but, with the recent developments in mobile OSs such as Android, there are more options capable of delivering high system performance for an embedded system.

Linux

Linux and the Linux kernel are covered in detail in Chapter 22 and Chapter 23 respectively, so we shall skip over it for now.

Android

Android is an OS which is mainly intended for use on touchscreen mobile devices, i.e. mobile phones and tablet computers. Originally created by Android, Inc. — a company financially backed by Google — Android was later bought, and is now developed and maintained, by Google. Due to its open source status, Android has since been customised for use on non-mobile devices such as smart TVs, cameras, media players, laptop computers, and wrist watches.

Google released the source code for Android under the Apache V2 open source license which means that anyone, be it a mobile phone manufacturer or a smart TV developer, who innovates using the Android platform has no requirement to share those additions with the open source community [4]. This makes Android a very commercially-friendly platform to work with.

The Android OS comprises of a kernel derived from the Linux kernel v2.6 for all versions up to and including Android 3.2, after which Android 4.0 and onwards are based on Linux kernel 3.x [12]. The Android software architecture, however, is largely different from that of a traditional Linux system, including some changes to the fundamental kernel functionality. Due to Android initially being targeted at mobile devices, a number of aggressive power management policies were introduced to minimise power consumption by forcing the kernel to go into sleep mode whenever possible. This is in contrast to the traditional desktop Linux variations which largely tend to never allow the kernel to enter sleep mode. Other changes include the introduction of timed GPIOs, alarm timers, paranoid network security, and the binder Inter-Process Communication (IPC), amongst others. The overall software architecture of Android is detailed in Figure 21.1.

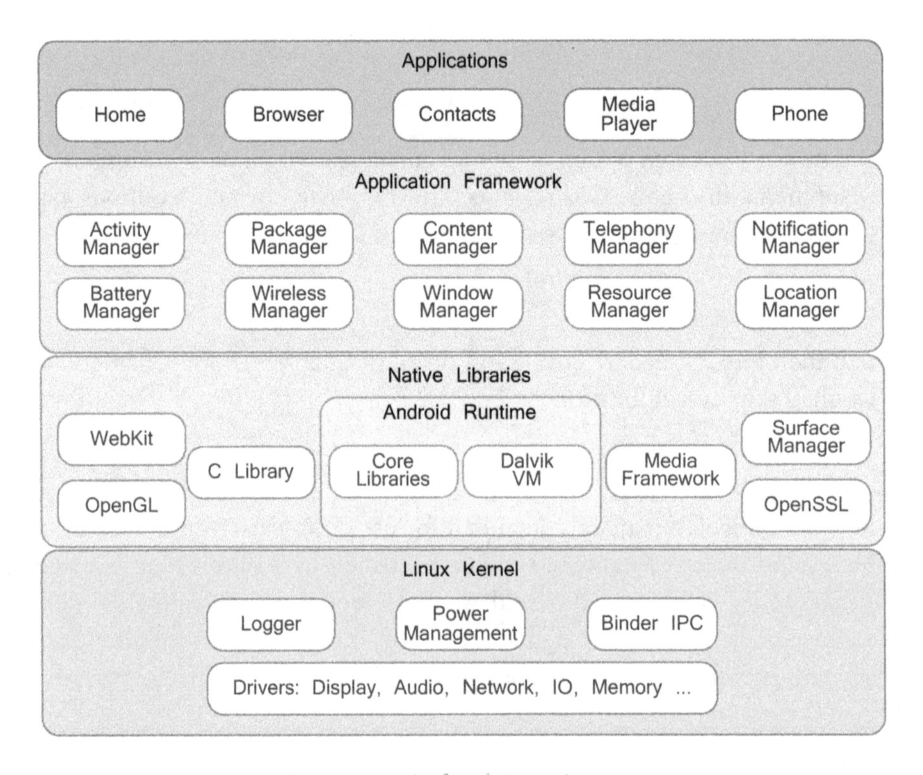

Figure 21.1: Android OS architecture

It is the more recent adoption of Android for non-mobile devices that is of interest here. There are a number of reasons — aside from its open source status — which make Android an appealing platform on which to build embedded systems. We will now take a look at few of those.

One of the things that makes Android desirable for developers is the fully featured SDK which provides a regular framework to work with through the use of a standardised API. Although Android is an ever-evolving platform, that has gone through a large number of releases in recent years, the API remains generally constant across releases; this allows developers to make a reasonably safe long-term investment as they can make savings by only designing and compiling applications once for multiple targets [12].

Android has out-of-the box support for a wide array of sensors (including GPS, accelerometer and camera), networking (WiFi, Bluetooth, NFC, 2G/3G) and also a large number of common multimedia formats. If your embedded application needs to make use of one or more of these features, then choosing Android could well improve your development time. Furthermore, due to its widespread use in mobile phones and tablets, the Android user interface is familiar to many potential users, thus reducing the learning curve.

21.2.4. Further Considerations

Other questions to consider when choosing an embedded OS:

- How much does it cost?
- How experienced is your design team in its use?
- How secure is it?

21.3. Applications

As mentioned in the previous section, the type of OS used for an embedded system depends on the type of application the system will be used for. If the intended application requires the system to respond to events within a strict time period, then the typical choice would be to user a RTOS. A simple example of this is the anti-locking braking system in a car, which requires that the brakes be freed within a specified time in order to prevent the car from skidding. However, other applications for embedded systems which are required to provide a high-level of system performance, without the need for the predictable response times of an RTOS would require an embedded OS such as Linux or Android. Such an example of this would be a TV set top box recorder that is required to provide a file system for the storage of television recordings, video processing, and a GUI for user interaction. The levels of embedded OS performance are depicted in Figure 21.2.

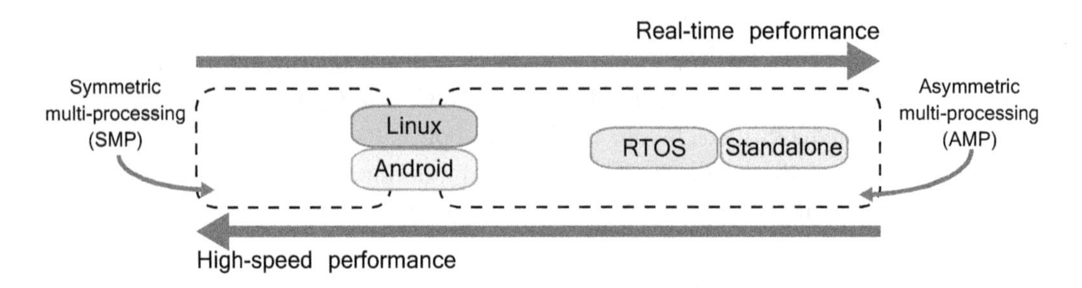

Figure 21.2: Embedded operating system performance

21.4. Multi-Processor Systems

The choice of OS for an embedded application may also depend on the number of processors in the system. As is the case with the Zynq platform, the system may include a multi-core processor, in which case you must make the decision whether to use a single OS which will run on both processing cores, or multiple OSs running on individual cores.

Figure 21.2 introduces the terms *Asymmetric Multi-Processing (AMP)* and *Symmetric Multi-Processing (SMP)* in terms of OS support. It is, however, useful to understand their definitions in terms of the system architecture as well.

Asymmetric multi-processing can be used on a system that utilises multiple CPU cores, which may be of different architectures. Each CPU, or CPU core, can run its own instance of an OS, which can either be homogeneous or completely different. An example of this would be a system running an RTOS on one CPU while running a Linux based GUI on another CPU. Communication between the CPU cores is facilitated through the use of shared memory, which provides a level of software abstraction.

Symmetric multi-processing on the other hand requires that all CPUs in the system are of matching architectures. A single OS instance is run across all CPUs which divides and coordinates the processing tasks between them. As with AMP, the shared memory space is used for communication between CPUs, as well as for the coordination of task execution.

The differences between AMP and SMP with reference to the dual-core ARM architecture of the Zynq platform is depicted in Figure 21.3.

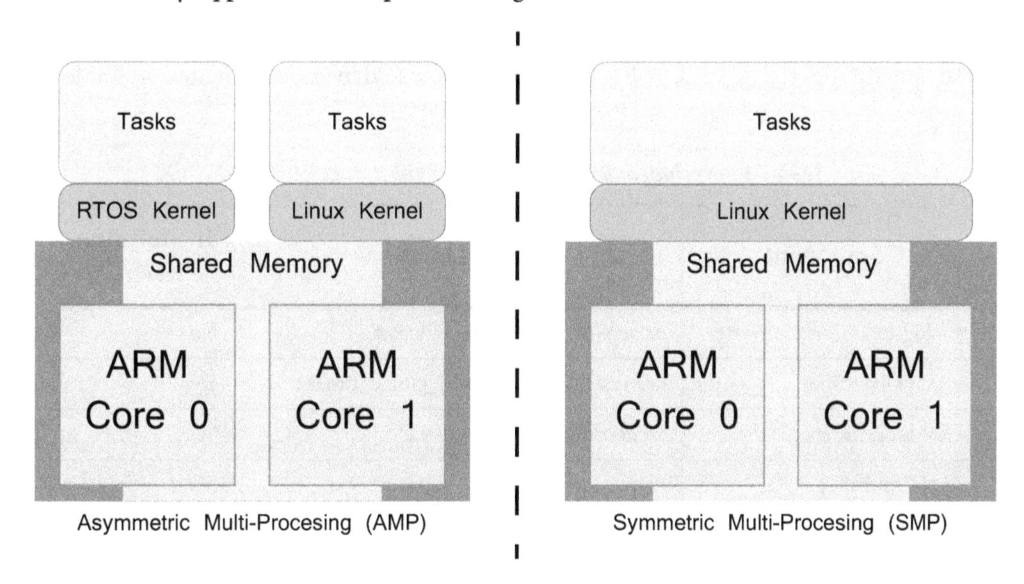

Figure 21.3: Asymmetric vs. symmetric multi-processing

21.5. Zynq Operating Systems

Now that the various types of OS, and their corresponding uses, have been introduced it would be useful to take a more detailed look at a couple of specific OSs from each category. The aim here is to provide a closer look at a number of OS choices and provide information on specific features that are available; how and where to find them; whether they are open source or commercially available; and the support that is available for each.

As was the case previously, this section will be split into two parts with one detailing a number of available Linux options, while the other will cover RTOS.

21.5.1. Linux

This section introduces a number of embedded Linux OSs and environments which are compatible with Zynq.

Xilinx Zynq-Linux

Zynq-Linux is an open source OS freely available from Xilinx. It is based on the 3.0 Linux kernel from kernel.org and includes a number of additions from Xilinx, such as a BSP and specific device drivers. A list of included device drivers is provided in Table 21.1 [9].

Table 21.1: Included device drivers for Zynq-Linux kernel [9]

Component	Driver Location	In Mainline Kernel
Analog-to-Digital Converter	drivers/hwmon/xilinx-xadcps.c	No
ARM global timer	drivers/clocksource/arm_global_timer.c	Yes
ARM local timers	arch/arm/kernel/smp_twd.c	Yes
CAN Controller	drivers/net/can/xilinx_can.c	No
DMA Controller (PL330)	drivers/dma/pl330.c	Yes
Ethernet MAC	drivers/net/ethernet/xilinx/xilinx_emacps.c	No
	drivers/net/ethernet/cadence/macb.c	Yes
GPIO	drivers/gpio/gpio-xilinxps.c	No
I2C Controller	drivers/i2c/busses/i2c-cadence.c	Yes
Interrupt Controller	arch/arm/common/gic.c	Yes
L2 Cache Controller (PL310)	arch/arm/mm/cache-l2x0.c	Yes
QSPI Flash Controller	drivers/spi/spi-xilinx-qps.c	No
SD Controller	drivers/mmc/host/sdhci-of-arasan.c	Yes
SDIO WiFi	drivers/net/wireless/ath/ath6kl/sdio.c	Yes
SPI Controller	drivers/spi/spi-xilinx-ps.c	No
Triple Timing Counter	drivers/clocksource/cadence_ttc.c	Yes
UART	drivers/tty/serial/xilinx_uartps.c	Yes
USB Host	drivers/usb/host/xusbps-dr-of.c	No
USB Device	drivers/usb/gadget/xilinx_usbps_udc.c	No
USB OTG	drivers/usb/otg/xilinx_usbps_otg.c	No

It should be noted that not all drivers are included in the mainline kernel, and can only be found in the Xilinx tree. Those included in the mainline kernel are indicated in Table 21.1. Drivers that are only found in the Xilinx tree may be old, and could be removed at any time.

Support is also provided for SMP, allowing the kernel to use both CPUs, and can also be configured to use a single CPU.

As well as providing access to all source code from the Xilinx repository, prebuilt releases are also available from the Xilinx Getting Started website.

http://www.wiki.xilinx.com/Getting+Started

Petalogix® - Petalinux

PetaLinux is provided in the form of a SDK that provides the combination of a fully-functional embedded Linux distribution and a development environment which integrates with the Xilinx hardware design flow [10]. PetaLinux provides a complete package that consists of everything required to build, test, develop and deploy embedded Linux systems.

PetaLinux comprises of three constituent parts:

- A fully customisable embedded Linux OS for Xilinx devices.
- Pre-built binary images (which are bootable out-of-the-box).
- PetaLinux SDK

Xillybus - Xillinux

Xillinux is a desktop distribution of Linux that will allow you to run a full graphical desktop environment on the Zedboard; a keyboard and mouse can be attached via the USB On-The-Go port of the Zedboard, while a monitor can be connected to the provided VGA port [7]. As well as providing a full Linux distribution, Xillybus also supply a development kit for interaction between the Linux host and peripherals running on the programmable logic. This is in the form of FIFOs on the logic side, and native Linux file operations on the host side. The Linux distribution, which is based upon the popular desktop distribution of Ubuntu 12.04 Long Term Support (LTS), and the development kit are available for free download from the Xillybus website [7].

21.5.2. RTOS

FreeRTOS

FreeRTOS is a lightweight real-time OS that is available for a wide range of devices and processor architectures. The core of the FreeRTOS kernel is made up of only 3 C files, making it very simple, and it has a minimal ROM, RAM and processing overhead — in many cases the kernel image size will be between 4 to 9 KB [5]. Xilinx provides an already developed version of the OS which is freely available from the FreeRTOS website [5].

21.5.3. Further Operating Systems

There are a large number of OSs for Zynq which are provided by Xilinx partners — far too many to cover in the context of this book. A list of solutions is provided in Table 21.2. Further details of any of the products listed can be found on the Zync-7000 SoC Ecosystem website [11].

Table 21.2: Zynq operating systems available from Xilinx partners

Xilinx Partner	Operating System/Software
Adeneo Embedded	Windows Embedded Compact 7, Linux, Android and QNX
Discretix	Security-centric software and IP
ENEA Software AB	OSE RTOS and ENEA Linux
eSOL	uITRON 4.0 RTOS, T-Kernel RTOS and IDE
Green Hills Software	INTEGRITY RTOS
Express Logic	ThreadX RTOS
iVeia	Android for Zynq
Mentor Graphics	Nucleus RTOS
Micrium	uC/OS RTOS
MontaVista Software	MontaVista Carrier Grade Linux
Open Kernel Labs	OKL4 Microvisor
QNX	QNX RTOS

Table 21.2: Zynq operating systems available from Xilinx partners

Xilinx Partner	Operating System/Software
Quadros	RTXC RTOS
Real Time Engineers Ltd	FreeRTOS
Sierraware	Open Source Hypervisor and Trusted Execution Operating System
SYSGO	Safe and Secure Virtualisation and Operating System
Timesys	LinuxLink
Wind River	VxWorks, Linux and Workbench IDE

21.6. Chapter Review

In this chapter the concept of embedded OSs has been introduced along with the reasoning behind their use. The various types of embedded OS were described and a number of examples provided along with example applications of possible products and devices. The concept of multi-processor systems was also introduced. In the next chapter we will take a more in-depth look at the Linux OS.

21.7. References

NOTE: All URLs last accessed June 2014.

[1] ARM, "Real-Time Operating Systems (RTOS)" webpage.
Available: http://community.arm.com/docs/DOC-2764

[2] Embedded Linux Wiki, "Android Kernel Features", webpage.
Available: http://elinux.org/Android_Kernel_Features

[3] P. A. Laplante and S. J. Ovaska, "Fundamentals of Real-Time Systems" in *Real-Time Systems Design and Analysis:Tools for the Practitioner*, 4th Ed. Wiley-IEEE Press, 2012, pp. 1 - 25.

[4] Open Handset Alliance, "Android Overview", webpage.
Available: http://www.openhandsetalliance.com/android_overview.html

[5] Real Time Engineers Ltd, "FreeRTOS", webpage.
Available: http://www.freertos.org/

[6] The Open Group, "The Open Group Base Specifications Issue 7, IEEE Std 1003.1™, 2013 Edition".
Available: http://pubs.opengroup.org/onlinepubs/9699919799/

[7] Xillybus, "Xillinux: A Linux distribution for the Zedboard", webpage.
Available: http://xillybus.com/xillinux

[8] Xilinx, Inc, "OS and Libraries Document Collection", UG643, June 2014.
Available: http://www.xilinx.com/support/documentation/sw_manuals/xilinx2014_2/oslib_rm.pdf

[9] Xilinx, Inc, "Linux Drivers".
Available: http://www.wiki.xilinx.com/Linux+Drivers

[10] Xilinx, Inc, "PetaLinux Software Development Kit".
Available: http://www.xilinx.com/tools/petalinux-sdk.htm

[11] Xilinx, Inc, "Zynq-7000 AP SoC Ecosystem".
Available: http://www.xilinx.com/products/silicon-devices/soc/zynq-7000/ecosystem/index.htm

[12] B. Zores, "The Growth of Android in Embedded Systems", The Linux Foundation Training Publication, 2012.
Available:
http://training.linuxfoundation.org/free-linux-training/download-training-materials/growth-of-android-in-embedded-systems

22

Linux: An Overview

This chapter will provide an overview of the Linux Operating System, beginning with a whistle-stop tour of the events leading up to the creation of Linux. A general system overview will be provided to pave the way for in depth analysis in future chapters. In particular these chapters will cover the Linux kernel and file system, device drivers and boot sequence.

Some of the tools and resources available for those with an express interest in Linux and application development will be presented.

22.1. A Brief History

Linux was conceived in April 1991, with the first announcement of the project given in August that year. A Finnish computer science student by the name Linus Torvalds made the following announcement on the Usenet group *comp.os.minix:*

> *"I'm doing a (free) operating system (just a hobby, won't be big and professional like gnu) for 386(486) AT clones. This has been brewing since april, and is starting to get ready. I'd like any feedback on things people like/dislike in minix, as my OS resembles it somewhat (same physical layout of the file-system (due to practical reasons) among other things)."*
>
> —Linus Torvalds (August 25th, 1991)

The Minix he was referring to is a variant of the UNIX OS, which was used as a guideline for the development of a free OS targeted towards the x86 consumer PCs of the time.

If we go back in time to 1983, Richard Stallman had previously embarked upon a project to create a free (as in free speech, not free beer), UNIX-like operating system known as GNU, a recursive acronym for "GNU's Not UNIX". In fact, the origins of Linux go even further back with the creation of the UNIX OS in AT&T's Bell Laboratories back in the early 1970s!

The major motivation for the GNU Project was the domination of proprietary software during the 80s. A free OS was seen as the fundamental step in paving the way for 'free' computing. However, writing an OS is no small undertaking, and by 1990 (and with funding from the Free Software Foundation in October 1985) all major components of the OS had been developed, except one. This final component, the core of the OS, is known as the *kernel* [1]. Using the GNU tools provided by Stallman, Torvalds created the Linux kernel (which will be discussed in detail in Chapter 23), marking the origins of Linux as it exists today.

Linux is a collaborative development, with many developers from many different companies cooperating on development labour, research and costs to form an ecosystem that is estimated to have grown from $12.3 billion in 2008 to $35.5 billion in 2013[2]. Today, Linux is more popular than ever and can be found in a vast range of devices, from mobile phones and satellite navigation systems to supercomputers and servers.

22.2. Linux System Overview

Figure 22.1 represents a generalised high-level architecture of a GNU/Linux System. In reality, the kernel space is a much more complicated concept, containing many more components. For our purposes, however, we shall consider only the most important ones.

The applications and system programs, as well as the GNU C library run on top of the kernel in the user-space. Applications refer to programs with a useful function, such as word processing, games, or C applications developed to run on the processor of a Zynq chip. System programs, however, are necessary in implementing various OS services that ensure the system actually works! The physical hardware resides at the opposite end of the chain and as such is abstracted from the user-level by the kernel space; this includes any system storage, network cards or the GPIO on a development board, for example. The indirect nature of access between the user-level and the hardware provides security as it ensures regulations through use of the kernel tools[3].

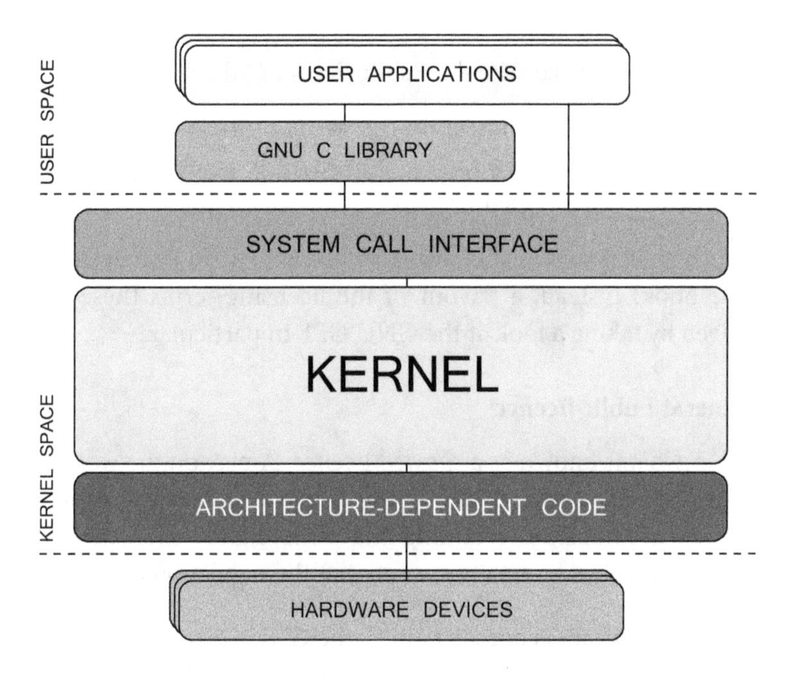

Figure 22.1: A High-level GNU/Linux system architecture

The System Call Interface (SCI) facilitates function calls from the user space to the kernel of the system. As previously discussed, the Linux kernel forms the heart of the OS and provides a set of tools with which the user space can interact with hardware. The Linux kernel itself can be broken up into its own layered subsystems such as memory and process management, the virtual file system and device drivers; however, these will be covered in further detail in Chapter 23. The kernel code is considered architecture-independent as this code is common to all processor architectures supported by Linux. Below this is the architecture-dependent code; that is, code which is processor and platform-specific and is often referred to as a BSP.

22.3. Licensing

The issue of licensing with Linux is a confusing one, replete with different standards and different acronyms. Remembering the GNU Project mentioned in Section 22.1, we recall that the basis for the project was the creation of an OS that was 'free'. In order to release free software, a free licence must be utilised. For GNU, the tools for which spawned Linux, the GNU General Public Licence (GNU GPL) is typically used. Other free licences exist (and in different versions), compatible with GPL such as the GNU Lesser GPL (LGPL), the GNU Affero GPL (AGPL), the GNU All-Permissive licence, Modified BSD licence, Apache licence, Intel Open Source licence, Mozilla Public licence (MPL)...the list goes on[4]!

The choice of licence used depends upon the needs of the developer, as the rules and clauses accompanying each will vary, so it is important to research which licence is best suited to the software being developed.

Clearly, with such an expansive list of available licences, to go into detail regarding each could fill a whole book! Instead, a flavour of the licensing terms these various options provide will be given by taking a look at the GNU GPL in particular.

22.3.1. GNU General Public licence

The GNU GPL is what is known as a *copyleft* licence. A play on the word copyright, this refers to the practice of using copyright law in the distribution of software and its modifications but with each modification retaining the same rights. This therefore ensures a program is free[5]. The following details just some of the regulations regarding GPL[6]:

- When a piece of software is released under the GNU GPL as free software, all subsequent updates or modifications must also be released as free software.
- Modifications can be made without publishing if they are for private use, including that of organisations and companies using modified code internally, however, if it is released publicly in any way the source code must also be made available.
- Since 'free' does not refer to 'free as in free beer', GPL still allows developers to charge for their software.
- However, once someone pays the fee on a piece of GPL'd software, they are then permitted to redistribute it with or without fees.
- All copies of software under GPL must carry appropriate copyright notices giving credit to authors where due. A copy of the GPL must accompany every copy of the program in question to ensure everyone knows their rights.
- Licences compatible with GPL allow code under such a licence to be included as part of a larger program covered by GPL. If a program features code under a non-compatible licence it is no longer considered fully free.

But what does this mean for the average developer? Well, a developer wishing to publish their program to a wider audience must take into account existing licences for any original or modified code they have included, and release it under a suitable licence that adheres to all the pre-existing terms. Put simply, it is advisable to read the terms of any licences relevant to a piece of code or its constituent parts to ensure the published program doesn't violate the terms of its licence.

The GNU Operating System homepage discusses the GPL (and other free licences) in detail and in fact you can download a full copy for yourself to read and gain a full understanding of what this means for a developer:

http://www.gnu.org/licences/gpl.html

22.4. Development Tools and Resources

22.4.1. Virtual Machines

Those lacking a machine running a Linux distribution but wishing to develop for Linux have the option of using a Virtual Machine (VM). This is a software implementation of an OS, emulated within the existing primary OS of a target machine provided it has enough processing power and memory.

There are many VM options available, at different price points and offering different features, to tailor to anyone from the casual user to more advanced solutions for enterprise. Table 22.1 highlights a few of these options and their relative advantages and disadvantages. A developer wishing to use a VM for the sole purpose of developing applications for Linux may find one of the free options more than suitable for their needs. Someone wishing to deploy a Linux based VM in a business scenario, however, may be required to look at paid alternatives.

Table 22.1: Comparison of VMs.

Virtual Machine	Free/Paid	licence	Advantages	Disadvantages
Virtual Box	Free	Proprietary/ GPL	Multicore support. Features a GUI. Supports USB. Supports snapshots. Support for Direct3D and OpenGL.	Cannot emulate other architectures.

Table 22.1: Comparison of VMs.

Virtual Machine	Free/Paid	licence	Advantages	Disadvantages
VMware Workstation	Paid	Proprietary	Multicore support. Features a GUI. Supports USB. Supports snapshots. Support for DirectX 9 and OpenGL.	Cannot emulate other architectures.
VMware Player	Free (personal use)	Proprietary	Multicore support. Features a GUI. Supports USB. 3D Acceleration with VMGL.	Less features than paid version (Workstation). No snapshot support.
KVM	Free	GPL	Emulates multiple architectures (ARM, MIPS, SPARC). Supports USB.	No proprietary GUI (third party alternatives available).
XEN	Free	GPL	Features a GUI. 3D Acceleration with VMGL.	No USB Support. No snapshot support.
Windows Virtual PC	Free	Proprietary	Features a GUI. Supports snapshots.	No multicore support. Linux support is unofficial/lacking some features. Cannot emulate other architectures. Partial USB support. No 3D Acceleration support.

22.4.2. Version Control

The collaborative nature of development for Linux necessitates the utilisation of some form of version control. Let us consider a rather contrived scenario in which an argument for version control is presented. Suppose you have invested several hours coding a particularly impressive application to run on a Zynq development platform. You've programmed the FPGA, downloaded the application to the processor and run it and everything functions as expected! With a spring in your step, you head to the kitchen to make a celebratory cup of coffee and grab a piece of cake. As the coffee brews, you discuss your success with a colleague, who has an idea on how you could improve the functionality of the system. Being a trusting person, you give them the all clear to implement the improvements. You come back to the project some time later, and sure enough it no longer works, the 'helpful' colleague is nowhere to be seen and your code is a mess! The only option is to try and remember what you had before and invest even more time in correcting things.

Utilising some form of version control is a way to prevent just this kind of disaster. There are three main types of version control that you could utilise. The first is simply making a local backup of all your work at regular intervals. This is highly susceptible to human error, and things can become rather confusing when you have many versions of many different files, created by many different colleagues, stored across many different directories. Maintaining a database of changes would help with this, but there exists a more elegant solution.

Centralised Version Control Systems (CVCSs) utilise a single server that hosts all versioned files. Every user can login to the server and check-out a local copy of the file for editing. When they are finished, they can commit any edits and the server maintains a record of who authored which version of which file and when. However, suppose the basic example of a case where regular full backups of the server have not been made and the hard disk is corrupted; the entire project is lost beyond repair. A Distributed Version Control System (DVCS) mitigates this. Instead of checking-out a single file, each user mirrors the entire repository on the server allowing for recovery of the entire repository should something go wrong[7]. The metadata utilised in DVCSs also facilitates operations such as merges. A DVCS is also of use on collaborative projects where progress is being made by more than one developer on a regular basis. If you want to utilise a colleague's latest advancements in unison with your own development you can simply take the latest snapshot on the server and continue working,

22.4.3. Git

Git is one such DVCS, designed by Linus Torvalds for the initial purpose of managing Linux kernel development. It was released in 2005 as a free alternative to a previous, proprietary Version Control System (VCS) that ceased to provide free use.

Similar in deployment to a DVCS, Git has one significant difference. A DVCS 'thinks' of the project as a set of original files and tracks the changes made to them over time. Git, however, takes a snapshot of all files at each commit which it then stores and references as required. To avoid redundancy, unchanged files are not re-stored and a link back is made to the previous file saved. This is visualised in Figure 22.2, where each version represents a snapshot of the project at that time.

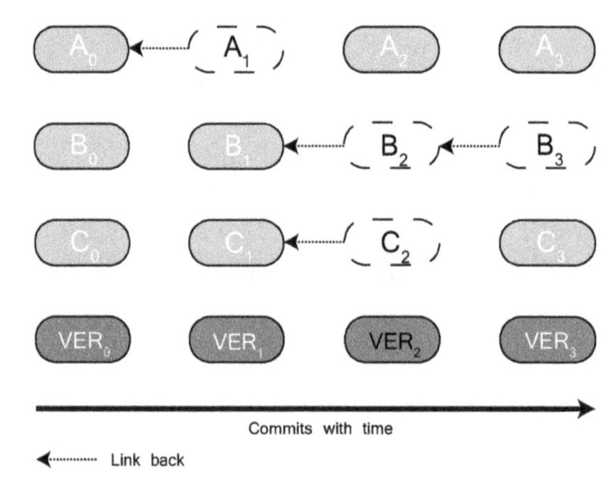

Figure 22.2: Exemplary Git version history

There are three main states in which any file can exist; these are referred to as *committed, modified,* and *staged.* A file is first modified in the working directory, then staged, marking it for the next snapshot to commit to the repository. This sequence of events is visualised in Figure 22.3. The working directory contains a checkout of a single version of the project for local modification. The staging area is really just file indexing of what will be entered into repository on the next commit. The repository is a store of the object database and metadata for the project.

Git operations occur locally, rather than on a remote server as with CVCS, mitigating any network latency. Any information regarding version history can be gleaned locally rather than through pulling information on older versions from the remote server. This also facilitates offline use when a connection to the repository may not be available, as commits can be made offline and uploaded when a connection is available.

Finally, since modification of a project within Git is generally considered an additive procedure to the database, it is considerably difficult to modify a project beyond the bounds of being recovered.For these reasons, Git is the preferred method of version control for Linux development[8].

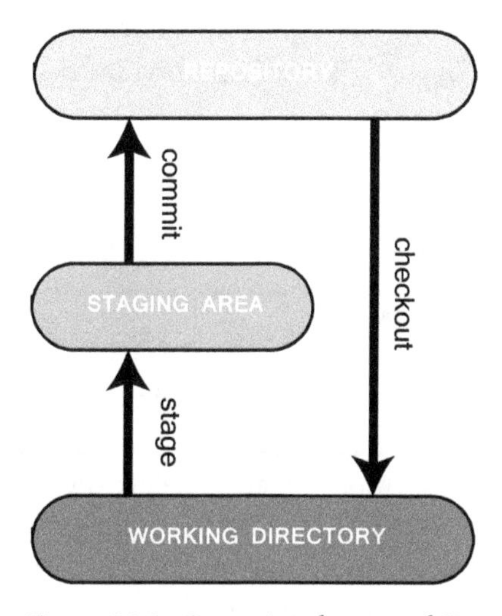

Figure 22.3: Operational stages of Git

22.4.4. Debugging Linux

Code, at some iteration of its development, will likely contain bugs; this is a statement that holds true for all application development, and Linux is no exception. Despite our best attempts to create perfect, logical code, often certain arguments will return an unexpected outcome, a subroutine may not receive the correct parameters, or something even stranger may occur due to a bug. In Linux, several tools are at our disposal to identify the nature and location of any bugs.

Problems with memory allocation are a common cause of program crashes; these could include memory leaks, where memory is never released with the appropriate calls, or buffer overruns, where writing occurs beyond the allocated memory. Examples of memory debugging tools include *MEMWATCH* and *Yet Another Malloc Debugger* (YAMD)[9].

Often, a program in the user-space can trigger a system call which can cause unexpected behaviour. One tool used for debugging such errors is *strace*, which monitors all the arguments to a particular call and its corresponding return value directly from the kernel to identify which call is at fault. If the return value is unexpected for the given arguments it can be deduced there is a problem with the function being called[9].

Tools are also available for bugs originating in the Linux kernel itself. The *GNU debugger* (gdb) is provided in the user-space in the form of a command line tool or GUI with the purpose of seeking out bugs in user-space programs, or for use in the kernel as the *remote host Linux kernel debugger through gdb* (kgdb). Other tools are provided in the kernel for sourcing these bugs, such as *Oops*[9]. Table 22.2 provides details of some of the availabile debugging tools.

Table 22.2: Linux debugging tools.

Source of Bug	Examples of Debugging Tools	Notes
Memory Allocation	MEM-WATCH	Memory error detection tool in C. Added to the program as a header file. Tracks memory leaks/corruptions. Provides a results log.
	YAMD	Locates problems with dynamic memory allocation in C/C++. An external tool that must be installed and run for a file under test. Provides a detailed analysis of code to ascertain memory leaks etc.
System Calls	strace	Command line tool. Details system calls issued by user-space programs including arguments to said calls and the return values. Information is taken directly from the kernel and requires no modifications to the kernel.

Table 22.2: Linux debugging tools.

Source of Bug	Examples of Debugging Tools	Notes
Kernel	gdb	Command line/graphical tool provided by the Free Software Foundation. Debugs user-space programs and the Linux kernel itself. When using gbd, the program to be debugged is run and details are provided about program termination.
	kgdb	Debugs the Linux kernel using gbd. Requires a development machine and a test machine connected via serial ports. Provided as an extension of the kernel allowing connection to a remote machine running the same extension during boot. This allows the setting of break points and data examination of the remote kernel.
	Oops	A message containing details of system failures sent to system console upon a crash. Message is sent to *ksymoops* utility to convert code to instructions and map stack values to kernel symbols to provide a meaningful reasoning for the cause of the failure.

There of course exists many other debugging tools available for getting to the root of a bug in a program, and the suggestions above are merely an introduction to the types of bug you may encounter and the debugging options available.

22.5. Chapter Review

An overview of a generalised Linux architecture has been presented, as well as a discussion of some of the issues surrounding Linux development, including licensing and development tools. As such the following chapters will focus on the key components of the kernel individually. Chapter 23 will delve deeper into the Linux kernel including the file system and drivers. Following this, we will discuss the Linux boot procedure.

22.6. References

Note: All URLs last accessed June 2014.

[1] GNU Operating System, "Overview of the GNU System", webpage
Available: http://www.gnu.org/gnu/gnu-history.html

[2] IDC white paper, "The Opportunity for Linux in a New Economy", April 2009
Available: http://www.linuxfoundation.org/sites/main/files/publications/Linux_in_New_Economy.pdf

[3] L. Wirzenius, J. Oja, S. Stafford and A. Weeks, "Overview of a Linux System — Important Parts of the Kernel" in *The Linux System Administrators Guide,* v0.9
Available: http://www.tldp.org/LDP/sag/html/overview.html

[4] GNU Operating System, "Various GNU licences and Comments about Them", webpage
Available: http://www.gnu.org/licences/licence-list.html#Softwarelicences

[5] GNU Operating System, "What is Copyleft?", webpage
Available: https://www.gnu.org/copyleft/

[6] GNU Operating System, "Frequently Asked Questions about the GNU licences", webpage
Available: http://www.gnu.org/licences/gpl-faq.html

[7] Scott Chacon, "Getting Started — About Version Control" in *Pro Git,*
Available: http://git-scm.com/book/en/Getting-Started-About-Version-Control

[8] Scott Chacon, "Getting Started — Git Basics" in *Pro Git,*
Available: http://git-scm.com/book/en/Getting-Started-Git-Basics

[9] IBM Developer Works, "Debugging Tools and Techniques for Linux on Power", 02 October 2013
Available: http://www.ibm.com/developerworks/systems/library/es-debug/index.html?ca=drs

23

The Linux Kernel

The previous chapter introduced the concept of the Linux kernel, and this chapter seeks to elaborate on this key part of the Linux operating system. The hierarchy of the kernel itself will be examined, with discussion of some of its main aspects: memory management, process management, and the file system.

23.1. Linux Kernel Hierarchy

Until now the Linux kernel has been presented as a mysterious but crucial part of a Linux based system. We will now investigate the kernel further and take a look at some of the core operations it is responsible for.

A *kernel* is effectively the core of an operating system. It is the part of the operating system that is first to load on boot, and therefore resides in main memory.

Figure 23.1 is similar to Figure 22.1 from the previous chapter, except the kernel has been expanded to show some of the most important components. In reality, if you were to search for a Linux kernel 'map' you would be presented with a complex web of many different components and paths, but this is beyond the scope of this book.

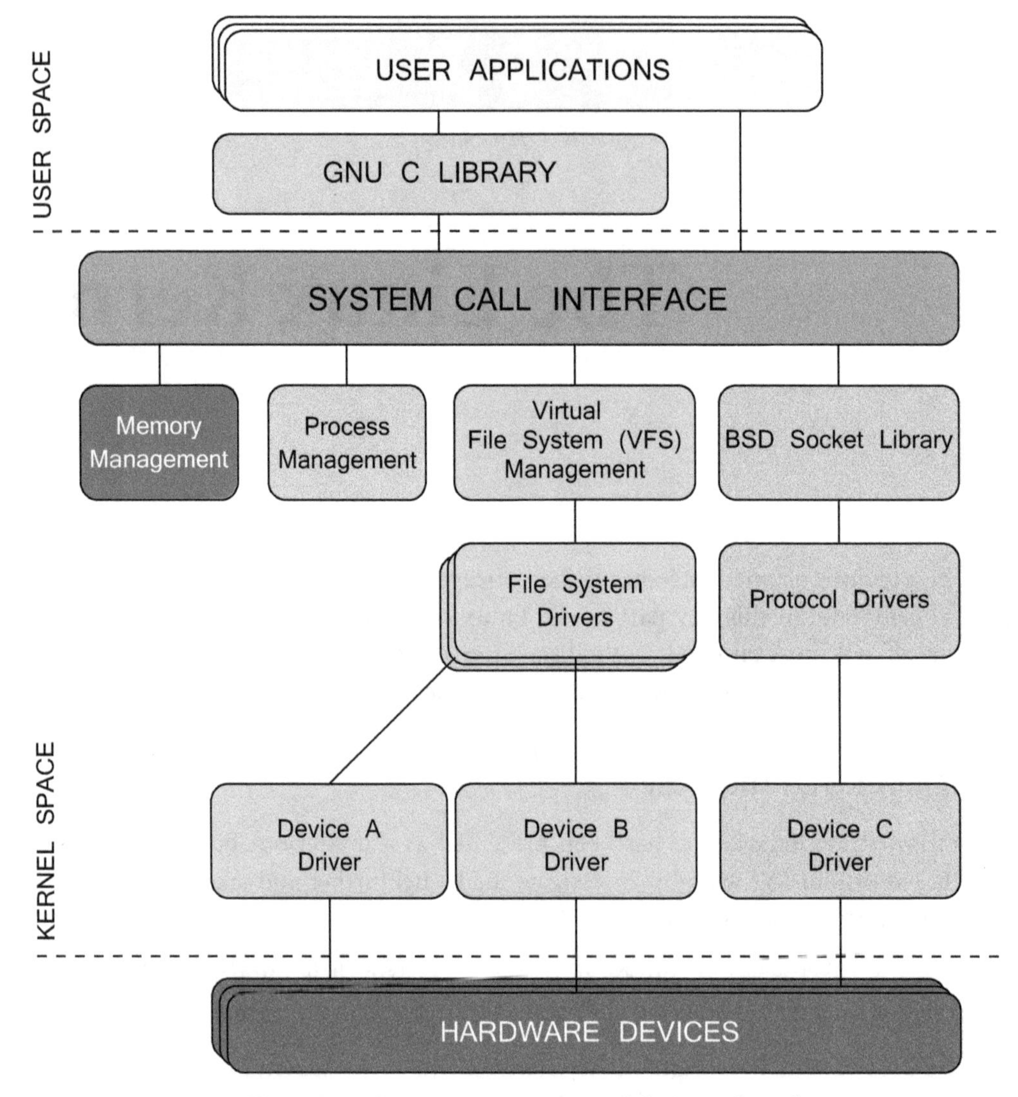

Figure 23.1: Important components of the Linux kernel

23.2. System Call Interface

A system call is an interaction between an application in the user space and the required service provided by the kernel. The SCI provides this link between the boundary of the user space and kernel space, where direct calls are not possible. The implementation of a

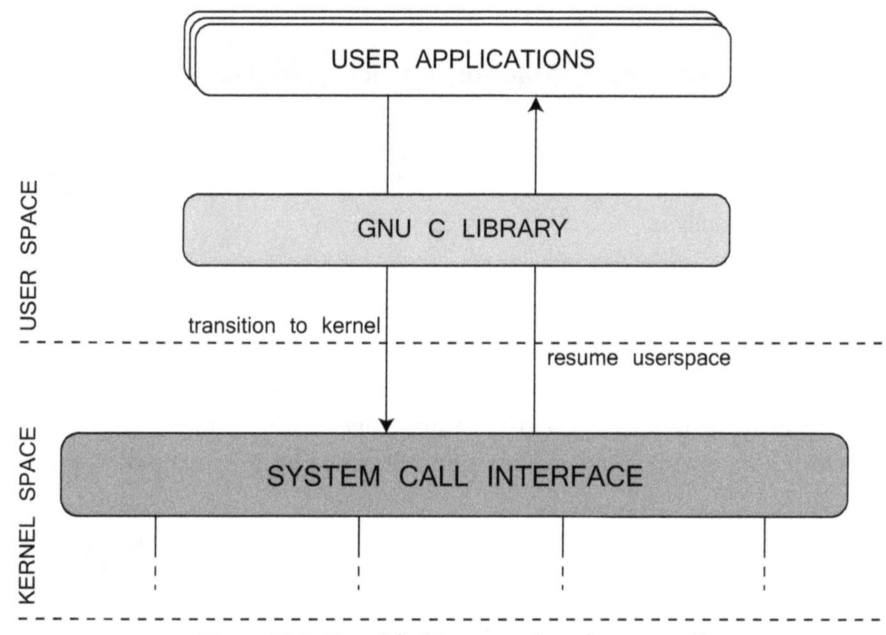

Figure 23.2: Simplified interrupt based system call

Linux system call can vary depending on the processor architecture. Figure 23.2 represents one such simplified approach based on processor interrupts[1].

- All system calls are multiplexed through a single kernel entry point, identifiable by a register as specified in and loaded from the C library.
- A software interrupt is triggered and the interrupt handler executes the *system_call* function.
- The identifying register indexes the *system_call_table* to trigger the correct system call via the SCI.
- Returning from the system call, there is a transition back to the user space where execution continues in the C library and original user application.

System calls therefore provide a level of abstraction between the programs in the user-space and the kernel itself. This also gives greater consistency to the interactions between all user applications and the kernel code.

23.3. Memory Management

Despite the rapid advancements in memory technology used in computing systems, the software used is always hot on its heels in terms of memory requirements. The truth is that there is a need for more memory than the finite amount available. Virtual memory can help with this problem, making it appear as though there is more memory available in the system than there actually is.

23.3.1. Virtual Memory

Figure 23.3 represents a simplified visualisation of virtual memory. Two processes, Process A and Process B, each have their own virtual address space and page table. A page is simply a segment of memory, given a unique Page Frame Number (PFN). The page tables therefore contain mapping information allowing the processor to resolve virtual

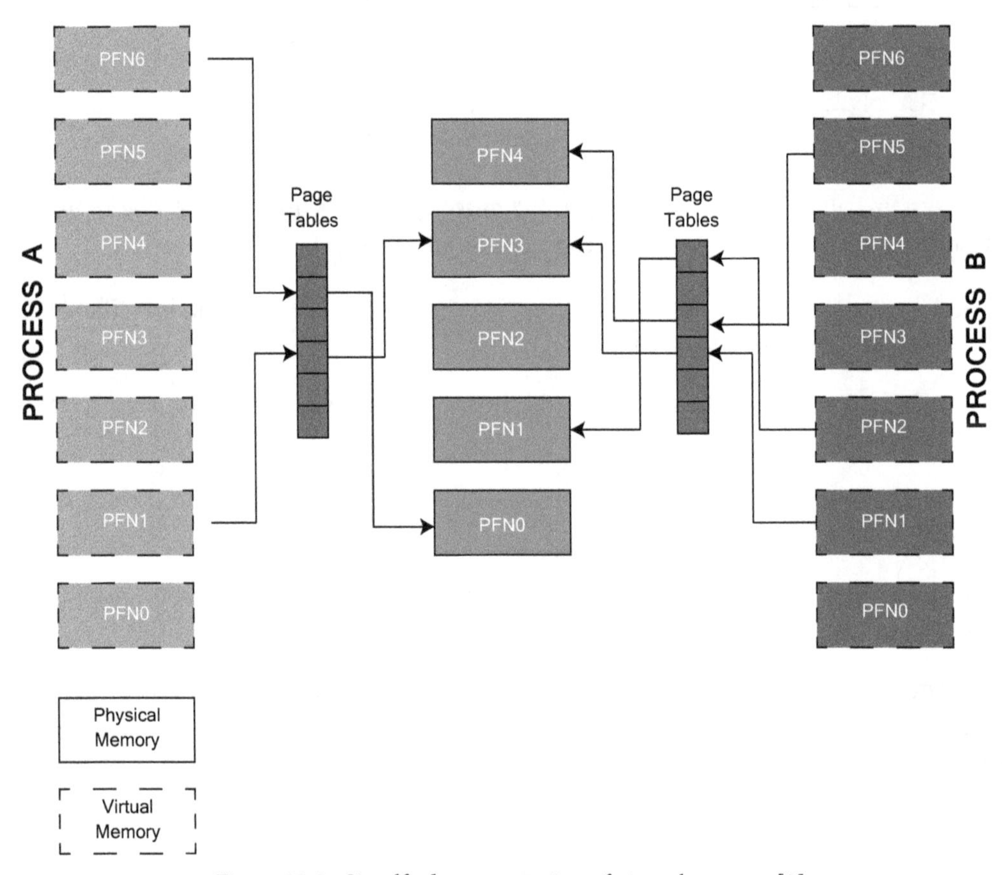

Figure 23.3: Simplfied representation of virtual memory[2]

memory addresses into physical ones. Note that Process A has virtual PFN1 mapped to physical PFN3. In order to translate this virtual PFN1 into physical PFN3 the processor will use an offset within the virtual page as an index to the page table. If the entry at that offset is valid, a physical PFN is retrieved. If not and the process is trying to access non-existent regions of memory then the address cannot be resolved and the operating system is notified of a page fault to be dealt with.

This subsystem for memory management offers a number of features; referring to Figure 23.3 once more, it can be observed that a virtual PFN does not have to match the physical PFN it is mapped to, and in fact several virtual PFNs from different processes can be mapped to the same physical PFN, as is observed with entries from both Process A and B mapped to PFN3 in physical memory. This allows the system to operate with a large address space, where the system appears to have more memory than is physically present. Virtual address spaces are independent for each process and so are offered a level of protection through separation, preventing code or data from undesired overwrites. A further benefit of this approach is that all running processes in a system receive a fair share of the physical memory available [2].

23.3.2. High and Low Memory

Historically, 32-bit Linux systems have only been able to support 4GB of memory due to 32 bits providing a kernel virtual address range of 0xC0000000 to 0xFFFFFFFF. The default mapping of this 4GB is with the 3GB in the upper address space given to the user space with the remaining 1GB beginning at 0xC0000000 set aside for the kernel code itself (in actuality this is slightly less than 4GB due to the presence of I/O).

To mitigate this, the Linux kernel now divides the virtual address space into two regions known as low and high memory. Low memory refers to the memory for which logical addresses exist in the kernel space. High memory, on the other hand, refers to memory for which logical addresses do not exist as it lies beyond the address range defined for kernel virtual addresses. Special page tables are required to map high memory into the kernel's address space; this can be an expensive operation with a limit on the number of high-memory pages mapped at any time. For this reason the core of the kernel is kept in low memory, while high memory can be used for some kernel tasks and process pages only[3].

23.4. Process Management

As with any other operating system, tasks to be carried out are performed by processes. Consider the fact that any program on an operating system is actually just some instructions, in the format of machine code, stored as an executable image; a process is therefore such a computer program in operation.

- As machine code instructions are executed by the system's processor, processes change and as such are dynamic.
- Linux is a multiprocessing OS, and each process has its own rights and responsibilities, meaning a crash of one process will not impact the other processes.
- Processes use both a variety of system resources from the CPU for running instructions and physical memory for holding the process and any corresponding data.

23.4.1. Process Representation

In Linux, a structure known as *task_struct* is used to represent a process. Within this structure is all the data required to represent that process as well as other a large amount of additional data used in associating the process with any parent/child processes. Some of the information relating to the process contained within *task_struct* is as follows[4]:

- *Process state of execution* — running, sleeping, sleeping but not interruptible, stopped, etc.
- *Flags* — process creation, exiting, memory allocation.
- *Process priority* — a lower value priority equates to higher process priority. This is dynamic and based on several factors.
- *Process address space.*

23.4.2. Process Creation, Scheduling and Destruction

Processes are created in the user space, either through execution of a program (which creates an entirely new process), or within that program itself using the *fork* system call (creating a child process). Each process has its own process identifier (PID) which is basically a number used to uniquely identify each process in the system. These numbers do not change throughout the life of the process, but upon process termination its PID is released for use with a new process. The process is passed back and forward between several different functions, including[4]:

- *copy_process* — creates a new process as a copy of the parent, consults the Linux Security Model to ensure it has permission to create new tasks, and initialises the *task_struct*

- *dup_task_struct* — allocates a new *task_struct* for the process and copies process descriptors to this structure.
- *wake_up_new_task* — initialises scheduler information and places the created process in a run queue

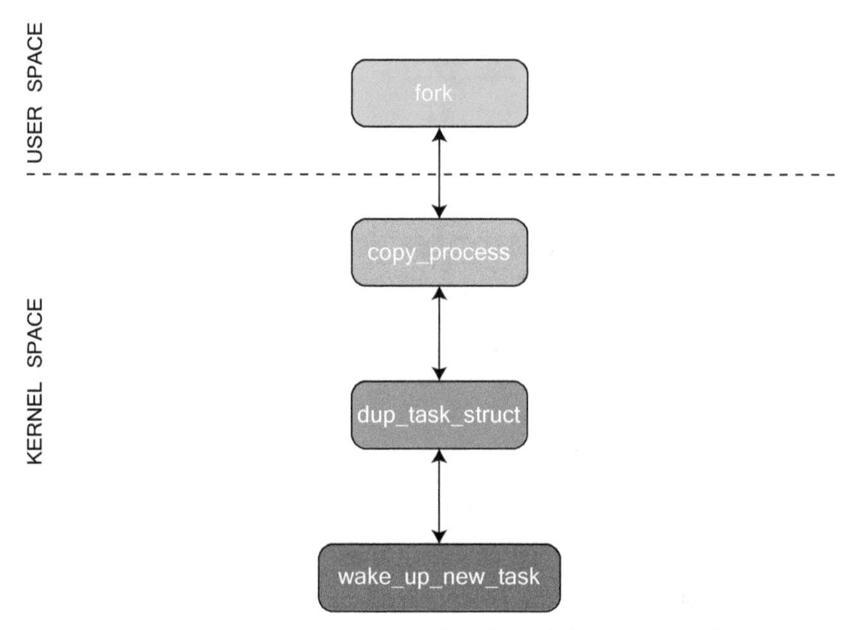

Figure 23.4: Process creation flow through fork *system call*

The Linux process scheduler is responsible for managing the various processes existing in the system based on their priority. It holds a list of all processes identified by their corresponding *task_struct* at each priority level. The *schedule* function selects the most appropriate process to run based on execution history and loading.

Several different events can initialise process destruction: normal process termination at the end of a run, termination through transmission of a special signal, or via the *exit* system call. All of the aforementioned methods produce a call to the kernel function *do_exit*, which eradicates all references to the current process from the OS. The process to be exited is indicated via the *PF_EXITING* flag, at which point a series of calls detaches the process from any resources used during its life. The *exit_notify* call provides a notification before the process state is altered to *PF_DEAD*, allowing the scheduler to select the next process for execution[4].

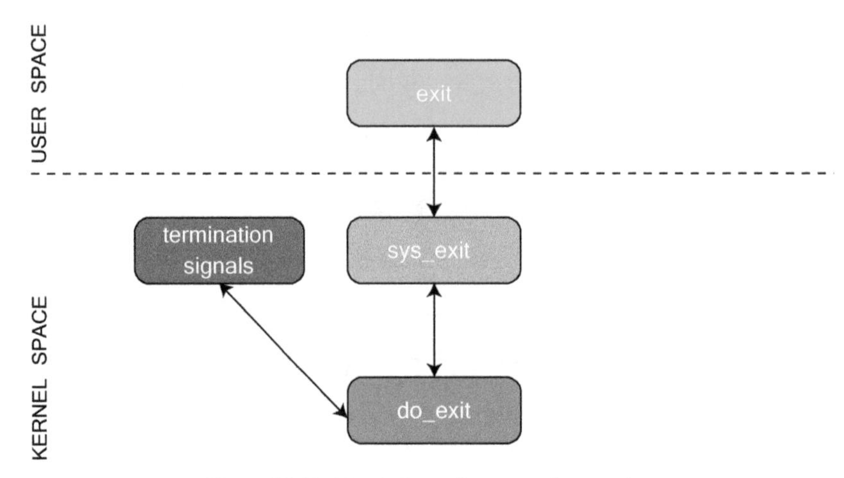

Figure 23.5: Operation of process destruction

23.5. File System

23.5.1. Linux File Systems

Linux is incredibly versatile in that it is capable of supporting a large number of different file systems including, but not limited to:

Table 23.1: Some supported Linux file systems

Linux File Systems	
ext	proc
ext2	smb
xia	ncp
minix	iso9660
umsdos	sysvs
msdos	hpfs
vfat	affs
ufs	

The numerous file systems are collated in a hierarchical tree structure and viewed by the system as a single file system entity. New file systems are mounted (that is, a file system is associated with a storage device in Linux) to this tree as they are added to the system[5].

23.5.2. Virtual File System

Referring back to Figure 23.1, the Virtual File System (VFS) plays a key role in facilitating the inclusion of the many file systems. It provides a level of abstraction, creating a unified interface between the individual file systems and the rest of the kernel. It is the root level of the file-system interface and is responsible for tracking all currently-supported and mounted file systems. Figure 23.6 represents the interactions between the VFS and the real file systems (in this instance we are only demonstrating the *vfat* and *ext2* file systems).

The buffer cache is used as a common data buffer from the hardware devices, speeding up file system access to the physical devices holding the file systems. It is independent of the file systems and as such keeps these file systems independent from the supporting device drivers and media.

During the navigation of mounted file systems inodes are continuously read or written, so the VFS maintains an inode cache to speed up access to the mounted file systems. (Note — *inode* stands for index node and is a data structure found within Unix file systems, containing all the information about a file system object).

The VFS also maintains a cache of directory entries to speed up access to those most commonly in use. When a directory is looked up by a file system, its details are added to the cache and can be accessed more quickly the next time the same directory is looked up. The VFS uses a form of Least Recently Used (LRU) retention of cache entries. As a directory entry is added to the cache, those used previously are moved down a level. When the cache is full, the entry at the lowest level will be discarded as the next entry is added[5].

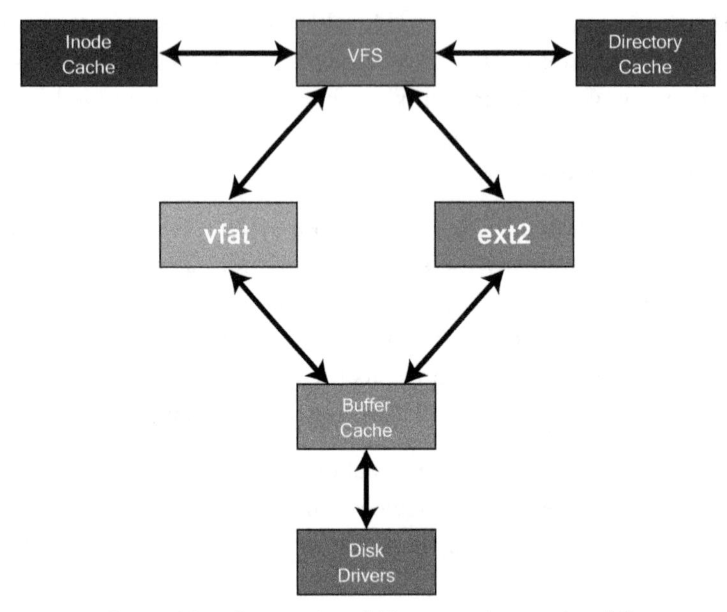

Figure 23.6: Linux virtual file system interactions[5]

23.6. Architecture-Dependent Code

The lowest layer of the Linux kernel stack is the architecture-dependent code. The Linux kernel supports a great variety of different hardware platforms, and so some platform specific code is required for each to work in unison with the more 'generic' architecture-independent code. This is commonly known as a BSP and includes source files for the architecture families and processors, common boot support files, DMA hardware interfaces, interrupt handling and other items relevant to a specific processor family[6]. When working through the associated ZedBoard-based tutorials you will notice that an important step in the design is building an appropriate BSP for the Zynq that allows the processor to communicate with the development board.

23.7. Linux Device Drivers

A Linux device driver, as in any OS, can be considered as a 'black box' that provides a level of abstraction between the hardware devices in a system and the programs running in the OS. By utilising device drivers, a standardised set of calls can be implemented across all programs which are independent of the specific device; these calls correspond to driver-specific functions that carry out the required operations. The result is a modular approach allowing the ability to build drivers outside the kernel and add them in as required[7].

23.7.1. A Note on Mechanisms Vs. Policies

An important distinction to make about device drivers is that they provide *mechanisms* and not a *policy* [7].

- A mechanism dictates what the code in question is capable of doing. A device driver therefore dictates the capabilities of a piece of hardware through the code contained within.
- Policies are defined at higher levels of the system and concern how these mechanisms should be used.

Ensuring that drivers are policy free gives flexibility, meaning the corresponding hardware is available to all users requiring it, regardless of their particular needs for the piece of hardware.

23.7.2. Module/Device Classification

In Linux, a *module* is a piece of code that is added to the kernel at runtime, a concept that will be clarified in the next chapter. Device drivers fall into this category, and file system types also exist as modules within the kernel.

Device drivers can exist as one of three fundamental types. These are[7]:

- *character devices* — devices that are accessed by a stream of bytes, such as a file. A driver responsible for such a device would perform open/close and read/write system calls.
- *block devices* — devices capable of hosting a filesystem, such as a hard disk. Linux allows the block device to act like a character device, reading and writing any number of bytes unlike traditional Unix systems which can only read/write in blocks of 512 bytes or more. The block devices differ from character devices in the way the kernel itself manages the data and hence have a different interface.
- *network interfaces* — devices concerning the transmission of packets of data through a network interface. Network drivers do not know about the connections regarding a transmission and instead only deal with the packets transmitted.

23.8. Chapter Review

This chapter has provided a high-level overview of some of the fundamental components of the Linux kernel. The system call interface was shown as a means of abstracting the user-level of a Linux system from the kernel. Within the kernel itself, memory and process management, the file system, and device drivers were discussed.

23.9. References

Note: All URLs last accessed June 2014.

[1] IBM Developer Works, "Anatomy of the Linux File System", 30 October 2007
Available: http://www.ibm.com/developerworks/linux/library/l-linux-filesystem/

[2] David A. Rusling, "Memory Management" in *The Linux Kernel*, v1.0
Available: http://www.tldp.org/LDP/tlk/mm/memory.html

[3] J. Corbet, G. Kroah-Hartman and A. Rubini, "Memory Mapping and DMA" in
Linux Device Drivers, 3rd Edition, O'Reilly, 2005, pp 412-463

[4] IBM Developer Works, "Anatomy of Linux Process Management", 20 December 2008
Available: http://www.ibm.com/developerworks/library/l-linux-process-management/

[5] David A. Rusling "The File System" in *The Linux Kernel*, v1.0
Available: http://tldp.org/LDP/tlk/fs/filesystem.html

[6] M. Tim Jones, "2. GNU/Linux Architecture" in *GNU/Linux Applications Programming*,
Cengage Learning, 2005, p17

[7] J. Corbet, G. Kroah-Hartman and A. Rubini, "An Introduction to Device Drivers" in
Linux Device Drivers, 3rd Edition, O'Reilly, 2005, pp 1-14

24

Linux Booting

Now that the Linux kernel has been introduced it is worth taking the time to consider the Linux boot process; the sequence of events that goes on behind the scenes when a Linux computer or embedded system is powered on. We will start with a high-level overview of the boot process of a Linux desktop computer ahead of covering the various individual stages. Next, we will look at how the boot process of an embedded Linux device differs from that of a Desktop variation by considering the Linux boot process specific to Zynq.

24.1. Overview

The first thing that happens when a Linux system is powered on, or reset, is that the processor will execute code in a predetermined location. For a desktop computer this location is stored in a section of flash memory on the motherboard, known as the Basic Input/Output System (BIOS). As modern day PCs offer such versatility in the range of boot devices, the first thing that the BIOS will do is determine which devices offer bootable options [1].

Once the boot device has been determined, the FSBL is loaded into RAM and executed by the processor. The FSBL is a very small section of code — — less than 512 bytes, or a single sector — and its sole purpose is to load the Second-Stage Bootloader (SSBL) into RAM.

The SSBLis the stage in the boot process where a boot menu is presented; if you have used Linux before you may have noticed this. This menu presents a list of possible boot options that are available. When a boot option from the list is chosen, the bootloader will load the corresponding OS, which will then decompress and load itself into memory before initialising the kernel. There are other functions that are carried out by the bootloader before the kernel is launched, but we will get to that later.

With the required kernel modules successfully launched, the final stage of the boot process is to invoke the first user-space function, *init*. This function initialises all high-level parts of the system.

This concludes the high-level overview of the Linux boot process. In the next section we will break down the boot process into its consecutive parts and take a look at each of them in greater detail.

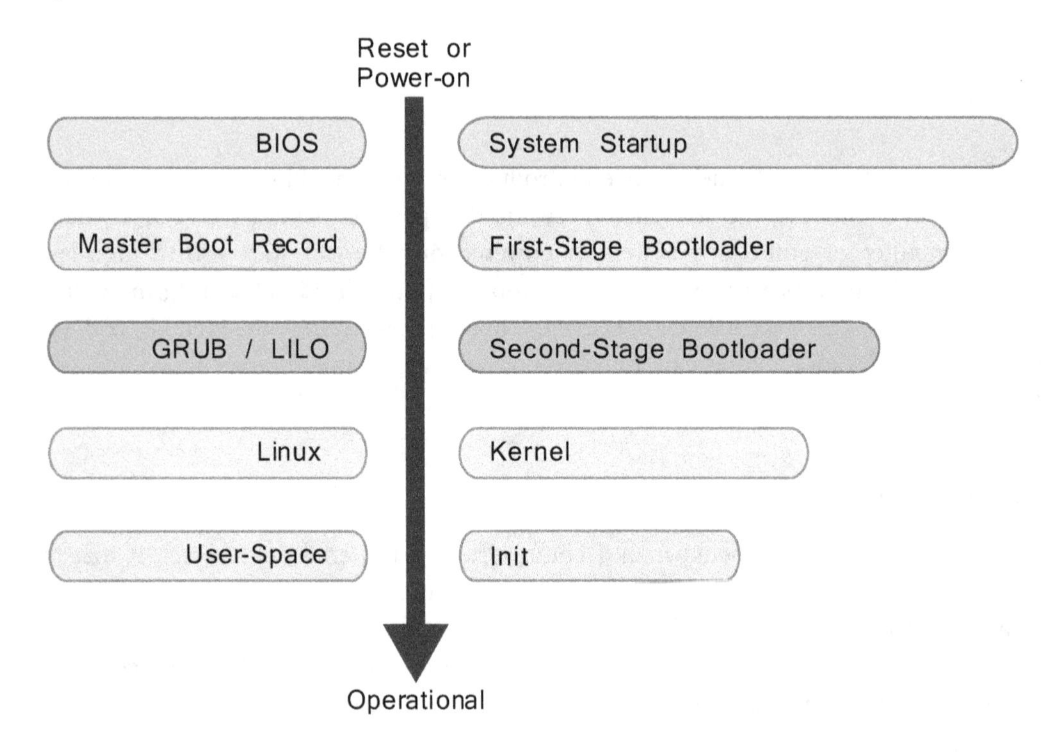

Figure 24.1: Stages of the desktop Linux boot process

24.2. Stages of the Desktop Linux Boot Process

24.2.1. BIOS

The BIOS is responsible for the initial system startup and locating the available bootable devices. When the system is powered on the processor will begin execution of the BIOS, which is split into two parts - the Power-On Self Test (POST) and runtime services.

The first operation carried out by the BIOS is dependent on whether the system was powered on (cold boot) or reset (warm boot). If the system has been powered on, the first operation is the initialisation and testing of the basic system hardware components - a process known as the POST [1]. Once all hardware devices have been initialised and verified, the POST function is flushed from memory. If, however, the system has been reset a special flag is raised in memory and the BIOS will not carry out the POST in order to save time.

Once the POST has been flushed from memory, the BIOS will advance to runtime services. This function searches for bootable devices that are available to the system; the devices must also be active to be found. The order in which the devices are located is defined in the Complementary Metal Oxide Semiconductor (CMOS) settings. Typically the Linux boot device is a hard disk; it can, however, be anything from a floppy disk, network device, USB memory or CD-ROM.

When booting Linux from a hard disk, the Master Boot Record (MBR), a 512 byte sector located in the first sector of the device, contains the primary bootloader. The last task carried out by the BIOS is the loading of the MBR into memory. Once loaded, the BIOS relinquishes control [1].

24.2.2. First-Stage Bootloader (FSBL)

The FSBL is a section of code contained in the MBR. The remainder of the MBR is a partition table and a validation signature. The largest section of the MBR (446 bytes) is known as the primary bootloader and contains executable code and error messages [1]. The partition table is contained in the next 64 bytes, which itself contains records of four primary partitions, of 16 bytes each. The final section of the MBR is a 2 byte boot signature — `0xAA55` — for validating the MBR. The structure of the MBR is provided in Figure 24.2.

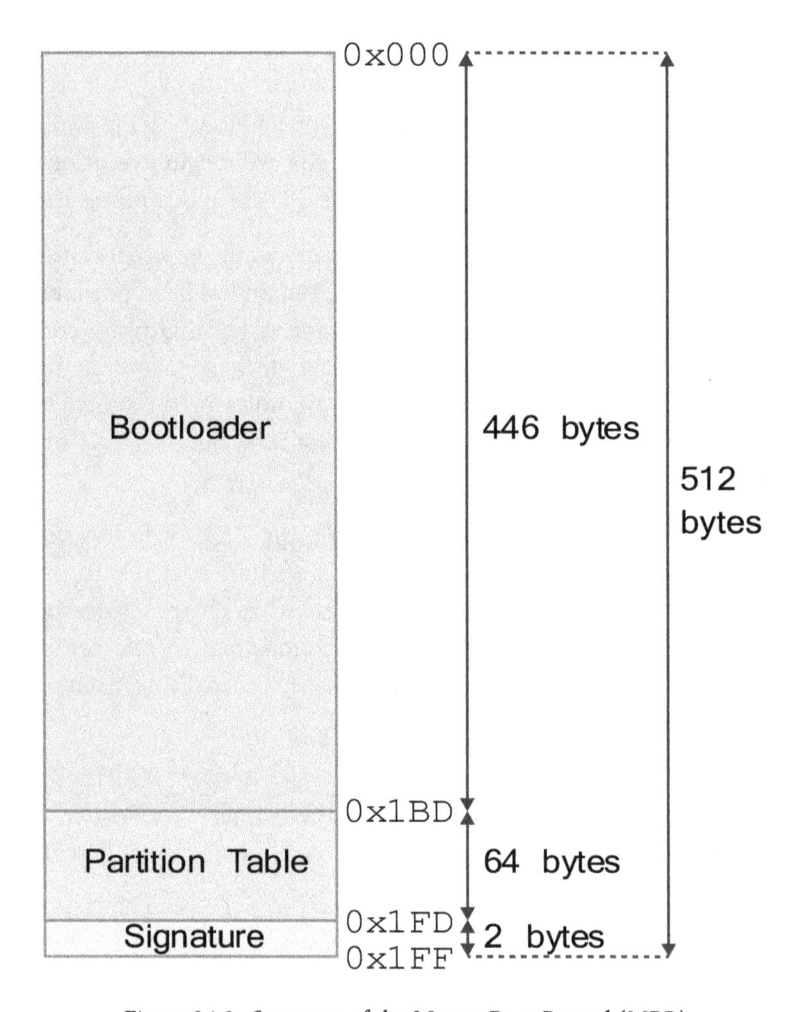

Figure 24.2: Structure of the Master Boot Record (MBR)

The primary bootloader's job is to locate, and then load, the SSBL by searching the partition table for an active partition. Once an active partition is located, all further partitions are scanned to ensure that they are inactive. The boot record from the active partition is then loaded into memory for execution.

24.2.3. Second-Stage Bootloader (SSBL)

The SSBL will present you with a list of operating systems that are available to be booted by the system. Once the desired operating system has been selected, the kernel image is decompressed and loaded into memory before the control of the processor is passed to the OS [4].

Two main bootloaders exist: Linux Loader (LILO) and Grand Unified Bootloader (GRUB). Both LILO and GRUB are made up of the combination of both the FSBL and SSBL. LILO has been around for long time and has, in most cases, been replaced with GRUB. While LILO requires that the kernel files be stored on raw disk sectors, GRUB is able to load kernels from ext2 or ext3 file systems [1]. GRUB is able to do this by adding an extra step to the bootloader process, between the first- and second-stages, which allows it to understand a specific file system.

Other methods of booting a Linux kernel exist, such as Syslinux or Loadlin which allow you to load Linux from, and replace, the currently running Windows/DOS environment.

24.2.4. Kernel

With control of the CPU handed over by the second-stage bootloader, a routine will run to perform a small amount of hardware setup before decompressing the kernel image. Once decompressed, the kernel image will be transferred to high memory - the part of physical memory not mapped directly by the kernel page tables. If a RAM disk image is present this will also be moved into memory and marked for later use [1]. At this point the kernel will be called for the first time and will boot.

During the kernel boot further hardware setup is carried out, including setting up the stack, configuring the page tables, enabling memory paging and the detection of the CPU and FPU type [1].

24.2.5. Init

The final stage of the Linux boot process is the initialisation of *init* - the first user-space application. Once invoked, *init* looks for the file **/etc/inittab** and determines whether it has an entry of the type **initdefault**, which details the initial runlevel of the Linux system. The various Linux distributions will have different configurations for the runlevels of the system. Taking the Linux standard base specification as an example, there are 7 runlevels, as detailed in Table 24.1 [3].

Init is the first program to be invoked that has been compiled with the standard C library. No standard C applications will have been executed prior to this point [1].

Table 24.1: Default runlevels for Linux Standard Base [3]

ID	Name	Details
0	Halt	Shuts down the system.
1	Single-user mode	Allows text mode access to a single user for administration purposes.
2	Multi-user mode without networking	Normal text mode operation allowing multiple users to access the system. All daemons run except Network File System (NFS).
3	Multi-user mode with networking	Same as runlevel 2 except with NFS enabled.
4	Reserved for local user	N/A.
5	Multi-user with a display manager	Full multi-user operation with the addition of graphical mode.
6	Reboot	Shuts down the system to runlevel 0 and reboots.

24.3. Booting Zynq

Now that we have looked at the traditional Linux boot process, we shall now focus on the changes that are introduced when booting Linux on a Zynq device. There are a number of key areas in which the embedded Linux boot process varies from that of a desktop distribution. We will start by taking a high-level look at the overall Zynq boot process, before taking a closer look at some of the individual stages.

Zynq devices boot over a number of stages, starting with the boot ROM which is initialised at power-on. The value of the boot mode strapping pins of the device determines the boot mode [5]. The boot mode defines from which of the supported interfaces — JTAG, NAND Flash, NOR Flash, QSPI Flash or SD card — the FSBL will be loaded from [2]. Once the boot mode has been determined, the boot ROM will read the boot header, and given the configuration parameters, will authenticate the image and load the FSBL image from the specified interface to the OCM. Once the image is transferred to the OCM, the control of the CPU is handed over to the FSBL.

Typically, the FSBL contains instructions for the CPU to further configure the PS using reads/writes, and to configure the PL using the DevC [5]. The Advanced Encryption Standard (AES) and Hash-based Message Authentication Code (HMAC) are used by the DevC to allow decryption of the FSBL and the PL bitstream, which are accessed via the DevC and the Processor Configuration Access Port (PCAP) [5].

The individual stages of the Zynq Linux boot process are detailed in Table 24.2 and visualised in Figure 24.3.

Table 24.2: Stages of the Zynq Linux boot process [5]

Stage	Description
Stage-0	On power-on reset, system reset or software reset, a hard-coded boot ROM is executed on the primary processor.
Stage-1	Typically this is the FSBL. It can, however, be any user-controlled code.
Stage-2	Typically this is the user design that will run on the processing system. It could also be the second-stage bootloader, and is completely within user control.

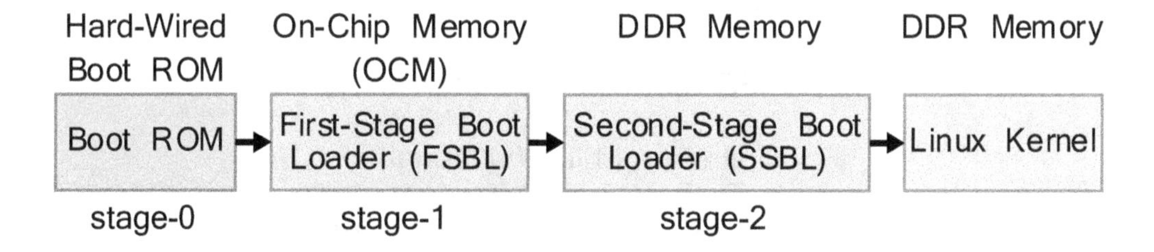

Figure 24.3: Zynq Linux boot process

Before we take a brief look at the individual Zynq boot stages in greater detail, it would be useful to first introduce the files that are required to boot Linux on a Zynq device. This will hopefully make things clearer for when they are mentioned later on in the section.

24.3.1. Zynq Boot Files

In order to boot Linux on a Zynq-7000 AP device, you need to have four files present on your boot medium:

1. BOOT.BIN
2. zImage
3. devicetree.dtb
4. ramdisk8M.image.gz

The first of these files, ***BOOT.BIN***, is the zynq boot image file. It is actually the combination of 2 compulsory files, the FSBL and SSBL executable and linkable format (*.elf*) files, and an optional bitstream (*.bit*) file. The FSBL and SSBL files, as their name suggests, contain the final stages of the bootloader which is uses to load Linux on the device. The bitstream is the file that is used to configure the programmable logic of the Zynq-7000 AP device.

The ordering of the files that are stitched together to form the boot image is important; the optional bitstream file, if required, must be placed after the FSBL file and before the SSBL file.

The required ordering of the boot image file is shown in Figure 24.4.

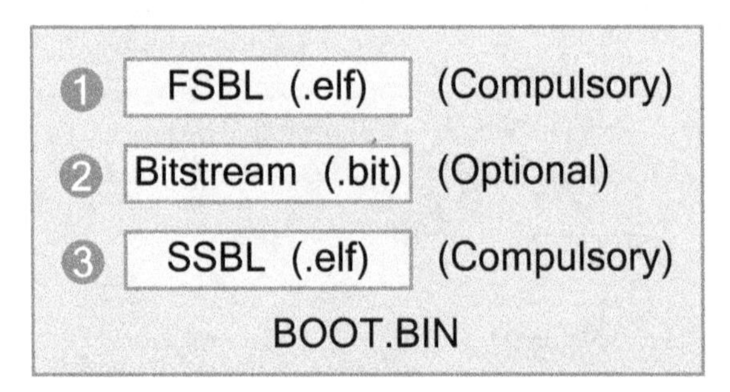

Figure 24.4: Order of boot image files

The ***zImage*** file contains the compressed Linux kernel. It will decompress itself once loaded into memory by the SSBL.

Information of the hardware that Linux is to be booted on is contained in the device tree blob (*.dtb*) file. The device tree is defined in human-readable text file, called the device tree source (*.dts*) file, and is then compiled into the binary form by the compiler to form the device tree blob which can be understood by U-Boot.

The final file, ***ramdisk8M.image.gz***, is a RAM disk image that, when loaded into memory, will allow a portion of RAM to be used as if it were a disk drive. This creates a temporary disk drive in the RAM which the Linux system can use as a file system to mount the root directory.

The overall content of the Zynq Linux boot medium is shown in Figure 24.5.

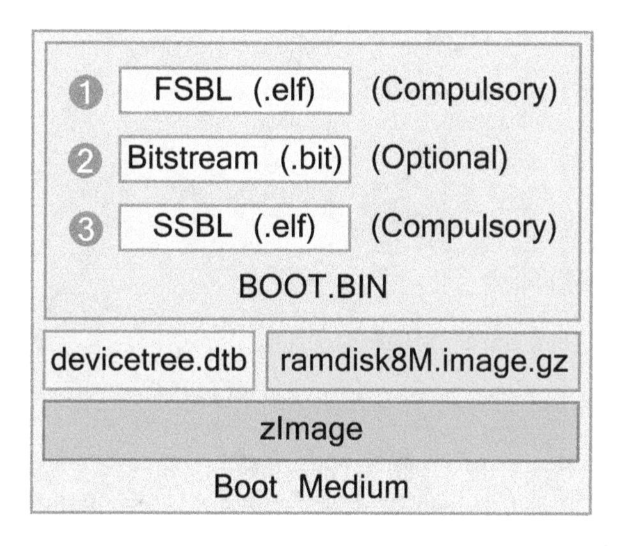

Figure 24.5: Required files for Zynq Linux boot medium

We will now take a more detailed look at the individual stages of the Zynq boot process.

24.3.2. Stage-0 (Boot ROM)

The function of the boot ROM is to load the stage-1 boot image. The location of the stage-1 boot image is determined at power on/reset by sampling the BOOT_MODE signals, which are applied to specific pins in the form of a weak pull-up or pull-down [5].

The boot ROM provides support for loading both encrypted (*secure*) and unencrypted (*non-secure*) boot images. The boot ROM also supports execution of the stage-1 boot image directly from linear flash sources - NOR and QSPI only - when using the eXecute-In-Place (XIP) function. This feature is only available when using non-secure boot images [5].

When in secure boot mode, and running secure boot ROM code, the CPU decrypts and authenticates the PS image before storing it in the OCM. The CPU then branches into the code in the OCM. When using non-secure boot, the CPU disables all secure boot features before branching to the boot image in the OCM or, if the XIP feature is in use, the non-volatile flash. Unless using the XIP feature, the boot image is limited to 192 KB [5].

As the PS uses hardware modules in the programmable logic for both the AES-256 and HMAC (SHA-256) decryption and authentication, the PL must be powered on during secure boot. The device encryption certificate is user-selectable from either the on-chip eFUSE unit or the Battery Backup RAM (BBRAM).

The five possible boot sources are outlined below:

1. NAND Flash
2. NOR Flash
3. Quad-SPI (QSPI) Flash
4. Secure Digital (SD) Card
5. JTAG

Boot sources 1 - 4 are used in master boot mode, whereby the external boot image is loaded by the CPU from non-volatile memory in the PS. JTAG, on the other hand, can only be used in slave boot mode and does not support secure booting. When booting from JTAG, a host computer acts as the secure master to load the boot image into the OCM via a JTAG connection to the device.

Once the stage-1 boot image has been loaded into OCM - or in the case of XIP, the boot image remains in non-volatile flash - the boot ROM releases control of the CPU to stage-1.

Figure 24.6 provides a high-level overview of the boot ROM flow.

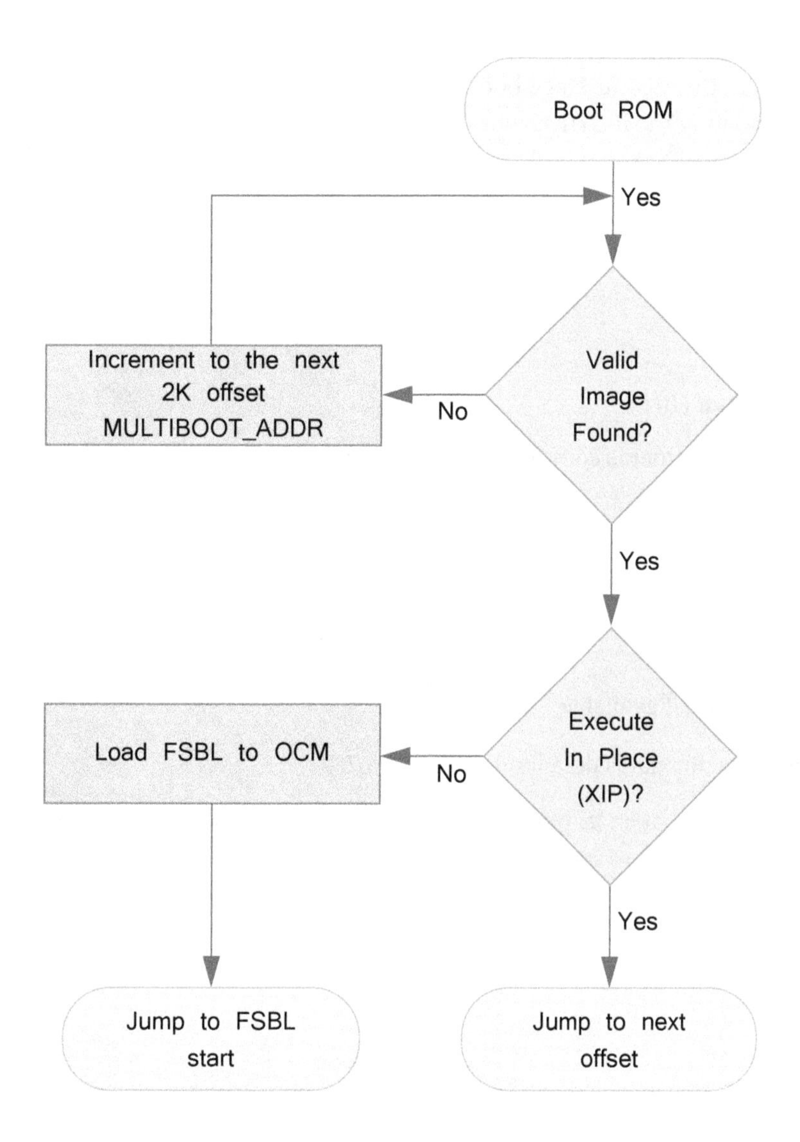

Figure 24.6: Boot ROM flow

24.3.3. Stage-1 (First-Stage Bootloader)

The FSBL is loaded into the OCM by the boot ROM after the initial boot period. The FSBL is responsible for a number of initialisation functions which include initialising the CPU with the PS configuration data, programming the PL using a bitstream, loading the second-stage bootloader or initial user applications into memory, and beginning execution of the second-stage bootloader/initial user application code. Before the control of the CPU is handed over to the second-stage bootloader, the FSBL disables the cache and the MMU, as well as invalidating the instruction cache, as U-boot assumes that they are disabled upon start [6].

Before exploring the functions of the FSBL further, it is useful to understand how the boot image is structured, as the bitstream for the PL and the SSBL, as well as any other code used by the SSBL, Linux or other operating systems, are grouped into partitions in the flash image.

Boot Image Format (BIF)

The boot image format is comprised of the following [6]:

- Boot ROM header
- FSBL image
- Partition image(s)
- Unused space, if available

The boot image format is detailed in Figure 24.7.

It is worth noting that encryption is not compulsory in the FSBL, but the option is made available for anyone who wishes to use it.

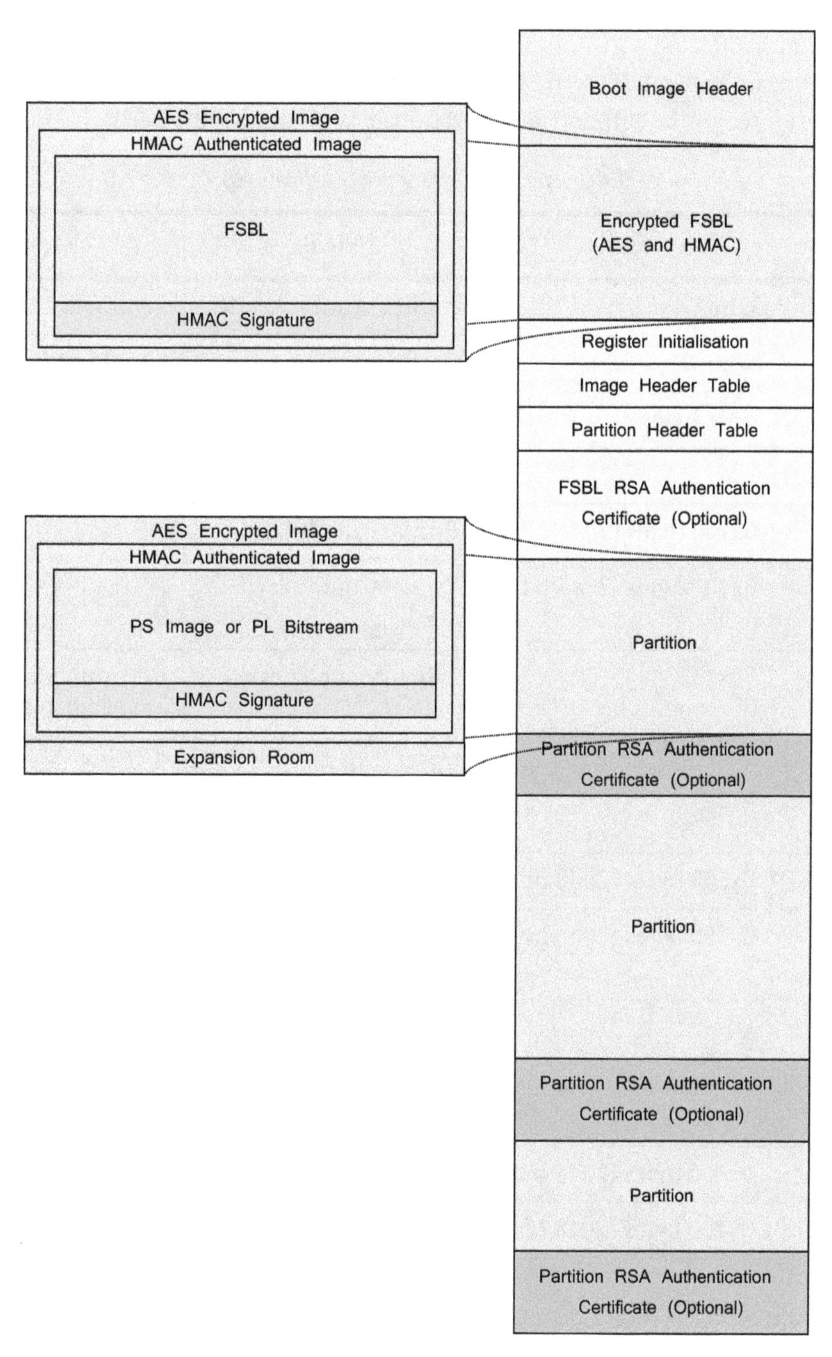

Figure 24.7: Zynq Boot Image Format (BIF) structure

Authentication Certificate

For each authenticated partition, an authentication certificate is appended at the end. All integers in the authentication certificate are stored in little-endian order [6]. Details of the information stored in the authentication certificate is given in Table 24.3.

Table 24.3: Authentication certificate [6]

Offset	Size	Field	Details
0x000	UInt32	Authentication header	See Table 24.4
0x004	UInt32	Certificate size	Should be 0x6C0
0x008	0x38 bytes	Reserved	Filled with zeros
RSA PPK			
0x040	0x100 bytes (2048 bits)	Primary modulus	----
0x140	0x100 bytes (2048 bits)	Primary modulus extension	----
0x240	UInt32	Primary public exponent	Recommended to be 0x00010001
0x244	0x3C bytes	Zero padding	----
RSA SPK			
0x280	0x100 bytes (2048 bits)	Secondary modulus	----
0x380	0x100 bytes (2048 bits)	Secondary modulus extension	----
0x480	UInt32	Secondary public exponent	Recommended to be 0x00010001
0x484	0x3C	Zero padding	----
0x4C0	0x100 bytes (2048 bits)	RSA SPK signature	----
0x5C0	0x100 bytes (2048 bits)	Partition signature	----
0x6C0	----	----	END of AC

The Xilinx BootGen tool (included with the SDK) precalculates the modulus extension used in the Montgomery reduction for modular exponentiation in order reduce the overhead on the FSBL [6]. These values are stored in the field directly following the modulus fields. The authentication certificate contents is detailed in Table 24.4.

Table 24.4: Bit authenticating certificate header [6]

Bit(s)	Field	Value
31 to 16	Reserved	Zeros
15 to 14	Authentication Certificate Format	00: PKSC #1 v1.5
13 to 12	Authentication Certificate Version	00: Current AC
11	Primary Public Key (PPK) Key Type	0: Hash Key
10 to 9	PPK Source	0: eFUSE
8	Secondary Public Key (SPK) Enable	1: SPK Enable
7 to 4	Public Strength	0: 2048
3 to 2	Hash Algorithm	0: SHA256
1 to 0	Public Algorithm	1: RSA

BootGen

The FSBL is combined with an optional PL bitstream and the SSBL/user application code using the provided Bootgen program. Bootgen is a standalone application for generating bootable images that are compatible with the Zynq-7000 processor. The tool assembles the boot image by prefixing a header block to a list of partitions - user ELF files, FPGA bitstreams and other binary files - where each partition can optionally be encrypted and authenticated. The resulting output is a single file which can be programmed directly into the boot flash memory of the Zynq system [6].

The BootGen tool can be used via the SDK to automatically generate a boot image, or can also be used via the command-line. The source code for the tool is also available, as well as in Windows and Linux 32- and 64-bit binary form.

When creating the boot image via the BootGen tool in the SDK, the first partition must be the FSBL ELF, followed by the optional bitstream partition and then the SSBL/application code ELF (it should be noted that if a customised FSBL is used, the boot sequence can be altered accordingly). If the optional bitstream is included in the boot image it must directly follow the FSBL. The SSBL/application ELF must have an execution address of greater than 1Mb, as during execution of the FSBL, DDR RAM is not remapped and any address lower than 1Mb is not accessible [6].

An execution flow of the FSBL is detailed in Figure 24.8.

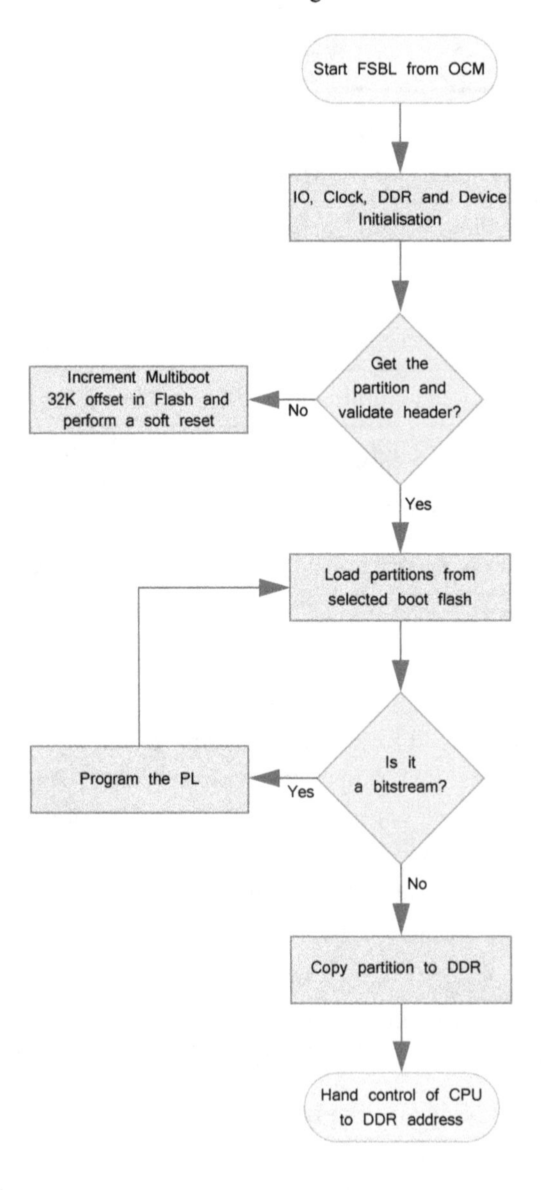

Figure 24.8: FSBL execution flow

24.3.4. Stage-2 (Second-Stage Bootloader)

Stage-2 of the Zynq boot process depends on the type of operating system being booted. For a standalone application, the application code will be loaded and executed at this stage. If, however, an operating system such as Linux or Android is being loaded, stage 2 would be a second-stage bootloader such as U-Boot.

Microprocessors execute code that resides in local memory. This is convenient for standalone or bare-metal applications, but not suited to larger operating systems such as Linux. This is because operating systems are stored on larger, permanent storage mediums such as hard disk drives, USB Flash memory or CD-ROMs. On processor power-on, the memory does not contain the OS; it is therefore the purpose of the bootloader to load the OS into memory from the permanent storage [6].

U-Boot is a popular open source universal bootloader within the Linux community, and is used by Xilinx for the Zynq-7000 AP processor. Xilinx provides the customised source code for U-Boot that is designed to run on Xilinx development boards via a Git tree located at the Xilinx Git repository [7]:

https://github.com/xilinx

As previously explained in this chapter, the Zynq boot process is split into three stages, with the Boot ROM loading the FSBL, and the FSBL loading the optional bitstream and SSBL. In this case, the SSBL is U-Boot and it is responsible for loading the compressed Linux kernel image, the system device tree and the ramdisk image into memory. Once these images are loaded into memory, U-Boot will initialise the execution of the Linux kernel.

24.4. Chapter Review

In this chapter we have taken a look at the traditional boot process of Linux on a desktop environment, including a description of the various stages such as the BIOS, the first- and second-stage bootladers, the kernel and init.

The process of booting embedded Linux on a Zynq-7000 AP device was then introduced and compared, where applicable, to the desktop boot sequence. Greater importance has been given to the Zynq process (this book is all about Zynq, afterall!), with care taken to detail the various files required to successfully complete the boot sequence, including the BOOT.BIN, zImage, devicetree.dtb and ramdisk8M.image.gz files. Further detail was also provided on the BIF and the required authentication certificates. Finally, the bootgen utility, used to assemble boot images for Zynq, was detailed.

24.5. References

NOTE: All URLs last accessed June 2014.

[1] IBM, “Inside the Linux boot process”, webpage.
Available: http://www.ibm.com/developerworks/linux/library/l-linuxboot/

[2] Digilent, Inc, “Embedded Linux Development Guide”, January 2013.
Available: https://www.digilentinc.com/Data/Products/EMBEDDED-LINUX/Digilent_Embedded_Linux_Guide.pdf

[3] Free Standards Group, “System Initialization - Run Levels” in *Linux Standard Base Core Specification 3.1*, 2005.
Available: http://refspecs.linuxbase.org/LSB_3.1.1/LSB-Core-generic/LSB-Core-generic/runlevels.html

[4] B. Ward, “How Linux Boots” in *How Linux Works*, 1st. Ed, No Starch Press, 2004, pp. 54 - 63.

[5] Xilinx, Inc, “Zynq-7000 All Programmable SoC Technical Reference Manual”, UG585, v1.7, February 2014.
Available: http://www.xilinx.com/support/documentation/user_guides/ug821-zynq-7000-swdev.pdf

[6] Xilinx, Inc, “Zynq-7000 All Programmable SoC Software Developers Guide”, UG821, v9.0, June 2014.
Available: http://www.xilinx.com/support/documentation/user_guides/ug821-zynq-7000-swdev.pdf

[7] Xilinx, Inc, “U-Boot”, webpage.
Available: http://www.wiki.xilinx.com/U-boot

Postscript

Glossary

123

7-series - A generation of Xilinx FPGAs, comprising the Artix-7, Kintex-7 and Virtex-7, and also including the Zynq-7000 SoC.

A

Accelerator Coherency Port (ACP) A direct connection between the *Application Processing Unit* and programmable logic parts of the Zynq, not routed through AXI interconnect blocks.

Advance Microcontroller Bus Architecture (AMBA) A family of on-chip bus architectures developed by ARM for use in system-on-chip designs.

Advanced eXtensible Interface (AXI) A high performance interconnect for use in high clock rate system designs. It is a member of the AMBA family

Android An operating system based on the open-source Linux kernel which is mainly intended for use on touchscreen mobile devices. In recent years Android has been customised for use on embedded devices.

Application Processing Unit (APU) A specific section of the Zynq processing system that includes the ARM processor cores, and their associated cache memories and NEON engines.

Application-Specific Integrated Circuit (ASIC) An integrated circuit which is designed for a specific use, rather than general-purpose use.

ARM A family of processor architectures. The hard processor type which forms the basis of the Zynq processing system is an ARM Cortex-A9 version. The term 'ARM' may also be used to refer to the developer of the processor, i.e. a company of the same name.

Asymmetric Multi Processing The use of different operating systems and software stacks on different cores within the same processor. In the context of Zynq, this corresponds to the two cores of the Zynq processing system's ARM Cortex-A9 processor.

Asynchronous Not aligned to a system clock.

AutoESL A tool for *high level synthesis,* superseded by *Vivado HLS.*

B

Basic Input/Output System (BIOS) The component of a computer system that is responsible for initial system startup and locating available boot devices. Also responsible for initialisation and testing of basic system hardware components.

Benchmarking The measurement and comparison of different processors or processing elements.

Binding The process of associating an operation with physical or logical resources to perform that operation.

Bitstream A file created as part of the Zynq (or FPGA) design flow, for programming the programmable logic with the desired configuration. The file extension is .bit.

Block devices In a Linux system, devices capable of hosting a file system.

Block RAM A dense, dedicated silicon memory resource located within the programmable logic section of the Zynq. Capable of being used in a variety of modes.

Board Support Package (BSP) The lowest layer of the software stack, enabling support of a specific hardware base system by a desired operating system.

Bootloader A computer program that is responsible for loading the main operating system on a computer system.

Bus An interconnection by which data and control signals are transmitted between processing components of a computer system.

C

Cache A fast memory to which frequently uses instructions and data is copied from regular memory for faster access. Fast access is enabled in the cache due to its high-speed and close proximity to the processor.

Centralised Version Control System (CVCS) Utilises a single server to host all versioned files in a project.

Character devices In a Linux system, a device which is accessed by a stream of bytes.

Configurable Logic Block (CLB) A functional grouping of logic resources in programmable logic. In the Zynq architecture, a CLB corresponds to two logic *slices.*

Constraint A user-supplied limit or prescribed value in respect of some aspect of the design. Examples include timing and placement constraints.

Coprocessor A special-purpose processing element that is used to supplement the functions of, and assist, the main processor.

Copyleft A play on the word copyright referring to the practice of using copyright law in the distribution of software and its modifications. Each modification retains the same rights.

CoreMark A standard benchmark used for evaluating the performance of processors.

Critical Path The longest combinatorial path between two clocked elements in a circuit.

D

Data flow A style of processing characterised regular passing of data through a chain of processing stages. May refer to hardware or software (usually used in the context of hardware in this book). Also a type of Vivado HLS directive for optimising the implementation of functions.

Debug To identify and resolve function problems in a software or hardware design.

Design Rule Check (DRC) A set of standard checks made by the development tools to find out whether a user-created design has any basic faults.

Digital Signal Processor (DSP) A processor which has been specifically designed for the task of digital signal processing, with an instruction set that is designed to carry out fast arithmetic operations.

Direct Memory Access (DMA) A method which alleviates the processor from performing memory transactions between coprocessing cores and memory through the use of an intermediate DMA controller.

Directive A mechanicm in *Vivado HLS* for influencing the behaviour of the synthesis process, with respect to a particular criterion.

Distributed Version Control System (DVCS) All versioned files in a repository are mirrored on the clients host machine.

DMIP (Dhrystone Millions of Instructions Per second) A measure of the performance of a processor when running the Dhrystrone standard test application.

Driver A software component that facilitates interaction with a hardware element of the system, or a peripheral.

DSP48x / DSP48E1 A dedicated silicon resource for high-speed arithmetic operations. Located within the programmable logic section of the Zynq. DSP48x is a general term, while the specific block used on the Zynq-7000 is the DSP48E1.

E

Embedded system A processing system which is embedded within a larger system or product to perform a single, dedicated function.

Extended Multiplexed Input/Output (EMIO) The mechanism for extending the fixed *Multiplexed Input / Output (MIO)* facilities of the processing system, by extending into the programmable logic.

F

Field-Programmable Gate Array (FPGA) An reconfigurable integrated circuit that contains a very large number of identical configurable logic blocks, connected by programmable interconneccts.

Firmware The components of an embedded system which integrate hardware and software to form a complete system, for instance the *drivers* and *bootloader*.

Fixed point A binary arithmetic representation of a number using a specified number of digits, and comprising integer and fractional components.

Flip-flop The fundamental sequential logic element, capable of storing one bit of data and implementing a 1 clock cycle delay. Found within the programmable logic section of the Zynq.

Floating point A method of representing numeric values as an approximation of the real value. The number is coded as a fraction and an exponent in a way which supports a large range of values.

FPGA Mezzanine Connector (FMC) A standard, high throughput interface for interfacing compatible extension cards to a development board. Such cards are usually referred to as FMC cards.

G

General Purpose Input Output (GPIO) A collective term for the basic user input and output facilities on a development board, usually including slide switches, push buttons, and LEDs.

General-purpose processor A processor which is designed to support a broad spectrum of applications, whilst not being optimised for any specific task.

Git A form of DVCS developed by Linus Torvalds.

GNU Project A mass collaboration project seeking to provide software that is free to run, distribute and modify.

GTX Transceiver A hard IP block located in the programmable logic section, for supporting various protocols for high-speed communications.

H

Hardware base system The custom designed hardware forming the basis of an embedded system.

Hardware Description Language (HDL) A category of programming language that describes the behaviour of hardware.

Hard IP block A functional unit that is implemented as a dedicated, high performance and fixed resource on the device. For instance, *DSP48x slices*, *GTX Transceivers*.

Hard processor A processor which exists as a dedicated silicon resource, with fixed configuration. See also *soft processor*.

High Level Synthesis (HLS) A process which converts a design expressed in a C-based language into a Register Transfer Level (RTL) description.

I

Input/Output Block (IOB) A unit on the programmable logic part of the Zynq which connects a signal to a pin, for interfacing with external signals. Each IOB accommodates a single 1-bit signal.

Instruction set The collection of operations and data types which a processor supports natively.

Integrated circuit A collection of electronic circuits together on a single chip.

Integrated Development Environment (IDE) A software application which provides developers with comprehensive features for the development of hardware and/or software systems.

Intellectual Property (IP) A hardware specification that can be used to configure the logic resources of an FPGA or physically manufacture a silicon device.

Interconnect A switch-like module which manages and directs traffic between interfaces within a system. The connects between two or more interconnects are formed using interfaces.

Interface A point-to-point connection for passing data, addresses and hand shaking signals between master and slave clients within a system.

Interrupt A signal sent to a processor from hardware/software which indicates that an event needs instantaneous attention. An interrupt signal tells the target processor to halt and save the status of its current task, before processing a specific set of instructions.

IP-XACT A standardised specification which provides a metadata description of IP that is compatible with a number of Integrated Development Environments.

ISE (Integrated System Environment) A Xilinx development tool suite for FPGAs and Zynq development. Superseded by *Vivado*.

J

JTAG (Joint Test Action Group) In common usage, the term is used to refer to a connection made for programming or testing a design on an FPGA or Zynq device. Officially refers to a standardised test protocol.

K

Kernel The core of an operating system. It is the part of the operating system that is first to load on boot, and therefore resides in main memory.

L

Latency A measurement of the delay between applying an input to a system or subsystem, and observing the corresponding output. Usually expressed in clock cycles.

Linux A UNIX-like, open source operating system based on the Linux kernel by Linus Torvalds.

Lookup Table (LUT) One of the fundamental logic resources in the programmable logic section. A LUT can implement a combinatorial logic function, a small ROM or RAM, or a shift register.

M

Machine code The low-level, binary stream of binary data relating to a computer program which a processor is able to interpret and process.

Mechanism When referring to Linux device drivers, a mechanism dictates the capabilities of the driver code.

Memory Management Unit (MMU) A component used to translate between virtual and physical addresses. In the Zynq architecture, located within the Application Processing Unit.

MicroBlaze Xilinx' standard *soft processor* type. Capable of being configured to user requirements.

Microcontroller An entire computer system on a single chip. Comprises of a processor, a fixed amount of memory, and peripherals.

Microprocessor A single integrated circuit which that can be programmed to perform a wide range of functions based on the instruction stored within its memory.

MicroZed A small and low cost Zynq development board with a limited number of peripherals, capable of being used as a carrier module on a larger board.

Minix A variant of the UNIX OS.

Multiplexed Input/Output (MIO) The set of peripheral outputs from the Zynq processing system, which are multiplexed to a dedicated IO bank.

N

NEON A resource within the ARM application processing unit that specifically caters for single-instruction-multiple-data operations, which are typically required in media processing and DSP type applications.

Netlist A netlist is a file representing connections and circuit elements. It is the primary output from the process of *synthesising* an input design description.

Network interface In a Linux system, devices concerned with the transmission of data packets.

O

Opcode A machine code representation of a processor instruction.

Operating system The software program that is initiated by the bootloader, and via which the management of all other software tasks and applications is performed.

OPMODE An instruction supplied to a DSP48x slice to configure the functionality it implements.

P

Package The physical outer housing of the Zynq device. The package dictates the set of IOBs that may be bonded to pins and thus made available for use.

Partitioning The process of dividing the operations of system between multiple implementations (i.e. hardware and software implementations).

Peripheral A functional block within an embedded system that is logically and physically separate from the main processor.

Phase Locked Loop (PLL) An intergral part of the clock management circuitry of a Zynq device / FPGA. Used to generate synchronised clock signals with low phase noise.

PicoBlaze A free-to-use, soft core 8-bit microcontroller, developed and maintained by Xilinx.

Pipelining The process of segregating a physical or logical processing stage (i.e. a set of operations grouped by a data dependency) into a set of smaller sections. Accomplished via the insertion of registers.

Place and Route A process undertaken to transfer a synthesised design to the physical resources of an FPGA or Zynq programmable logic.

Pmod A de facto standard interface type for interfacing small add-on modules to a development board.

Policy When referring to Linux device drivers, a policy dictates how the driver mechanisms are carried out.

PowerPC A type of *hard processor* that was used in selected previous generations of Xilinx FPGAs.

Pragma In software development, a method of providing direction to the compiler beyond that which is embodied within the code itself. In HLS, a method of integrating *directives* into the source code.

Processing System The part of the Zynq device that includes an ARM processor and associated facilities, and which is capable of running software applications.

Processor Hardware that executes instructions in a sequential manner.

Profiling A form of dynamic program analysis that aids the optimisation of a software application by measuring properties of the application code.

Program Software written in a specific programming language to represent desired machine behaviour.

Programmable Logic The section of the Zynq device which comprises reconfigurable logic and interconnects. It has the same architecture as 7-series FPGA devices.

Q

R

Random Access Memory (RAM) A category of modifiable storage medium.

Read Only Memory (ROM) A category of non-modifiable storage medium.

Real-Time Operating System (RTOS) A category of operating systems defined by their ability to respond quickly and predictably for a given task.

Register Transfer Level (RTL) A digital circuit design abstraction which models data flow between hardware registers and the logical operations performed on that data.

Repository The database of all versioned files in a project utilising version control.

S

Simulation The process of imitating the execution of hardware on a general-purpose processor.

Single Instruction Multiple Data (SIMD) A type of operation in which the same instruction is applied to several independent sets of input data simultaneously.

Slice A low-level grouping of programmable logic resources. In the Zynq architecture, a slice comprises 4 *Lookup Tables* and 8 *Flip-Flops*, and associated circuitry.

Soft processor A processor which is formed from logic slices in the programmable logic section of a Zynq device, or in general FPGA logic fabric. See also *hard processor.*

Snoop Control Unit (SCU) Part of the Zynq *Application Processing Unit.* Responsible for maintaining memory coherency between cache memories and processor cores.

Solution A term used in the context of HLS. Refers to a particular design iteration based on a standard set of source code, which may be unique in terms of applied directives and constraints, resulting in a unique set of results.

Switch Matrix A resource for configuring the routing of signals between different parts of the programmable logic.

Synchronous Being aligned to a reference clock signal.

Synthesis The process of translating an RTL source into a logic configuration which is targeted for a specific programmable logic device.

System Call Interface An abstracting layer between the user-space and kernel-space in Linux that handles all communications between the two.

System-on-Chip (SoC) An integrated circuit which combines all components of a system or computer on to a single chip. SoCs can contain analog, digital and mixed-signal components.

SystemC A programming language (or, strictly, C++ with specific extensions) which is used for modelling hardware at various levels of abstraction. Can be used as an input language for *high level synthesis.*

System Generator A block-based design tool for the creation of DSP systems. Hosted in the MATLAB/Simulink environment.

T

TCL script ('tickle script') A file containing a set of commands written in an industry standard format, usually used for directing and controlling the behaviour of design tools.

Terminal A program run on a host PC to facilitate communication with software operating on a Zynq device. Used for debugging and interaction.

Testbench A harness for testing a design. Usually refers to an HDL file, although in the context of HLS, may also refer to a C-based harness.

Trip Count Used in Vivado HLS to represent the number of times a loop is executed.

U

Universal Asynchronous Receiver / Transmitter (UART) A connection that facilitates communication over a serial link. Often used in conjunction with a Terminal application for debugging / interaction with software.

UNIX A multitasking computer OS developed by AT&T at Bell Labs.

Unroll The process of expanding a described set of operations into a fully parallel hardware implementation.

V

VHSIC Hardware Description Language (VHDL) A standardised hardware description language.

Verilog A standardised hardware description language.

Version control The management of changes to source code and other files, usually involving the retention of previous versions.

Virtual file system A level of abstraction, creating a unified interface between the individual file systems and the rest of the kernel.

Virtual machine An OS implemented in software through emulation in the system's primary OS.

Virtual memory A memory management technique whereby the storage available to a process or task is virtualised as a contiguous global address space.

Vivado The Xilinx software suite used for developing Zynq systems. Also supports FPGA development.

Vivado HLS Xilinx tool for creating hardware designs from C/C++/SystemC descriptions.

W

Webpack A version of the Xilinx design tools available free of charge from the Xilinx website. Some limitations apply, in terms of device targets and functionality.

X

Xilinx Design Constraints (XDC) A file type used for defining constraints as part of the design process. File extension is .xdc.

Xilinx Inc. The programmable logic company and pioneer of FPGAs, and the creators of Zynq.

Xilinx University Program A department within Xilinx devoted to supporting academic uses of Xilinx design tools and development boards.

Y

Z

ZedBoard A low cost development board featuring a Zynq-700 0 SoC, and a number of peripherals.

ZedBoard.org A website supporting the above development board.

ZyBo A smaller and lower cost Zynq development board than the ZedBoard, but retaining a number of peripherals.

Zynq-7000 A umbrella term for the first generation of Zynq devices.

List of Acronyms

123

2G	2nd Generation (mobile cellular networks)
3G	3rd Generation (mobile cellular networks)
3GPP	3rd Generation Partnership Project
3GPP LTE	3rd Generation Partnership Project Long Term Evolution

A

ACP	Accelerator Coherency Port
AES	Advanced Encryption Standard
AFI	AXI FIFO Interface
AGPL	Affero General Purpose License
AHB	Advanced High-performance Bus
ALU	Arithmetic and Logic Unit
AMBA	Advanced Microcontroller Bus Architecture
AMP	Asymmetric Multi Processing
ANSI	American National Standards Institute
AP	All Programmable
APB	Advanced Peripheral Bus

API	Application Programming Interface
APSoc	All Programmable System on Chip
APU	Application Processing Unit
ASIC	Application Specific Integrated Circuit
ADC	Analogue to Digital Converter
ATP	Authorised Training Provider
AWDT	Watch Dog Timer
AXI	Advanced eXtensible Interface
AXI_HP	Advanced eXtensible Interface - General Purpose
AXI_HP	Advanced eXtensible Interface - High Performance

B

BBRAM	Battery Backup Random Access Memory
BIF	Boot Image Format
BIOS	Basic Input / Output System
BOM	Bill of Materials
BSD	Berkeley Software Distribution
BSP	Board Support Package
BTAC	Branch Target Address Cache

C

CAN	Controller Area Network
CCTV	Closed Circuit Television
CDK	C/C++ Development Kit
CDMA	Code Division Multiple Access
CFR	Crest Factor Reduction
CLB	Configurable Logic Block
CLI	Command Line Interface
CMOS	Complementary Metal Oxide Semiconductor

CORDIC	Co-Ordinate Rotation DIgital Computer
CPU	Central Processing Unit
CTT	Concepts, Tools and Techniques
CVCS	Centralised Version Control System

D

DAP	Debug Access Port
DDR	Double Data Rate (referring to memory)
DDRC	DDR Controller
DDRI	DDR Interface
DDRP	DDR PHY
DevC	Device Configuration (Unit)
DMA	Direct Memory Access
DMAC	Direct Memory Access Controller
DMIPs	Dhrystone Millions of Instructions Per Second
DOS	Disk Operating System
DPD	Digital Pre-Distortion
DPR	Dynamic Partial Reconfiguration
DRAM	Dynamic Random Access Memory
DRC	Design Rule Check(s)
DSP	Digital Signal Processing / Digital Signal Processor
DTS	Device Tree Source
DUT	Device Under Test
DVCS	Distributed Version Control System
DVI	Digital Visual Interface

E

EABI	Embedded Application Binary Interface
ECC	Error Correction Coding / Error Correcting Code

EDA	Electronic Design Automation
EDK	Embedded Development Kit
EEMBC	Embedded Microprocessor Benchmark Consortium
ELF	Executable Linkable Format
EMIO	Extended Multiplexed Input / Output *(equivalent to* Extended MIO)
ESD	Electro-Static Discharge
ESL	Electronic System Level

F

FF	Flip Flop
FFT	Fast Fourier Transform
FIFO	First In First Out
FIQ	Fast Interrupt reQuest
FIR	Finite Impulse Response
FMC	FPGA Mezzanine Card
FPGA	Field Programmable Gate Array
FPU	Floating Point Unit
FSBL	First Stage Boot Loader
FSF	Free Software Foundation
FSM	Finite State Machine

G

GCC	GNU Compiler Connection
GDB	GNU Debugger
GIC	General Interrupt Controller / Generic Interrupt Controller
GigE	Gigabit Ethernet
GHB	Global branch History Buffer
GNU	GNU's Not Unix (recursive acronym)
GP	General Purpose

GPIO	General Purpose Input / Output
GPL	GNU General Purpose License
GPP	General Purpose Processor
GPS	Global Positioning System
GPU	Graphics Processing Unit
GPV	Global Programmers View
GRUB	Grand Unified Boot Loader
GTX	High speed serial transceiver (note: not an acronym as such!)
GUI	Graphical User Interface

H

HD	High Definition
HDL	Hardware Description Language
HDMI	High Definition Multimedia Interface
HIL	Hardware In the Loop
HLS	High Level Synthesis
HMAC	Hash-based Message Authentication Code
HP	High Performance
HR	High Range
HTML	Hyper Text Markup Language

I

IC	Integrated Circuit
ICAP	Internal Configuration Access Port
IDE	Integrated Design Environment
IEEE	Institute of Electrical and Electronics Engineers
IF	Intermediate Frequency
II	Initiation Interval
IO	Input / Output

IOB	Input / Output Block
IOP	Input / Output Peripheral(s)
IOSERDES	Input / Output Serialiser / Deserialiser
IoT	Internet of Things
IP	Intellectual Property
IPC	Inter-Process Communication
IPI	Inter-Processor Interrupt
IPv4	Internet Protocol version 4
IPv6	Internet Protocol version 6
IRQ	Interrupt ReQuest
ISE	Integrated System Environment
ISO	International Organization for Standardization

J

JTAG	Joint Test Action Group
JTNC	Joint Tactical Networking Centre
JTRS	Joint Tactical Radio System

K

kgdb	Remote host Linux kernel debugger through gdb

L

L1	Level 1 (Cache)
L2	Level 2 (Cache)
L3	Level 3 (Cache)
LCD	Liquid Crystal Display
LED	Light Emitting Diode
LGPL	Lesser General Public License
LILO	LInux LOader
LMS	Least Mean Squares

LRU	Least Recently Used
LTS	Long Term Support
LUT	Lookup Table

M

M2M	Machine to Machine
MAC	Media Access Control
MBR	Master Boot Record
MCS	Micro Controller System
MIO	Multiplexed Input / Output
MIPs	Millions of Instructions Per Second
MMU	Memory Management Unit
MPE	Media Processing Engine
MPL	Mozilla Public License
MSPS	Mega Samples Per Second

N

NCO	Numerically Controlled Oscillator
NFC	Near Field Communications
NFS	Network File System
NIC	Network Interface Controller
NMI	Non-Maskable Interrupt

O

OCM	On-Chip Memory
OEM	Original Equipment Manufacturer
OFDM	Orthogonal Frequency Division Multiplexing
OLED	Organic Light Emitting Diode
OMP	Orthogonal Matching Pursuit
ONFI	Open NAND Flash Interface

OpenCL	Open Computing Language
OpenCV	Open source Computer Vision
OPMODE	Operation Mode
OS	Operating System
OSCI	Open SystemC Initiative
OTG	On The Go

P

PC	Personal Computer
PCAP	Processor Configuration Access Port
PCB	Printed Circuit Board
PCI	Peripheral Component Interconnect
PCIe	Peripheral Component Interconnect express
PFN	Page Frame Number
PHY	PHYsical Layer
PID	Process Identifier
PL	Programmable Logic
PLL	Phase Locked Loop
PMU	Performance Monitor Unit
Pmod	Peripheral module
PPI	Private Peripheral Interrupts
POSIX	Portable Operating System Interface
POST	Power On Self Test
PR	Partial Reconfiguration
PS	Processing System
PPK	Primary Public Key

Q

QoS	Quality of Service

QPSK	Quadrature Phase Shift Keying
QSPI	Queued Serial Peripheral Interface

R

RAM	Random Access Memory
RF	Radio Frequency
RISC	Reduced Instruction Set Computer
RM	Reconfigurable Module
ROM	Read Only Memory
RP	Reconfigurable Partition
RTL	Register Transfer Level
RTOS	Real Time Operating System

S

SATA	Serial Advanced Technology Attachment
SCI	System Call Interface
SCSI	Small Computer System Interface
SCU	Snoop Control Unit
SD	Secure Digital
SDF	Standard Delay Format
SDK	Software Developer's Kit
SDIO	Secure Digital Input Output
SDR	Software Defined Radio
SERDES	Serialiser / Deserialiser
SFP	Small Form factor Pluggable (type of connector)
SGI	Software Generated Interrupts
SIMD	Single Instruction Multiple Data
SLM	System Level Model / Modelling
SMA	SubMiniature version A (type of connector)

SMC	Static Memory Controller
SMP	Symmetric Multi Processing
SoC	System on Chip
SoPC	System on Programmable Chip
SPI	Serial Peripheral Interface
SPI	Shared Peripheral Interrupts
SPK	Secondary Public Key
SRAM	Static Random Access Memory
SRL16	Shift Register Length 16
SRL32	Shift Register Length 32
SSBL	Second Stage Boot Loader
SSL	Secure Socket Layer
SWDT	System Watch Dog Timer

T

TCL	Tool Command Language
TLB	Translation Look-aside Buffer
TLM	Transaction Level Model / Modelling
TRM	Technical Reference Manual
TTC	Triple Timers / Counters

U

UART	Universal Asynchronous Receiver Transmitter
UCF	User Constraints File
UG	User Guide
USB	Universal Serial Bus

V

VCD	Value Change Dump
VCS	Version Control System

VFP	Vector Floating Point
VFS	Virtual File System
VGA	Video Graphics Array
VHDL	VHSIC Hardware Description Language (see VHSIC below!)
VHSIC	Very High Speed Integrated Circuit
VLSI	Very Large Scale Integration
VM	Virtual Machine

W

WFE	Wait For Event
WFI	Wait for Interrupt
WiFi	Wireless Fidelity

X

XADC	Xilinx Analogue to Digital Converter
XDC	Xilinx Design Constraints
XIP	eXecute In Place
XMD	Xilinx Microprocessor Debugger
XML	eXtensible Markup Language
XPS	Xilinx Platform Studio
XST	Xilinx Synthesis Technology
XUP	Xilinx University Program

Y

YAMD	Yet Another Malloc Debugger

Z

ZED	Zynq Evaluation & Development

Index

123

3GPP LTE 357
7-series 1, 66
- Artix-7 40, 80
- Kintex-7 80

A

Abstraction 7, 116, 261
- levels of 256

Academic 133
- curriculum 149
- licenses 158
- project 147, 150, 154
- research 147, 156
  - Dynamic Partial Reconfiguration (DPR) 157
  - embedded vision 157
  - hardware virtualisation 157
  - image processing 156–157
- software support 158
- teaching 147, 149, 154, 158
  - computer science 150
  - electronic engineering 150
  - embedded systems design 150, 154
  - FPGA 150–151
  - materials 160
  - processor architecture 151
- technical support 160

Academic development boards
- ZedBoard 159

AccelDSP 261
Accelerator Coherency Port (ACP) 19, 33
ACP 193–195
Address
- address range 401
- address space 400
- space 402
- virtual addresses 401

address
- address phase 354

address channel 354
Advanced Driver Assistance Systems (ADAS) 102
Advanced eXtensible Interface 353
Advanced Microcontroller Bus Architecture 353
AMBA 30
Android 59, 375–377
Application Processing Unit (APU) 17, 31
Application Programming Interface (API) 372
Application Specific Integrated Circuit (ASIC) 2
Applications 101
- aerospace 103
- defence 103
- GPS 102
- high performance computing 105
- image and video processing 104
- Image processing 19
- medical 105
- motor control 104
- RADAR 102
- satellite 102
- Signal processing 19, 28

Video processing 19
Zynq 77, 98
Architecture-dependent code 406
Arithmetic 26–27
DSP48E1 26
fixed point 289, 291
wordlengths 26
ARM 353
ARM v7 architecture 20–21
Cortex-A9 1, 83–84, 89, 92
Documentation 20
Processor IP 21
ARM Cortex A9 16, 20
Artix-7 40
Assembly language 262
atomic operations 356
Authorised Training Providers (ATPs) 160
AutoESL 65, 262
Avnet 133
AXI 5, 54, 96, 120, 197, 203, 208, 353, 360
Accelerator Coherency Port 33
AXI channels
Address Channels 356
Read Address Channel 362
Read Data Channel 356, 362
Write Address Channel 362
Write Data Channel 356, 362
Write Response Channel 356, 362
AXI Interconnect 362
AXI interconnect 354
AXI interface 192
AXI masters 354
AXI slaves 354
AXI4 30, 353–354
AXI4-Lite 31, 354
AXI4-Stream 31, 354
AXI_ACP 202
AXI_GP 198–199, 202
AXI_HP 198–200
bus protocol 298
buses 30
General purpose interface 33
High Performance Ports 33
Interconnect 31
Interface 31
interface 1
PS-PL Interfaces 32
AXI ports (general purpose) 33
AXI ports (high performance) 33
AXI4 30
AXI4-Lite 31
AXI4-Stream 31

B

Bare-metal 59, 374
Basic Input/Output System (BIOS) 409, 411
Bill of Materials (BOM) 77, 79, 95, 107
BIT (*.bit) file 59
Bitstream 59
Black box 62
Block RAM 25, 40
Block RAM (BRAM) 55
Board Support Package (BSP) 58, 406
Boot sequence 385
Booting
Boot Image Format (BIF) 420
bootloader 410
First-Stage Bootloader (FSBL) 411–412, 414, 419–420, 423–425
ZedBoard 139
BRAM 357
Buffer
buffer overruns 393
cache 405
burst-based transactions 354
Bus 5, 184
arbitration 186
AXI 30
bandwidth 188
bridge 186
master 186
peripheral bus 176, 185
slave 186
system bus 176, 185

C

Cache 178
Level 1 (L1) 17–18, 21, 179
Level 2 (L2) 17, 180
Level 3 (L3) 180
cache 17
Cache memory 31
Carry logic 25
C-based languages 261, 264
C (programming language) 262
C++ (programming language) 263
SystemC 243, 245, 261, 263–264
Channel Decoder 357

ChipScope 66, 357
Clocks 29
Cloud 115
Cognitive radio 110–111
Command line 395
Communication overhead 93, 96
Complementary Metal Oxide Semiconductor (CMOS) 411
Computer vision 115–119, 126, 153
    example 120
Configurable Logic Block (CLB) 23
Constraints 34, 67
    timing 274
Controller Area Network 357
Controller Area Network (CAN) 22, 55
Co-processing 94
    co-processor 54, 177
Copyleft 388
CORDIC 358
CoreMark 82–83, 86, 89
CPU Private Peripheral Interrupts (PPI) 214
Crest Factor Reduction 357
Critical path 274, 310

D

Data
    big data 115, 149
    conversion 71
    flow 54
        backpressure 302
        downstream 302
        upstream 302
DDR 203–204
    DDRC 203
    DDRI 203
    DDRP 203
Debug Access Port (DAP) 198
Debugging 52, 393
    gbd 395
    gdb 394–395
    kgdb 394–395
    MEMWATCH 393–394
    Oops 394–395
    strace 393–394
    Yet Another Malloc Debugger (YAMD) 393–394
Deinterlacer, 357
Deinterleaver 358
Design flow 9, 47, 53, 96
    Zynq SoC 8
Design reuse 7–8, 62, 255
    education 152
Design Rule Check (DRC) 55
Development Boards
    ZC702 Evaluation Kit 157
Development boards 52, 150, 154, 159
    MicroZed 142
    ZC702 Evaluation Kit 67, 69, 71
    ZC706 Evaluation Kit 69
    ZedBoard
        See also ZedBoard
    ZyBo 159
Device Configuration Unit (DevC) 198, 415
Device drivers 385
Devices
    block devices 407
    character devices 407
    drivers 59, 406
        ZedBoard USB-UART 138
    network interfaces 407
    selection critera 78
Dhrystone 82
Digilent 133, 155
Digital Predistortion 357
Digital Signal Processing (DSP) 19, 28, 109
Digital Signal Processor (DSP) 89, 177
Direct Memory Access 353, 357
Direct Memory Access (DMA) 34, 187
Distributed RAM (DRAM) 25
DMIPs 82, 86
Documentation 249
Documentation Navigator 48
Drivers 395
DSP48E1 25–26, 40
    OPMODE 27
DSP48x 274
    DSP48E1 55
Dynamic Partial Reconfiguration (DPR) 110, 121, 123–124, 157
Dynamic reconfiguration 109
Dynamic spectrum access 110

E

Ecosystem (Zynq) 101
Electronic System Level (ESL) design 263
ELF (*.elf) file 59
Embedded system
    hardware system 5

software system 5
Embedded systems 173, 371, 373
embedded vision 115
See also computer vision
networked 149
operating system 371
Android 375–377
embedded Linux 373
Real-Time Operating System (RTOS) 373, 382
standalone 59, 373–374
processor 80, 177
Emulated 389
Error-Correcting Code (ECC) 203
Ethernet 22, 357
Executable image 402
Expansibility 114
Extended MIO (EMIO) 21, 34

F

Fast Fourier Transform (FFT) 19
FFT 358
Field Programmable Gate Array (FPGA) 1, 4, 237
FPGA Mezzanine Connector (FMC) 71
FIFO 358, 360
File system 385, 395, 397, 404
affs 404
ext 404
ext2 404
hpfs 404
iso9660 404
minix 404
mounted 405
msdos 404
ncp 404
proc 404
smb 404
sysvs 404
ufs 404
umsdos 404
vfat 404
xia 404
Finite Impulse Response (FIR) filter 19, 28, 92
Finite State Machine (FSM) 322
FIR 358
First In First Out (FIFO) 25
First In First Out (FIFO) buffer 33
First-Stage Bootloader (FSBL) 409
Flip-Flops (FFs) 24
Floating Point
Floating Point Unit (FPU) 17, 20
IEEE 754 standard 20
Floating point 19, 84
Floating Point Unit (FPU) 21, 80
FPGA 391
Free Software Foundation 386, 395

G

Gamma Correction, 357
General Purpose Input/Output (GPIO) 21–22
General Purpose Processor (GPP) 89
Generic Interrupt Controller (GIC) 196, 211–213, 215, 218
Git 392
GNU 385–389, 394
GPIO 55
GPL 387–390
Grand Unified Bootloader (GRUB) 413
Graphical User Interface (GUI) 373
GTX Transceiver 29
GTX transceiver 69

H

Hardware
acceleration 89, 93–94, 98
Hardware in the Loop (HIL) 57, 59
system design 47
Hardware Base System 58
Hardware Description Language (HDL) 63, 240–241
behavioural 257
simulation 57
structural 257
Verilog 150, 239, 243
VHDL 150, 239, 243, 330
Hardware platform 58
Hardware/software partitioning 92, 97, 259
HDL Coder 241, 247, 261
Hierarchy 295
Hierarchy (system) 54
High Level Synthesis (HLS) 8, 78, 96, 153, 255, 281
algorithm synthesis 265–266
design iterations 269
directives 258
education 152
golden reference 268
interface synthesis 265
ports 266

protocols 266
scheduling and binding 273
High Performance (HP) Input/Output 28
High Range (HR) Input/Output 28

I

I2C bus 22
ICAP 357
Image Edge Enhancement 357
Image processing 19, 89, 115–116, 118
abstraction 117, 119
description 118
education 152
features and objects 118–119
implementation 118
IP 120
pixel level processing 117, 119
iMPACT 66
Implementation
metrics 259, 266, 286
clock frequency 271
I/O requirements 271
latency 271
power consumption 271
resources/area 271, 274
throughput 271, 274
Inode 405
Input/Output Block (IOB) 25, 28
Instruction sets 19
Intellectual P 244
Intellectual Property (IP) 7, 56, 62, 126, 152, 237, 239–240, 343–345, 347–350
IP subsystem 239, 349
packaging 281
Vivado HLS 330
Interconnect 31, 198–199, 345
Central interconnect 198
Interconnect master 199
Interconnect slave 199
Master 31
Master interconnect 198
Memory interconnect 198
OCM interconnect 198
Slave 31
Slave interconnect 198
interconnect 353
Interconnect master 199
Interconnect slave 199
Interface 31
Interfaces
Communications 21
Communications transceivers 40
Controller Area Network (CAN) 22
Ethernet 22
External 21
GPIO 21–22
GTX Transceiver 29
I2C bus 22
PCI Express 29, 69
PS-PL 21, 32
SD card 22
Serial Peripheral Interface (SPI) 22
UART 22, 55
USB 22, 52
Interleaver 358
Internet 148
connected devices 148
Internet of Things (IoT) 115, 148–149
protocol
IPv4 148
IPv6 148
traffic 148
Interrupts 183, 211
CPU Private Peripheral Interrupts (PPI) 213
interrupt handler 399
interrupt interface 196
Shared Peripheral Interrupts (SPI) 213
Software Generated Interrupts (SGI) 213
software interrupt 399
IOSERDES 28
IP block
Hard IP block 29
IP Integrator 353
IP library 346
IP-XACT 7, 62, 152, 243, 330, 346, 351
ISE Design Suite 47, 64
ISim 65
System Generator 50, 55, 65, 152, 229, 241, 247, 261, 330
Xpower Analyzer 65

J

Java 261
JTAG 29
modes 139
ZedBoard programming 138

K

Kernel 385–387, 392, 394–395, 397, 403
- code 399, 401
- flags 402
- kernel-space 386, 398

Kintex-7 40, 85
KVM 390

L

Latency 286, 310–311
LEON4 86
Licensing
- free 386
- GNU Lesser General Public License (LGPL) 348
- management tools 48
- Modified-BSD license 348

Linux 59, 157, 375, 379, 381, 385–387, 389–390, 394–395, 397, 409–410, 413–415
- Boot ROM 417–418, 420
- Grand Unified Boot Loader (GRUB) 413
- Init 413
- Kernel 413, 425
- Linux booting 410, 413–415
- Linux Loader (LILO) 413
- security model 402

Linux Loader (LILO) 413
Lookup Tables (LUTs) 24

M

Machine code 402
Machine vision
- See Computer vision

Machine-to-Machine (M2M) 115, 148
Master Boot Record (MBR) 411–412
MATLAB 51, 57, 67, 120, 241, 247, 261
Media Access Controller 357
Memory
- access 187
- cache coherence 18
- controller 176
- Distributed RAM (DRAM) 25
- high memory 401
- low memory 401
- main memory 397
- management 397, 401
- memory allocation 393
- Memory Management Unit (MMU) 17
- On-Chip Memory (OCM) 17–18
- physical memory 402
- Programmable Input/Output (I/O) 187
- Random Access Memory (RAM) 25, 409
- Read Only Memory (ROM) 25

Metadata 392
MicroBlaze 16, 80, 83–85, 87
- Micro Controller System (MCS) 84

Microcontroller buses 353
Minix 385
Multiplexed Input/Output (MIO) 21
Multi-processing 402
- Asymmetric Multi-Processing (AMP) 378
- Symmetric Multi-Processing (SMP) 378

N

NEON 17, 19, 21, 55, 89, 92, 94, 119
Numerically Controlled Oscillator (NCO) 109

O

On-Chip Memory (OCM) 31, 193, 196–200, 202, 208–210, 215, 414
OpenCL 153
OpenCV 120, 126, 153, 156
OpenRISC 86
OpenSparc 86
Operating system 385
Operating System (OS) 58–59, 403
Optimisation 307
OS 386

P

Page 400
- page fault 401
- Page Frame Number (PFN) 278, 400
- page table 400

Parallelism 54
Partitioning
- See Hardware/software partitioning

PCI Express 29
Peripheral 5–7
PetaLinux 381
Phase Locked Loop (PLL) 29
Physical layer (PHY) 109
PicoBlaze 85
Pipelining 310
PlanAhead 65
Pmod 71, 154
Power consumption 124
Power-On Self Test (POST) 411
PowerPC 86, 88

Printed Circuit Board (PCB) 2
Process
child process 402
creation 402
destruction 403
fork 402
management 397, 401
scheduler 403
termination 402
Processing
mechanisms 407
policies 407
Processing System (PS) 4, 15
Processor 176
co-processor 177
Digital Signal Processor (DSP) 177
embedded 177
execution 90
execution cycles 89–90, 180
hard 80, 98
hard processor 16
instruction set 84
microcontroller 176
microprocessor 176
multi-core 90
operation 90
operation (non real-time) 90
priority 90
processor architecture 398
processor interrupts 399
routine 90
soft 80, 88, 98
soft processor 16
subordinate 84
Professor Workshops 160
Profiling 224–225
Programmable Logic (PL) 4, 15, 23, 80, 89, 109, 265
Block RAM 25
carry logic 25
Configurable Logic Block (CLB) 23
DSP48E1 25
Flip-Flops (FFs) 24
Input/Output Block (IOB) 25
Lookup Tables (LUTs) 24
slices 24
switch matrix 24
Programming and debug 29
JTAG 29
security 139

R

Radio
portability 108
standards 107–108
transceiver 107
Radio Frequency (RF) 109
spectrum 110–111
Random Access Memory (RAM) 25
Read Only Memory (ROM) 25
Real-time
processing 116
Real Time Operating System (RTOS) 59, 126, 374
firm real-time 374
hard real-time 374
soft real-time 374
systems 92
hard real-time 92
soft real-time 92
Real-Time Operating System (RTOS) 382
Register Transfer Level (RTL) 8, 243, 245–248, 257
simulation 245–246
Remote machine 395
Repository 391–393
Research (academic) 147

S

SD card 22
Second-Stage Bootloader (SSBL) 409
SelectIO Resources 28
Serial Peripheral Interface (SPI) 22
Shared Peripheral Interrupts (SPI) 215
Simulink 51, 57, 67, 120, 241, 247
Single Instruction Multiple Data (SIMD) 19, 27
Slices 24
Smart
agriculture 111–112
building 111–112
city 111–112, 114, 148
grid 111–112
grids 148
home 111, 113, 148
network 111, 114
system 111
transport 111–112
Snoop Control Unit (SCU) 18, 31, 197–198
Software

system design 47
Software Defined Radio (SDR) 71, 107–109, 121–122, 156
education 152
military 108
Software Development Kit (SDK) 19, 50, 55, 57, 65, 228, 381
Software Generated Interrupts (SGI) 217
Static Memory Controller (SMC) 206–208
Switch matrix 24
Synthesis
logic synthesis 257
physical synthesis 257
System bus 176
System call 399, 403
System Call Interface (SCI) 398
System integration 60
System Level Model (SLM) 257
System-level design 96
System-on-Chip (SoC) 2, 4, 147, 149, 263
System-oriented design 47
Systems-on-Chip (SOC) 239

T

TCL scripting language 63, 286
commands 284
Team-based design 54, 60
Throughput 310
Time-to-market 107
Training 147, 160
See also Academic
online 161
QuickTake 161
SpeedWay 143
videos 161
Transaction Level Modelling (TLM) 263

U

UART 22
UCF file (*.ucf) 67
Universal Serial Bus (USB) 22
UNIX 386
UNIX-like 386
User-space 393–394, 398, 402

V

Verification 8, 62, 263
Verilog
See Hardware Description Language (HDL)
Version control 391
CVCS 391–392
DVCS 391–392
VCS 392
VHDL
See Hardware Description Language (HDL)
Video & Imaging Kit 69
Video processing 19, 91, 115–116, 118
compression 116
Virtual address space 401
Virtual Box 389
Virtual File System 405
Virtual machine 389
Virtual memory 400
Vivado 353, 360
IP Integrator 360
ZedBoard support 136
Vivado Design Suite 10, 47, 64, 79, 149, 227, 241, 281, 343–344, 350
board support 136
documentation 72
IP Catalog 241, 343, 345–346, 349–350
IP Integrator 55, 57, 62, 65–66, 227, 330, 349
IP Packager 350
System Generator 50, 55, 65, 152, 229, 241, 247, 261, 330
Vivado Device Programmer 66
Vivado Integrated Development Environment (IDE) 50, 55, 65
Vivado Logic Analyser 50
Vivado Logic Analyzer 66
Vivado Power Analyzer 65
Vivado Simulator 49, 59, 65
Vivado Synthesis 65
WebPACK 10, 158
Vivado HLS 8, 50, 55, 63, 65, 98, 121, 243, 255, 264, 281
algorithm synthesis 309
data types 311
loops 319
arithmetic
floating point 294
arrays 327
directives 327–328
resources 327
C/RTL cosimulation 268–270
clock period 310
clock uncertainty 310
command line interface 286

concurrency 317, 319
constraint 310, 328–329
  interface 329
  latency 328
  resource utilisation 328
constraints 275–276
control component 322
data dependency 312–313
data types 287
  arbitrary precision 287, 289, 294
  boolean 289
  C and C++ (native) 287
  complex 289
  extended precision floating point 289
  fixed point 291
  floating point 294
  overflow 291–292
  quantisation 291–292
  saturation 291–292
  SystemC 292–293
  validation 294
dataflow (pipelining) 316–319
datapath component 322
design flow 267
directive 328
directives 96, 266, 269, 276, 284, 307, 319, 321–322, 327–329
export 330
export from 269, 276
function return statement 297
functional verification 268
Graphical User Interface (GUI) 284
  perspectives 284
hardware implementation 309
hierarchy 295
HLS process 272
implementation metrics 309
interface 295
  AXI 303
  block-level protocol 295, 302–304
  directive 304–305
  manual specification 308–309
  port-level 296
    protocol 295
  protocol 296, 298, 300–301
  specification 295
  synthesis 295
  SystemC 308
  top level function 295
IP 277
Iteration Interval (II) 311
latency 310–311, 313, 315, 317
loops 286, 319, 325
  flattening 324–325
  hardware 321
  latency 319–320, 326
  merging 321–322
  nesting 323, 325
  partially unrolled 321
  pipelining 326
  rolled 319, 321
  throughput 321, 324, 326
  unrolled 321, 326
optimisation 275, 307, 316, 319, 328
output languages 276
perspective 286
pipelining 310–312, 314–316, 326
pragma 284, 307, 329
scheduling and binding 261
solutions 266, 276, 284, 309, 328–329
synthesis report 286, 329
SystemC 308
throughput 310, 313, 315, 317
video libraries 120
VMware Player 390
VMware Workstation 390
VOIP 357

W

WebPack 10, 49
Windows Virtual PC 390
Wireless connectivity 107
write channel 355
write response channel 354

X

XADC 71
Xcell Daily Blog 106
Xcell Journal 106
XDC (*.xdc) file 67
XEN 390
Xilinx
  Documentation 20
Xilinx Alliance Program 125
Xilinx Microprocessor Debugger (XMD) 59
Xilinx Platform Studio (XPS) 65–66, 227, 330
Xilinx Synthesis Technology (XST) 65
Xilinx University Program (XUP) 72, 147, 154, 156,

158–159
contact details 160
membership 160
Professor Workshops 160
teaching materials 160

Z

ZedBoard 10, 133, 154, 159
booting 139
community 144
constraints 143
documentation 142
hardware configuration 137
jumper settings 140
kit contents 137
layout 135
memory 134
oscillator 134
programming 138
reference designs 144
support forums 145
training 143
tutorials 143
USB connections 138
USB-UART drivers 143
Vivado support for 136
Zynq device 134
Zynq 4, 66, 353, 360, 391
applications of 101
Architecture 15, 41
architecture 6
booting 414–415, 425
design flow 107
ecosystem 101, 125
physical size 96
power consumption 107
Zynq-7000
Kintex-7 40
Zynq-7000 devices 39
Automotive grade 40
Defence grade 40

CPSIA information can be obtained
at www.ICGtesting.com
Printed in the USA
LVHW010309241122
733904LV00002B/23